Climate Change and Water Governance

ADVANCES IN GLOBAL CHANGE RESEARCH

VOLUME 54

For further volumes:
http://www.springer.com/series/5588

Margot Hill

Climate Change and Water Governance

Adaptive Capacity in Chile and Switzerland

Springer

Margot Hill
Research Group on Climate Change
and Climate Impacts
Institute for Environmental Sciences
University of Geneva
Carouge, GE, Switzerland

ISSN 1574-0919
ISBN 978-94-007-5795-0 ISBN 978-94-007-5796-7 (eBook)
DOI 10.1007/978-94-007-5796-7
Springer Dordrecht Heidelberg New York London

Library of Congress Control Number: 2012952092

Portion of text reprinted from Environmental Science and Policy, 14/7, Beniston, M., Stoffel, M., Hill, M. (s), Impacts of climatic change on water and natural hazards in the Alps: can current water governance cope with future challenges? Examples from the European "ACQWA" project, 731 740, Copyright (2012), with kind permission from Elsevier."

Printed on acid-free paper

Springer is part of Springer Science+Business Media (www.springer.com)

Foreword

The contents of this volume of *Advances in Global Change Research* represent several years of research conducted by Dr. Margot Hill, focusing on adaptive capacity and water governance in two widely-separated regions of the globe, namely the Swiss Alps and the Chilean Andes. While there are clearly different institutional frameworks in the two countries in terms of the water policies that are implemented, there are close similarities in both regions in terms of the response of hydrology and water resources to a changing climate. These include shifting precipitation patterns, highly variable winter snow packs, and receding glaciers, ultimately resulting in changing seasonality and amounts of runoff that will subtly modify water availability and water use.

As climate change is likely to amplify already observable trends in surface runoff, the question is posed as to whether adaptive capacity in the regions studied is sufficiently robust to respond to a situation which has never been experienced to date. Indeed, because of the presence of snow and ice in the Alps and the Andes, the runoff from the melting cryosphere has up till now largely buffered the negative impacts of hot, dry seasons on water availability. For example, during the 2003 heat wave in Western Europe, rivers such as the Rhine or the Rhone saw large increases in discharge as a result of enhanced glacier melt. However, as long-term global warming will inevitably accelerate glacier melt and shorten the winter snow season, there is a very likely risk of seeing a major change of paradigm by the middle of this century, in particular very low flows from spring to autumn that will be in sharp contrast to the peak flows that occur in today's climate during these very same seasons. Because up till today there is no precedent for the situations projected to occur in coming decades, there has been little thought dedicated to the manner in which water-dependent economic sectors (e.g., hydro-power, agriculture, or tourism) may respond to significant water shortfalls at certain critical times of the year.

It is thus in the context of complex and interlinked environmental and socio-economic issues that Margot Hill has focused her attention. By looking specifically at adaptive capacity and exploring possible avenues for new water governance, she has assessed the robustness of current water policies in the contrasting case-study regions and, whether in a changing environmental context, existing water policies will be sufficient to cope with the large changes in water resources that are expected over the course of the twenty-first century. The thought-provoking narrative, upheld by very clear tables, graphics, and an abundant literature, suggests that technology and changes to infrastructure will not in themselves resolve all future problems that a changing climate will impose upon hydrological resources. Nor will these totally resolve the problems faced by a number of key economic sectors that depend directly or indirectly upon water in the right amounts and at the right times of the year for their revenue. Margot Hill emphasises that there is instead a genuine need for "developing a stronger focus and understanding of institutional adaptation and adaptability".

The innovative ideas outlined in this volume come at a timely moment for national and supra-national authorities, in particular the European Commission which is monitoring the Water Framework Directive and will need to progressively adapt its texts to incorporate the changes that are now becoming apparent. The contents of the book will certainly provide some essential guidance for the decision-making process that will need to be initiated fairly rapidly if we are to avoid disruptions to many key economic sectors where water is an essential element for their business, and the potential and sterile rivalries between sectors that will inevitably arise if no forward-planning is envisaged.

Professor and Head of the Institute of Environmental Sciences Martin Beniston
University of Geneva, Switzerland

I met Margot Hill in 2010 at World Water Week in Stockholm, as I scrambled for a seat in a seminar on climate adaptation and water governance. She was presenting a comparison of the institutional, ecological, hydrological, and legal challenges of two snowpack-mediated regions in Chile and Switzerland. Her talk was as exciting and thoughtful as it was sobering about the shifting landscape that we all face as a result of accelerating climate change. When she proceeded from the podium to the chair next to me, much animated discussion followed. She convincingly articulated that climate change adaptation was not a "science" or "policy" problem but an institutional issue, exposing weaknesses in our governance and operating rules. She remains in a small, if growing, coterie of insightful observers and this volume distils much of her experience from Chile and Switzerland.

When we met, I had just returned from the Tibetan plateau, where traditional herders reported what Margot here refers to as "transformations" of their grasslands, soil, and wetlands and rivers. Over the span of about a decade, the plateau grasslands were becoming something unrecognisable to families that had lived there for millennia. As an ecologist, the rate of change was occurring on a scale that I had never seen outside of regions of intense industrial development such as Eastern Europe or coastal China. Similar rates of climate-induced ecological change are occurring elsewhere–the Andes, the cloud forests of Central America, the Himalayas and their flanks, many coral-rich marine zones, and of course the latitude and boreal zones. Given such dramatic ecological shifts, the social, political, and cultural systems of the plateau were stressed beyond the experience of many generations.

Most of us are headed to the same unfamiliar place. Hydrology is destiny on some level, and the water cycle has proven to be exquisitely sensitive to climate. Moreover, current impacts are not simply shifts in the frequency or severity of extreme events. These impacts are essentially geological-scale leaps, occurring in less than a single human lifetime. They are largely unidirectional and irreversible, and they are hard to predict with confidence.

What Margot's talk confirmed for me was that one of the most crucial components for how well we deal with transformative ecological change is to take back resource management decisions from the kingdom of the engineers and economists. We need to understand that while individuals (and often technical specialists) make most of the *direct* decisions about managing water resources, these individuals also reflect broader intra- and inter-institutional arrangements. Individuals are the faces of governance, but they are also expressions of larger forces. And by extension, resilience comes from adjusting the operating rules for whole governance systems to promote many of the qualities enumerated here.

Can we cope with unknown and hard to predict climate conditions? I have a great deal of faith in humans from our long evolutionary and ecological history, but that history also provides many concerning examples. What I take away now from Margot's insights in this volume is that our future security will emerge from our ability to realize that resilience is a shared, governed quality that reflects learning, memory, imagination, and creative anticipation. She is right in particular to focus our attention on the centrality of both water and institutional, regulatory, and legal frameworks to our social and ecological well-being. And she redefines this landscape of decision making in a useful, exciting manner.

We can expect transformation. Can we prepare by engineering flexibility?

Director, Freshwater Climate Change
Conservation International

John H. Matthews

Preface

Despite gridlock in the supra-national climate governance regime and continuing uncertainty in climate modelling outputs, regional climate impacts are being observed with quickening pace from the Alps to the Andes. The stresses on linked social and ecological systems (SES) from shifting precipitation patterns, glacial retreat and associated changes in run-off regimes are exacerbating a number of underlying governance and management challenges that suggest present water governance regimes may not be robust or resilient enough to cope. While SESs have long adapted to climate influences, the speed and magnitude of change in future climatic and hydrological conditions pose serious challenges, and are increasingly recognised as potentially lying beyond human experience and the coping ranges of social and natural systems.

This book is for all those interested in the growing theoretical and management challenges surrounding climate change adaptation, adaptive capacity and resilience in the governance of linked social-ecological systems. This book looks beyond the technology, modelling, engineering and infrastructure so often associated with water resources management and climate change adaptation, to the decision making environment within which these water and adaptation decisions are made.

Climate change will not only impact on the function and operation of existing water infrastructure, but also the institutions (government agencies, ministries, river basin authorities and user group associations) that manage valuable water resources and water courses. The focus on governance looks to the broader sets of rules, norms and policy frameworks, within which institutions operate. Not only will institutions and water governance frameworks need to respond and shape adaptation responses (through the legislation, operations, policies, decisions) but they will also need to become more adaptable to better manage increasing uncertainty and change as climate change impacts become increasingly prevalent.

In order to achieve this, it is vital to go beyond the technical and hard infrastructural solutions for climate change adaptation that have so far been the corner stone of climate change adaptation. It is vital to better understand the adaptive processes that allow the regimes that govern water resources to respond to new shocks and changes in the hydrological system, in order to build more resilient water governance systems that can bend, but not break, in the face of new and unexpected challenges. This increasing focus on adaptation has signalled a shift to focus on the need for more flexible and adaptive processes in water governance regimes, to manage uncertainty. Over the past decade, the concept of adaptive capacity, its identification and characterisation, has received increasing attention, but primarily through work relating to other related fields, such as adaptive governance and adaptive management approaches.

Despite the increasing amount of attention more recently paid to adaptive capacity and adaptive processes, the understanding of how adaptive capacity to respond to climate change may be developed within water governance regimes is still in its relative nascence. Moreover, even with the advances in the conceptualisation of adaptive capacity, there still are considerable gaps in understanding the role of different governance regimes in building adaptive capacity and challenges in mobilising proactive and reactive capacity at different scales as well as the mechanisms that allow transformation to more sustainable water resources management. To date there still has been relatively little empirical verification of indicators of adaptive capacity at local and regional levels, as well as across different scales.

This book aims to contribute to the conceptualisation and operationalisation of adaptive capacity, as well as proffering new case studies to the empirical body of evidence on adaptation and adaptive capacity. It attempts to bridge the conceptual gap by contributing a more nuanced conceptualisation and operationalisation of adaptive capacity, through better understanding how the governance context and mechanisms within those frameworks contribute to an enabling environment for adaptive capacity. It also seeks to better understand the challenges in generating adaptive capacity across temporal and spatial scales by drawing heavily on resilience based approaches.

Evidence in this book highlights the challenge of balancing out proactive and reactive responses, as well as responses to multiple forms of stress at different magnitudes of physical change and scales of governance to ensure that responses to one kind of risk do not undermine the capacity to address others. Recently, there has been a growing recognition of the challenges in ensuring that short term adaptation actions do not undermine long term social-ecological resilience, by limiting the adaptive capacity to cope with shocks at different magnitudes of change.

Adaptation and long term adaptability are not therefore one and the same thing, and this needs to be better understood in the process of developing adaptation and broader environmental policy, plans and projects that address the impacts of climate change. The framework developed in this book is therefore intended to improve the assessment of different forms of adaptation outcome in the context of transformation to more adaptive water governance frameworks for coping with climate change impacts. Closer attention is now needed to better identify and understand the nature of the trade-offs between adaptation policies, plans and adaptability across multi-scale contexts.

The two case studies presented in this book come from the highly contrasting cases of Chile and Switzerland, namely the Rhône Basin in the Canton Valais,

Switzerland, and the Aconcagua Basin in Valparaiso, Chile. Despite their many differences, both regions do represent mountain watersheds, nivo-glacial regimes, in which observed impacts of climate change on glacial melt and elevation of the snow line have been documented.

Conclusions drawn from these two geographies do encompass broader implications for other regions. Both countries have repercussions outside their national boundaries for broader water, economic and political issues. To date, most academic and practitioner studies on Chile have focussed either on issues concerning the water market (for which there is broad international interest, in terms of reports by the World Bank and the Global Water Partnership) or physical impacts of climate change. This book bridges those questions and looks at the implications of climate change for the broader governance context, and the adaptability of that context to the impacts of climate change.

Understanding the adaptability of the Chilean case is particularly relevant in the broader context of Latin American. The style of water governance in Chile has long been held as a potential model by international institutions such as the World Bank for other Latin American countries seeking to reform their own water governance frameworks. Closer inspection of the Chilean water governance context in relation to its adaptive capacity to climate change is warranted not only for water managers and policy makers in the country itself, but also for many of the international experts who often cite Chile as one potential model of water governance for other countries (often, but not limited to Latin America).

The case of Chile also has important repercussions for global economic issues, considering its important role as an exporter of water intensive/polluting commodities to the global marketplace (copper, avocado, table fruit, vegetables, and wine). Chile can also potentially serve as a "canary in the coal mine", for a context that is more advanced in terms of global change impacts and closer to tipping points (reduced glacier melt contribution etc.) in the physical system. On the other hand, the case of Switzerland, as the water tower of Europe, has high relevance for the neighbouring European countries that its headwaters eventually flow into. The adaptability of the governance context and the impacts of climate change in the headwaters of the Alps are of high interest and relevance to those countries further downstream.

Acknowledgements

The contents of this volume represent the outcome of several years of research, over the course of which a number of people were instrumental both professionally and personally. Professor Martin Beniston from the Research Group in Climate Change and Climate Impacts at the University of Geneva and Andrew Allan from UNESCO Water, Science and Policy Centre at the University of Dundee were both vital presences in their unerring support, encouragement and guidance throughout the course of the research. A number of people both at the Institute of Environmental Sciences in Geneva as well as in other organisations provided insightful feedback and enlightening conversations. I am therefore extremely grateful for the support and advice of Dave Huitema, Markus Stoffel, Geraldine Pflieger, Alexandre Babak Hedjazi, Roderick Lawrence and Bich Le. Furthermore, I retain fond memories of vital debate and building valuable friendships with Beatrice Mosello, Eva Lieberherr and Nate Engle.

I would also like to thank all those who were interviewed for the research or gave their time for my questions and enquiries. While I cannot name all those who agreed to take time out of their day to be interviewed by me, or answer questions via email, your knowledge, expertise and contribution are greatly valued, and I can only hope that this book and additional research outputs provide some recompense for your own time and efforts. Furthermore, the field work in Chile was supported by the ACQWA Mobility Fund, which significantly helped me to make the most of the limited amount of time in the Chilean region. I would also like to thank the hospitality and support of Veronica Torres and Michael Hantke-Domas in Santiago. In addition, I would like to thank Professor Alejandro Leon, from the University of Chile, my research assistant, Manuela Bressi and my translation assistants Ana Maria Daza and Isabelle Pereira. I am also incredibly grateful to Lars Ribbe and his group at the Institute for Technology and Resource Management at the University of Köln, who were so kind as to include me in their field trip to the Aconcagua that was so vital in providing introductions to the case area and its actors.

Finally, I would like to thank my family for their encouragement and support from near and far. I thank my husband Stuart, for boundless support, reassurance and constant interest in my work, which cannot have always been as easy as he made it look.

Contents

Abbreviations

AIWM	Adaptive and integrated water resources management
AM	Adaptive management
ADB	African Development Bank
AVDE	Association Valaisanne des Distributeurs d'Eau
BAFU	Bundesamt für Umwelt (Federal Office for the Environment)
BFE	Bundesamt für Energie (Federal Office for Energy)
CCIAV	Climate Change Impacts, Adaptation and Vulnerability literature
CERISE	Cellule scientifique de crise (Scientific cell for crises)
CNE	Comisión Nacional del Energía (National Commission for Energy)
CNR	Comisión Nacional del Riego (National Commission for Irrigation)
CONAMA	Comisión Nacional del Medio Ambiente (National Commission for Environment – now MMA)
COP15	15th Conference of the Parties
CODELCO	Corporación Nacional del Cobre de Chile (National Copper Corporation of Chile)
COREPIL	Commission Régional de Pilotage
CPR	Common Property Resource Regime
CRA	Confederacion de Regantes de Aconcagua
EOS	Energie Ouest Suisse
EU	European Union
FP7 ACQWA	Framework Project 7 'Assessing Climate change impacts on water Quantity and Water quality in vulnerable mountain regions
DETEC	Department of the Environment, Transport, Energy and Communications
DFID	Department for International Development (UK)
DFSB	Dienstelle für Strass- und Flussbau (Administration for Road and River Building)
DGA	Direcion General de Aguas (General Directorate of Water)
DOH	Direcion de Obras Hidrologicas (Directorate of Hydraulic Works)

DWL – Valais	Dienstelle für Wald und Landschaft (Office for Forests and Landscape)
EAWAG	Eidgenössische Anstalt für Wasserversorgung, Abwasserreinigung und Gewässerschutz (Federal Institution for Water Provision, Sanitation and Protection)
EIA	Environmental Impact Assessment
ENSO	El Nino Southern Oscillation
EPFL	Ecole Polytechnique Fédéral Lausanne
EWZ	Elektrizitätswerk Zermatt (Electricity Utility Zermatt)
FOEN	Federal Office for the Environment
GCM	Global Climate Model
GWP	Global Water Partnership
GWP-TEC	Global Water Partnership – Technical Committee
HEID	Institut de Hautes Etudes Internationales et du Développement (Graduate Institute of International and Development Studies)
IISD	International Institute of Sustainable Development
IPCC	Intergovernmental Panel on Climate Change
ICWE	International Conference on Water and the Environment
IWRM	Integrated Water Resources Management
IUCN	International Union for the Conservation of Nature
JdV	Junta de Vigilancia
KEV	Kostendeckende Einspeisevergütung (Cost covering Compensation)
MIDEPLAN	Ministerio de Planificación (Ministry of Planning)
MINERVE	Modélisation des Intempéries de Nature Extrême, des Retenues Valaisannes et de leurs Effets (Modelling of Extreme Events, Valais Resevoirs and their Effects)
MLG	Multi-level governance
MMA	Ministerio del Medio Ambiente (Ministry of Environment)
MOP	Ministerio de Obras Publicas (Ministry of Public Works)
NeWater	New Approaches to Adaptive Water Management under Uncertainty
NGO	Non-governmental organisation
OcCC	Organe Consultatif sur les Changements Climatiques (Consultative Body on Climate Change)
OECD	Organisation of Economic Co-operation and Development
PDO	Pacific Decadal Oscillation
SAEFL	Swiss Agency for Environment, Forests and Landscape
SEA	Strategic Environmental Assessment
SES	Social Ecological System
SIB	Services Industriels de Bagnes (Industrial Services of Bagnes)
SRES	Special Report on Emission Scenarios
SSIGE	Société Suisse de l'Industrie du Gaz et d'Eau (Swiss Society of Gas and Water Industry)
TRC	Third Rhône Correction

UN	United Nations
UNDP	United Nations Development Programme
UNECE	United Nation Economic Commission for Europe
UN-ECLAC	United Nation Economic Commission for Latin America and Caribbean
UNFCCC	United Nations Framework Convention on Climate Change
UNESCO	United Nations Educational, Scientific and Cultural Organisation
UVEK	Eidgenössischen Departement für Umwelt, Verkehr, Energie und Kommunikation (Federal Department for Environment, Transport, Energy and Communication)
WA21	Wasser Agenda 21
WB	World Bank
WCC-3	World Climate Change Conference in Geneva
WEF	World Economic Forum
WWC	World Water Council
WWF	World Wildlife Fund

Part I
Addressing Water Governance Challenges in the Anthropocene

Chapter 1
Addressing Water Governance Challenges in the Anthropocene

Abstract Water governance, negotiation between actors and institutions for the effective implementation of acceptable water allocation and regulation, faces a plethora of challenges over the coming decades. The challenges arising from population growth, development, climate variability as well as climate change impacts. Concurrently, a crisis of governance has been recognised as one of the major issues facing global water resources over the past decades. The duality of essential role water governance plays in responding to these challenges and the recognised limitations and failures of governance regimes to adequately manage legacy issues predicates the value of closer investigation of both water governance challenges and solutions in the context of climate change and uncertainty. This chapter provides an introduction to the developments in both the challenges to and solutions from water governance over the past few decades.

Keyword Water governance challenges • Climate change uncertainty • Hydro-climatic pressures • Water governance solutions • Adaptive and integrative water management

1.1 Climate Change and Uncertainty: The Great Acceleration

The crisis of governance in the challenges facing global water resources is now well recognised (Gleick 2009; UNESCO 2006; WEF 2009). Governance reflects the negotiation between society and government for effectively implementing socially acceptable allocation and regulation by mediating behaviour through values, social norms and laws (Rogers and Hall 2003). Water governance therefore encompasses the laws, regulations, property rights, institutions, policies and actions, which manage and negotiate water resources as well as networks of influence, such as international market forces, the private sector and civil society (UNDP 1997). Population growth, development, and diminishing water supply from current climate variability

M. Hill, *Climate Change and Water Governance: Adaptive Capacity in Chile and Switzerland*, Advances in Global Change Research 54,
DOI 10.1007/978-94-007-5796-7_1, © Springer Science+Business Media Dordrecht 2013

are already stressing the availability of high-quality water resources. Water governance is essential to managing variability in water supply and delivery (due to seasonality and local variability), in part through the construction and management of regulating infrastructure, but also through the rules (permits, ownership rights, laws, regulations) that administer valuable water resources.

Even if greenhouse gas emissions cease tomorrow, the inertia of the climate system is committed to a likely increase in global temperatures of at least 2°C by the end of the century (IPCC 2007). The associated shifts in climatological patterns will require us all, but water managers in particular, to adapt in a timely and effective manner. The physical and environmental changes pose significant challenges to water infrastructure and management systems, despite the fact that water stakeholders have long dealt with changes and stresses relating to climate variability. The projected speed and magnitude of anthropogenic climate change is set to exacerbate underlying variation and stresses, rendering future situations less manageable (IISD 2006) unless our current institutional arrangements can become adaptive to the realities of future environmental situations.

The release of the fourth assessment report by the Intergovernmental Panel on Climate Change (2007) could have been seen as a tipping point for an increasing awareness of the linkage between climate change and related resource management issues, including water management. Significant progress was made, yet the subsequent years have seen a number of setbacks to significant traction being made by the scientific community on a number of resource related issues. Climate and water cannot be separated as independent issues, especially as water is the primary medium through which climate impacts will be experienced, through changes in local hydrological patterns (Parry et al. 2007). The significance of the water, energy, food nexus is so fundamental to economic development globally, that the intensification of hydrological cycle will impact on both rich and poor, whether through too much water, or too little. Moreover, mountainous areas, commonly considered 'Water Towers' of the world are at the forefront of these warming patterns (Häberli and Beniston 1998). Climate impacts on glacier retreat, precipitation patterns (seasonality and snow line) and associated changes in run off regimes are already observed in Alpine and Andean regions, and model projections suggest a continuation if not heightening of current trends (Viviroli et al. 2011).

In 2002, a Nature paper (Crutzen 2002) suggested that the advent of a new geological period was upon us, one defined by the fact that human actions were playing a dominant role in shaping biospheric processes. This period was called the 'anthropocene', and has fundamentally challenged our perception of human interaction with bio-physical processes. Humans can no longer view themselves as an observer of bio-physical or bio-chemical processes, but instead have become a major contributor and actor in them. This has significant consequences for how human actors should view their part in the 'management' of bio-spherical process and natural resources. Moreover, it prescribes a shift in how actors evaluate and design the management processes to cope in a less stable climatological period, and the increasing need to be aware of the planetary boundaries that we are rapidly approaching (Rockström et al. 2009). The Nature article on planetary boundaries suggested that the regulatory capacities of the earth maintained a safe operating space of natural environmental change within which humanity could thrive and develop (Rockström et al. 2009). It goes on to define a set of interlinked biophysical thresholds, or planetary boundaries, which if crossed, could lead to irreversible and abrupt environmental change with disastrous consequences for human development. These planetary boundaries are: climate change; rate of biodiversity loss; interference with the nitrogen and phosphorus cycles; stratospheric ozone depletion; ocean acidification; global freshwater use; change in land use; chemical pollution; and atmospheric aerosol loading.

The 15th Conference of the Parties meeting (COP15) in Copenhagen was seen as a major disappointment for the global change science research community on many fronts. The water community was one of many that came out of Copenhagen severely disenchanted, since all references to water were dropped entirely from the final text on adaptation, which represented a widening of the gap between the climate and water contingents when many had hoped a connection would be further fused.[1] COP15 showed that many were still not making the link between the climate and water agendas, or even the wider environmental issues at stake. It also raises the issue that many governance regimes focus on separate aspects of the social or ecological systems (e.g. climate, or forests, freshwater fisheries, marine fisheries, or even less coherently across sector specific legislation or different institutional combinations at ministerial level). However, there is an increasing focus from the global change community on the need for human society and the governance systems that moderate our actions and decisions to operate within multiple inter-connected earth systems. Since the climate negotiations centred purely on the climate system, those involved in carving out the climate regime fell short in recognising the need for human society to operate within the other earth systems (Rockström et al. 2009).

The link between tipping points in these planetary boundaries has been reflected in theories of environmental resource management and governance, as well as in the water disciplines, but has not yet been widely adopted by those outside of the research and scientific community (Rockstrom et al. 2009). The retreat of mountain glaciers is one of the indications that certain sub systems of the earth are moving out of their relatively stable Holocene state, and into the anthropocene (Crutzen 2002; Rockström et al. 2009). Global freshwater consumption has moved from a pre-industrial value of 415 km^3 per year to 2,600 km^3 per year, which while it may fall under its proposed planetary boundary, is tightly coupled with other boundaries in the system. Our ability to stay within the climate boundary may depend on stopping the transgression of the freshwater boundary and vice versa, since all of them are conceived as 'bio-physical preconditions for human development…and well-being' (Rockström et al. 2009, p 474).

[1] Co-operative Programme on Water and Climate (CPWC); Netherlands Commission for Environmental Assessment (MER); Institute for Environmental Studies (IVM); Netherlands Environmental Assessment Agency (PBL).

Additionally, it should be noted that uncertainty does not stem only from the increasing risks and hazards for a potentially warmer world, but also from the very nature of the knowledge system used to map out climate impacts. Despite significant advances in climate change science and modelling techniques, the uncertainty associated with such projections (rather than predictions) at either global or regional levels is likely to continue for the foreseeable future (Carter et al. 2007). Yet, decisions about how to adapt the governance and management of complex water resource systems to climate change impacts cannot just wait until climate model projections are more precise.[2] While models can project a range of futures or alternative scenarios of change, the complex nature of the bio-spherical processes that drive water hydrological patterns means that in the conceivable future short and long term management decisions about future water quality, security and availability will still be subject to a large range of uncertainty in both projected and unanticipated changes.

Social systems have tended to have rules or tools to cope with normal ranges of uncertainties, or moderate deviations from the norm (what Mathews et al. (2011) term 'predictable certainty'), such as wet years followed by dry years on an inter-annual or decadal timescale (Smit and Wandel 2006; Yohe and Tol 2002). For example, from a governance perspective, prioritisation rules may kick in when indicators suggest a dry year is underway. From a management perspective, reservoir storage could tie over water provision during dry years, or flood management strategies such as dykes and early warning systems might protect against high precipitation events (Herrfahrdt-Pähle 2010; Huntjens et al. 2010; Smit and Wandel 2006). However, climate change embodies a more unpredictable and indeterminate form of uncertainty (Matthews et al. 2011) or irreversible changes in state (reduced run off contribution from glacier and snow melt, shifts in seasonality, increasingly consecutive dry years) that may lie outside or beyond the boundaries of past and present coping ranges of water management and governance regimes[3] (Smit and Wandel 2006; Yohe and Tol 2002).

Climate change is therefore seen as exacerbating these broader challenges affecting water governance, acting as an overarching pressure that causes these underlying stresses on water institutions to become even more pronounced as impacts intensify (Lettenmaier et al. 2008). Since climate change is a systemic threat that will have significant interactions with other drivers of change (as discussed above), it will require fundamental shifts in how water governance regimes operate, and how they interact and coordinate across local, regional, national, and trans-boundary scales. More specifically, increasing uncertainty of future conditions, or 'non stationarity'

[2] Also refer to http://www.newater.info/index.php?pid=1045

[3] *Adaptive capacity has been analyzed in various ways, including via thresholds and "coping ranges", defined by the conditions that a system can deal with, accommodate, adapt to, and recover from (de Loe and Kreutzwiser 2000; Jones 2001; Smit et al. 2000; Smit and Pilifosova 2001, 2003). Most communities and sectors can cope with (or adapt to) normal climatic conditions and moderate deviations from the norm, but exposures involving extreme events that may lie outside the coping range, or may exceed the adaptive capacity of the community.* (Smit and Wandel 2006, p 287).

(Kiang et al. 2011; Milly et al. 2008) and possible bifurcations ("thresholds") in the climate system implies that water governance cannot approach the future based on the assumption that it will replicate the relatively stable conditions of the past. The resulting implication is that a shift is required in how we plan and manage water resources, which respects non stationary conditions and embraces (rather than seeks to remove) increased levels of uncertainty, transforming how water governance relates to ecosystems and communities over climate-relevant timescales.

Climate change impacts on hydrological resources and patterns will affect water governance and management primarily through alterations in the timing of hydrological patterns (seasonality), quantity of water resources (floods and droughts) and quality (suitability for consumption or use) (Matthews and Le Quesne 2009; Cook et al. 2011). Impacts include alterations in seasonality, a rise in the frequency or intensity of extreme hydrological events (increased drought and flood recurrence and duration), higher variability of precipitation patterns, increased hurricane intensity, changing trends in snow pack, and generally accelerating rates of glacier melt leading to changes in run-off (first increasing then decreasing) (IPCC 2007). These changes imply both a shift in the alteration (shifts in timing and averages) and intensification (increasing number and severity of extreme events) of the hydrological cycle. Changing seasonality, water temperatures and alterations in precipitation patterns affect water quality, in terms of dissolved oxygen levels, concentration of pollutants, as well as levels of toxic algae and sedimentation impacting aquatic species (Matthews and Le Quesne 2009) and infrastructure such as dams.

Therefore, governance processes that were designed in a context of 'stationarity' may not be equipped to address accelerated changes to the hydrological cycle and more unpredictable uncertainties in relation to future climate. Water rights, regulatory and policy contexts that do not take into account the ecological requirements for maintaining healthy, productive and protective waterways threaten to undermine the resilience of the socio-ecological system, at a time when it is needed most (i.e. as climate impacts mount). Likewise rights, plans, policies and regulation that do not acknowledge inherent uncertainties by allowing for revision if the bio-physical parameters, upon which they are based, change, are likely to become increasingly ineffective in managing the rivalries and negative impacts arising from climate change. Legislation and rules set now or in the past may impact decisions on investment and management paths for the next 10, 20 or 30 years, over which time these impacts will intensify. Simply scaling up past solutions to environmental challenges to tackle climate related issues may not be adequate to manage future challenges, because rules may not have taken unpredictable uncertainty into account, or solutions have been focussed primarily on enabling technical 'hard' adaptations that do not address the social reality in which they must be implemented, or because the timelines for re-assessment and the integration of new knowledge do not match increasing speeds of change.

However, water governance, and the institutions it effects, do not just experience climate change, but play a crucial role in developing an enabling environment for successful adaptation (Tompkins and Adger 2004), to anticipate and respond to a changing climate. Governance regimes define the context within which adaptation takes place (Adger et al. 2005), requiring the institutions these regimes define to be simultaneously both climate adaptive and yet able to drive sustainable adaptation efforts. To respond to this dual challenge, the water resources and research community have in recent years focussed more heavily on better understanding adaptive governance processes for sustainable water resources management.

The recognition of an anthropocene requires the water research community to focus more heavily on strategies that would effectively manage water resources in the context of a new epoch. Thus it signals the need to shift attention from assessing and shaping responses in order to avoid over-exploitation of resources to also include dealing with uncertainty under changing climatic conditions. Therefore, when investigating water resource issues, it is vital to recognise and take into account the complex inter-connected and multi-functional role that water resources serve for healthy ecosystems, societies and economies, and thus the ability for humankind to stay within the bio-physical preconditions that are necessary for our own development and well-being (Rockstrom et al. 2009).

1.2 Shifting Lens: Sustainability to Adaptability

In his seminal book 'On the Origin of Species', Darwin famously noted that "It is not the strongest of the species that survives, nor the most intelligent that survives. It is the one that is the most adaptable to change" (Darwin 1859). This observation perfectly elucidates how humans have always had to adapt to change, including climatic and meteorological variation. So what is different now? Why do we worry so much about society's ability to adapt to future variation in the twenty first century? The answer to this can be found by looking at the speed of current climatic change, and the complex geo-political-environmental context within which it is and will take place. Current rates of change and the increasing global, rather than local, drivers of concatenating shocks (Biggs et al. 2011) have meant that a more concerted effort must be placed on creating an enabling environment for adaptive capacity to accelerating rates of change in today's more complex and interconnected world.

Discussions around resource based institutions have held prominent place since Hardin argued in his seminal paper 'The Tragedy of the Commons' (Hardin 1968), that resource users in shared resource extraction and use systems, are inevitably locked into the trap of destroying the resource on which they depend. In the preceding 40 years, much of the debate around institutional arrangements for resource management has been pinned on whether or not this 'tragedy of the commons' prophesy is universally true, or if enough examples can be found to counter argue the proposition (Ostrom et al. 1999), identifying favourable institutional processes that resolved these shared resource problems. While Hardin proposed polarised solutions of either socialism or privatisation of free enterprise, Ostrom continues to chart a number of alternative methods of restricting access and creating incentives that resolve over-exploitation issues related to shared resources that are open to public consumption (e.g., fisheries catchment quotas, local forest management practices, and water allocation agreements to name a few).

Ostrom herself has noted that accelerating rates of change are a major challenge in establishing sustainable institutions to manage such shared resources that are open to public consumption (Ostrom et al. 1999). While this in part points to the historic exclusion of ecological requirements in the governance system leading to negative environmental impacts it also shows that the convergence of human induced global change processes, such as climate change, with diverse governance challenges (i.e. lack of clarity around existing water use rights and over-exploitation), is pushing institutions and humanity in general, past those environmental thresholds beyond which it becomes increasingly difficult to apply previous practices to future problems (Kane and Yohe 2000). This calls for a new lens through which to assess the appropriateness of governance frameworks in a rapidly changing environment of increasingly indeterminate risks. It also calls for suitably robust criteria to be established with which to shape fitting responses.

In response to these increasing stresses on global hydrological resources, increasing attention has been paid to the failure of governance in the water sector in the preceding two decades. Investigations of different governance regimes and outcomes have sought to pinpoint elements in a system which may produce more effective results in creating 'good governance' (Rieu-Clarke et al. 2008). Normatively the concept appeals to the democratic advantages of broadening the participation base and the durability of solutions which evolve through negotiation and cooperation by a greater number of stakeholders. The frameworks which have arisen out of these studies and research programmes have primarily centred on goal-specific approaches such as integrated water resources management (IWRM), as inspired by the Dublin Principles (Solanes and Gonzalez-Villareal 1999). While the focus on good governance and IWRM has provided a vital goal on which water managers could frame solutions (UNECE 2009), a better understanding is needed of how relevant these frameworks are in relation to the challenges induced by climate change.

Scholars and practitioners have therefore become increasingly critical of traditional command and control approaches for their rigidity and impracticable goal of decreasing uncertainty (Johnson 1999). Instead, approaches that focus on governance and management that is adaptive as well as integrative have been posited as being more suitable to managing uncertainty (Engle et al. 2011). This has led in recent years to a number of the water resources and research community to focus more heavily on better understanding adaptive processes, either in relation to how systems have coped with past variability as well as shocks outside past and present coping ranges (Engle 2010; Herrfahrdt-Pähle 2010; Huntjens et al. 2011; Pahl-Wostl 2007; Pahl-Wostl and Sendzimir 2005). In the past decade, there have been many more studies from the governance, adaptation and resilience discourses that have sought to improve the baseline understanding of adaptation and adaptive capacity in water governance regimes. Case evidence has been used to suggest an increasingly converging set of criteria required to foster adaptive processes (Dovers and Hezri 2010). Within the context of river basins, it has been noted that more attention needs to be devoted to understanding and managing the transition to more adaptive regimes that 'take into account environmental, technological, economic, institutional and cultural characteristics of the basin' (Pahl-Wostl et al. 2007, p 49).

Flexibility in governance systems is one key criterion in building adaptive capacity to react to the unanticipated conditions that may result from climate impacts (Hurlbert 2009). Empirical studies have also suggested that designs that focus on participatory, collaborative, and learning-based approaches can increase adaptive capacity and support the sustainability of water systems (Folke et al. 2005; Kallis et al. 2006; Pahl-Wostl et al. 2007; Tompkins and Adger 2004). Other studies have identified the important role that leadership plays in championing innovative approaches and strategies for adapting to climate change (Engle 2010) as well as steering social systems through transformative processes (Olsson et al. 2004). Better understanding how to identify and assess these governance mechanisms that foster adaptive capacity is an integral part of transitioning to more sustainable water governance regimes.

1.3 Converging Threats

Chile and Switzerland both face an interesting set of converging challenges. Both countries have OECD status and possess high levels of the classic determinants of adaptive capacity. Their citizens enjoy democratically elected legitimate governments, strong economies (even through current economic woes of the financial crash) and educated populations, despite recent events in Chile that have elucidated the disproportionate levels of education between economic elite and lower socio-economic levels. However, both case areas within the countries face multiple challenges driven by climate, economic, socio-political and ecological factors. In Chile, the neoliberal model implemented by the Pinochet regime validates strong deregulation, privatisation and market liberalisation in the interests of improving economic efficiency. While the particular market model pursued has been seen to be effective so far for the development of supply and sanitation[4] and export based economic growth, its limitations concerning effective protection of ecosystems, climate change and upstream-downstream rivalries have gradually been recognised (Vergara-Blanco 2004). The water rights market and Water Code do not take into account the diverse nature of the different sectoral stakeholders, yet assumes agriculture, mining, energy and industry could all compete for the same resource on equal terms. Carl Bauer has discussed at length the social and environmental consequences of the Chilean water model, and presented a detailed description of the major political challenges in reforming the 1981 Water Code to take better account of environmental and social externalities (Bauer 1997, 1998, 2004).

In March 2010, President Piñera took over the presidency from the Bachelet government, heralding the first *Alianza* government after 20 years of the *Concertacion* coalition. For the first time since the Pinochet regime fell, the right wing neo-liberal

[4] However, a recent UN-ECLAC study (Lentini 2011) has presented evidence that shows rising costs for domestic consumers due to increasing water losses because utilities have allowed infrastructure to deplete.

coalition is in power, with potential consequences for the development and direction of water governance. Field work in Chile took place about 7 months after the change of government, with many of the civil servants from the previous government recently out of their positions. The strong political influences of the neo-liberal dogma for water resources management in Chile, defines not just the new government, but also left its mark on policies followed during the period of the left wing Concertacion. For example, in the northern areas of Chile, in response to growing stresses from mining use and population growth, the previous government had attempted to pressure, unsuccessfully, the Superintendencia into forcing the regional utility to move to desalination.

As central and northern areas become drier, this policy could potentially imply a transference of the costs of industrial over-consumption onto the domestic customer, as the cost of moving to a desalination system would have increased water prices three- or four-fold. Similar levels of worry persist with regards to the hydropower sector, where concerns exist that Italian owned ENEL control 80% of non-consumptive water rights in Chile and 96% of non-consumptive rights in the Aysen area, which is the most water rich in Chile and one of the richest in the world (Patagonia 2011). The challenges in the case area Aconcagua Basin are presently not as highly contentious as northern or Patagonian areas of Chile, but increasingly recurring drought, changes in glacier and snow coverage, mounting pressures from mining and expanding agricultural coverage are all exerting mounting pressure on water resources.

In Switzerland, a very different set of drivers frame the challenges, particularly within the Alpine context of the Canton Valais. Traditional socio-economic structures in the alpine zone have undergone large upheavals over the last 50 years (Hill et al. 2010), with consequent challenges for resource management. Not only have alpine farmers played an important role in the governance of water through common property resource regimes, but they have been crucial in the development and maintenance of water infrastructure in the upper watersheds of the Rhône. As there are fewer full time and part time farmers, these traditional structures have suffered, with consequences for water management in crucial periods.

These transitions have also brought new rivalries for water resources. The convergent expanse in tourism in the major Valais ski resorts, with increasing requirements of water for artificial snow production as snow coverage becomes less predictable intensifies existing rivalries on the tributaries to the Rhone in the Valais. The on-going inclusion of environmental flows as a new 'user' of water resources adds further tension to water governance across multiple sectors. Despite the image of Switzerland, and in particular the Valais, as the Water Tower of Europe, the abundance of water resources are highly spatially dependant, and periodic rivalries exist not just during peak winter periods, but are increasing in the later periods of summer (e.g. La Reche; Saviése; Conthey).

Both regions represent mountain watershed nivo-glacial regimes, where climate change (as experienced through glacier melt and snow pack changes) will correspond with changes in the seasonality of river flows. In both areas impacts of climate change have already been observed on glacial melt and elevation of the snow line with associated impacts on the timing and amount of run off (Häberli and Beniston 1998; Pellicciotti et al. 2007) projected to increase (Christensen et al. 2007). As mountainous areas, climate change impacts will be keenly felt in both cases, mainly through alterations in seasonality (Viviroli et al. 2011). However, shifts in seasonality and decreases in glacier melt take on particular significance in the Andean region where dependence on glacier and snow melt run off is high for water availability during the dry summer months (Pellicciotti et al. 2007; Souvignet et al. 2008).

Global climate models show that warming and drying trends have already been observed and can be projected to intensify for the Andean region (Christensen et al. 2007). Temperature increases in the Alps have exceeded 1–1.5 °C since 1900 (about three times the global-average temperature rise), with corresponding implications for increased glacial melt and changes in snow pack (OcCC 2008; Solomon et al. 2007). Furthermore, in combination with the strong El Nino Southern Oscillation (ENSO) event currently occurring, the central-northern regions of Chile have been experiencing one of the worst drought periods in memory (DGA 2010). The convergence of climate change impacts with the complex political and economic issues poses significant challenges across the two case areas that will need to be navigated through effective water governance frameworks.

1.4 Summary

Effective adaptation and building adaptive capacity should therefore be seen as crucial to the sustainable management of water resources in the Anthropocene. Governance is recognised as being an issue at the heart of water resource challenges, and therefore strengthening adaptive capacity through governance frameworks is essential for responding effectively to future climatic uncertainty and stress (Folke et al. 2005; UNECE 2009) and shifting to means of managing freshwater in a way that incorporates climate change associated changes in timing, quantity and quality. Moreover, higher uncertainties and the increasingly indeterminate nature of water risks (e.g. years of drought followed by extreme flooding) from climate change challenge the fixed rules and regulations that define many water institutions, and may lie beyond current planning practices (Matthews and Le Quesne 2009).

As attention has shifted to better understanding adaptive processes, a set of assumptions and panaceas (single solution applied to wide range of problems) have arisen in the literature that address how to foster governance arrangements that are more adaptive, integrative and flexible. However, despite an upsurge in research into governance and adaptation and the water sector over the past decade, a lack of comparative analyses of the application of these approaches in river basins persists (Huntjens et al. 2011). Furthermore, there remain considerable gaps in the empirical exploration and understanding of the complex dynamics that effect the stimulation and mobilisation of adaptive capacity at different scales as well as the role of different governance regimes in building adaptive capacity.

This calls for a better understanding of how governance systems adapt to climatic stimuli. Other studies have shown that investigating how these systems have adapted (or not) to recent past stresses from extreme events may allow us to draw lessons about adaptive capacity to future climate change (Adger et al. 2007; Engle 2011; IISD 2006); allowing managers to learn from what has already been done, successfully or unsuccessfully, to inform their decisions about what should be done. The research presented in this book aims to contribute to the conceptualisation and operationalisation of adaptive capacity in order to help bridge these conceptual gaps. In so doing, it hopes to contribute a more nuanced conceptualisation and operationalisation of adaptive capacity, through better understanding how the governance context and mechanisms within those frameworks contribute to an enabling environment for adaptive capacity. It also seeks to better understand the challenges in generating adaptive capacity across temporal and spatial scales and in so doing, generate a framing of adaptive capacity that better serves policy and decision makers.

References

Adger WN, Arnell NW, Tompkins E (2005) Successful adaptation to climate change across scales. Glob Environ Chang 15(2):77–86

Adger WN, Agrawala S, Mirza MMQ, Conde C, O'Brien K, Pulhin J, Pulwarty R, Smit B, Takahashi K (2007) Assessment of adaptation practices, options, constraints and capacity. In: Parry ML, Canziani OF, Palutikof JP, van der Linden PJ, Hanson CE (eds) Climate change 2007: impacts, adaptation and vulnerability. Contribution of working group II to the fourth assessment report of the Intergovernmental Panel on Climate Change. Cambridge University Press, Cambridge

Bauer CJ (1997) Bringing global markets down to earth: the political economy of water rights in Chile, 1976–95. World Dev 25(5):639–656

Bauer CJ (1998) Slippery property rights: multiple water uses and the neoliberal model in Chile, 1981–1995. Nat Resour J 38(1):109–155

Bauer C (2004) Siren Song: Chilean water law as a model for international reform. Resources for the Future, Washington, DC

Biggs D, Biggs R, Vasilis D, Sholes RJ, Schoon M (2011) Are we entering an era of concatenated global crises? Ecol Soc 16(2):27. [online] URL: http://www.ecologyandsociety.org/vol16/iss22/art27/

Carter TR, Jones RN, Lu X, Bhadwal S, Conde C, Mearns LO, O'Neill BC, Rounsevell MDA, Zurek MB (2007) New assessment methods and the characterisation of future conditions. In: Parry ML, Canziani OF, Palutikof JP, van der Linden PJ, Hanson CE (eds) Climate change 2007: impacts, adaptation and vulnerability. Contribution of working group II to the fourth assessment report of the Intergovernmental Panel on Climate Change. Cambridge University Press, Cambridge

Christensen JH, Hewitson B, Busuioc A, Chen A, Gao X, Held I, Jones R, Kolli RK, Kwon RT, Laprise R, Magaña V, Mearns CG, Menendez CG, Raisanen J, Rinde A, Sarr A, Whetton P (2007) Regional climate projections. In: Climate change 2007: the physical science basis. Contribution of working group I to the fourth assessment report of the Intergovernmental Panel on Climate Change. Cambridge University Press, New York

Cook J, Hill M, Freeman S, Levine E (2011) Shifting course: climate adaptation for water management institutions. WWF US, Washington, DC

Crutzen PJ (2002) Geology of mankind. Nature 415:23
Darwin C (1859) On the origin of species, London: John Murray
DGA (2010) MOP declara zona de escasez hídrica en regiones de Coquimbo y Valparaiso
Dovers SR, Hezri AA (2010) Institutions and policy processes: the means to the ends of adaptation. Wiley Interdiscip Rev Clim Change 1(2):212–231
Engle NL (2010) Adaptation to extreme droughts in Arizona, Georgia, and South Carolina: evaluating adaptive capacity and innovative planning and management approaches for states and their community water systems. University of Michigan, Ann Arbor
Engle NL (2011) Adaptive capacity and its assessment. Glob Environ Chang 21(2):647–656. doi:10.1016/j.gloenvcha.2011.01.019
Engle NL, Johns OR, Lemos MC, Nelson DR (2011) Integrated and adaptive management of water resources: tensions, legacies, and the next best thing. Ecol Soc 16(1):19. [online] URL: http://www.ecologyandsociety.org/vol16/iss11/art19/
Folke C, Hahn T, Olsson P, Norberg J (2005) Adaptive governance of social-ecological systems. Ann Rev Environ Resour 30(1):441–473. doi:10.1146/annurev.energy.30.050504.144511
Gleick PH (2009) Launch of the world's water 2008–2009: the biennial report on freshwater resources. Paper presented at the Woodrow Wilson International Centre for Scholars, Washington, DC, 2 Apr 2009
Häberli W, Beniston M (1998) Climate change and its impacts on glaciers and permafrost in the Alps. Ambio 27(4):258–265
Hardin G (1968) The tragedy of the commons. Science 162:1243–1248
Herrfahrdt-Pähle E (2010) The transformation towards adaptive water governance regimes in the context of climate change. Universität Osnabrück, Osnabrück
Hill M, Wallner A, Furtado J (2010) Reducing vulnerability to climate change in the Swiss Alps: a study of adaptive planning. J Clim Policy 10(1):70–86
Huntjens P, Pahl-Wostl C, Grin J (2010) Climate change adaptation in European river basins. Reg Environ Change 10:263–284
Huntjens P, Pahl-Wostl C, Rihoux B, Schlüter M, Flachner Z, Neto S, Koskova R, Dickens C, Nabide Kiti I (2011) Adaptive water management and policy learning in a changing climate: a formal comparative analysis of eight water management regimes in Europe, Africa and Asia. Environ Policy Gov 21:145–163
Hurlbert M (2009) The adaptation of water law to climate change. Int J Clim Change Strat Manage 1(3):230 240
IISD (2006) Designing policies in a world of uncertainty, change and Surprise: adaptive policy-making for agriculture and water resources in the face of climate change. International Development Research Centre & The Energy and Resources Institute, Winnipeg/Manitoba/New Delhi
IPCC (2007) Climate change 2007: synthesis report. Contribution of working groups I, II and III to the fourth assessment report of the Intergovernmental Panel on Climate Change. Intergovernmental Panel on Climate Change, Geneva
Johnson BJ (1999) The role of adaptive management as an operational approach for resource management agencies. Conserv Ecol 3(2):8. [online] URL: http://www.ecologyandsociety.org/vol3/iss2/art8
Kallis G, Videira N, Antunes P, Pereira G, Spash CL, Coccossis H, Quintana SC, del Moral L, Hatzilacou D, Lobo G, Mexa A, Paneque P, Mateos BP, Santos R (2006) Participatory methods for water resources planning. Environ Plann C Govern Policy 24(2):215–234
Kane SM, Yohe GW (2000) Societal adaptation to climate variability and change: an introduction. Clim Chang 45(1):1–4
Kiang JE, Olsen JR, Waskom RM (2011) Featured collection on "Nonstationarity, Hydrologic Frequency Analysis, and Water Management". J Am Water Resour Assoc 47(3):433–570. doi:10.1111/j.1752-1688.2011.00551.x
Lentini E (2011) *Servicios de agua potable y saneamiento: lecciones de experiencias relevantes*. United Nations Economic Commission for Latin America and the Carribean, Santiago de Chile.
Lettenmaier D, Major D, Poff L, Running S (2008) Water resources. In: The effects of climate change on agriculture, land resources, water resources, and biodiversity in the United States.

A report by the U.S. Climate Change science program and the subcommittee on Global Change Research. Washington, DC

Matthews J, Le Quesne T (2009) Adapting water management: a primer on coping with climate change. WWF-UK, Godalming

Matthews JH, Wickel BAJ, Freeman S (2011) Converging currents in climate-relevant conservation: water, infrastructure, and institutions. PLoS Biol 9(9):e1001159. doi:10.1371/journal.pbio.1001159

Milly PCD, Betancourt J, Falkenmark M, Hirsch RM, Kundzewicz ZW, Lettenmaier DP, Stouffer RJ (2008) Stationarity is dead: whither water management? Science 319(5863):573–574. doi:10.1126/science.1151915

OcCC (2008) Das Klima ändert – was nun? Der neue UN-Klimabericht (IPCC 2007) und die wichtigsten Ergebnisse aus Sicht der Schweiz. Organe consultatif sur les changements climatiques, Bern

Olsson P, Folke C, Berkes F, Hahn T (2004) Social-ecological transformation for ecosystem management: the development of adaptive co-management of a wetland landscape in southern Sweden. Ecol Soc 9(4):2. doi:Artn 2. [online] URL: http://www.ecologyandsociety.org/vol9/iss4/art2/

Ostrom E, Burger J, Field CB, Norgaard RB, Policansky D (1999) Revisiting the commons: local lessons, global challenges. Science 284(9 April):278–282

Pahl-Wostl C (2007) Transitions towards adaptive management of water facing climate and global change. Water Resour Manage 21(1):49–62. doi:10.1007/s11269-006-9040-4

Pahl-Wostl C, Sendzimir J (2005) The relationship between IWRM and adaptive water management. NeWater working paper 3

Pahl-Wostl C, Kabat P, Möltgen J (eds) (2007) Adaptive and integrated water management: coping with complexity and uncertainty. Springer, Berlin/New York

Parry ML, Canziani OF, Palutikof JP (2007) Technical summary. Climate change 2007: impacts, adaptation and vulnerability. contribution of working group II to the fourth assessment report of the intergovernmental panel on climate change. In: Parry ML, Canziani OF, Palutikof JP, van der Linden PJ, Hanson CE (eds) Fourth assessment report of the Intergovernmental Panel on Climate Change. Cambridge University Press, Cambridge, pp 23–78

Patagonia (2011) Patagonia Chilena: Sin Represas: El Problema – Monopolio y Concentración. http://www.patagoniasinrepresas.cl/final/contenido.php?seccion=problema. Accessed 15 Sept 2011

Pellicciotti F, Burlando P, Van Vliet K (2007) Recent trends in precipitation and streamflow in the Aconcagua River basin, central Chile Glacier Mass Balance Changes and Meltwater Discharge. Selected papers from sessions at the IAHS Assembly in Foz do Iguaçu, Brazil, 2005, IAHS Publication 318

Rieu-Clarke A, Allan A, Magsig BO (2008) Assessing governance in the context of IWRM. Striver policy brief: strategy and methodology for improved IWRM – an integrated interdisciplinary assessment in four twinning river basins issue no. PB6.2. Available online: http://www.dundee.ac.uk/water/projects/striver/

Rockström J, Steffen W, Noone K, Persson A, Chapin FS, Lambin EF, Lenton TM, Scheffer M, Folke C, Schellnhuber HJ, Nykvist B, de Wit CA, Hughes T, van der Leeuw S, Rodhe H, Sorlin S, Snyder PK, Costanza R, Svedin U, Falkenmark M, Karlberg L, Corell RW, Fabry VJ, Hansen J, Walker B, Liverman D, Richardson K, Crutzen P, Foley JA (2009) A safe operating space for humanity. *Nature* 461 (7263): 472–475. doi: 10.1038/461472a

Rogers P, Hall AW (2003) Effective water governance. GWP TAC background papers no. 7. Global Water Partnership Technical Committee, http://www.gwpforum.org/servlet/PSP?iNodeID=215&itemId=197

Smit B, Wandel J (2006) Adaptation, adaptive capacity and vulnerability. Glob Environ Chang 16(3):282–292

Solanes M, Gonzalez-Villareal F (1999) The Dublin principles for water as reflected in a comparitive assessment of institutional and legal arrangement for integrated water resources management. Global Water Partnership: Technical Advisory Committee (TAC), Available online at: http://info.worldbank.org/etools/docs/library/80638/IWRM4%5FTEC03%2DDublinPrinciples%2DSolanes%26Gonzales.pdf. Accessed 12 Sept 2008

Solomon S, Qin D, Manning M, Chen Z, Marquis M, Averyt KB, Tignor M, Miller HL (2007) Summary for policymakers. In: Climate change 2007: the physical science basis. Contribution of working group I to the fourth assessment report of the Intergovernmental Panel on Climate Change. Cambridge University Press, Cambridge/New York

Souvignet M, Gaese H, Ribbe L, Kretschmer N, Oyarzún R (2008) Climate change impacts on water availability in the Arid Elqui Valley, North Central Chile: a preliminary assessment. Paper presented at the IWRA World Water Congress, Montpellier, France

Tompkins EL, Adger WN (2004) Does adaptive management of natural resources enhance resilience to climate change? Ecol Soc 9(2):10. doi:Artn 10. [online] URL: http://www.ecologyandsociety.org/vol19/iss12/art10

UNDP (1997) Governance for sustainable human development. United Nations Development Programme. http://mirror.undp.org/magnet/policy/

UNECE (2009) Guidance on water and adaptation to climate change. United Nations Economic Commission for Europe, Geneva

UNESCO (2006) Water: a shared responsibility – The United Nations World Water Development report 2. UNESCO, Paris

Vergara-Blanco A (2004) Discrecionalidad administrativa y nuevas limitaciones a los derechos de agua, vol 276. Centro de Estudios Publicos, Santiago

Viviroli D, Archer DR, Buytaert W, Fowler HJ, Greenwood GB, Hamlet AF, Huang Y, Koboltschnig G, Litaor MI, Lopez-Moreno JI, Lorentz S, Schadler B, Schreier H, Schwaiger K, Vuille M, Woods R (2011) Climate change and mountain water resources: overview and recommendations for research, management and policy. Hydrol Earth Syst Sci 15(2):471–504. doi:10.5194/hess-15-471-2011

WEF (2009) The politics of water. Paper presented at the World Economic Forum annual meeting: shaping the post-crisis World Davos, Switzerland 28 Jan–1 Feb 2009

Yohe G, Tol RSJ (2002) Indicators for social and economic coping capacity – moving toward a working definition of adaptive capacity. Glob Environ Chang 12(1):25–40

Chapter 2
A Starting Point: Understanding Governance, Good Governance and Water Governance

Abstract Governance has been a widely and deeply discussed concept in the political sciences. As global freshwater resources have become increasingly degraded and impacts of climate change begin to take hold on local hydrological systems, scholars and practitioners have increasingly recognised a crisis of governance. This chapter presents a broad overview of governance theories and discusses the shifts from state centric notions of 'government' to a wider range of governance modes and types, as a way of contextualising the shift from a 'command and control' paradigm in water governance to more decentralised, integrated and flexible approaches.

Keywords Developments in water governance • Institutional arrangements • Integrated water resources management • Scales of governance • Physical and human boundaries

2.1 Understanding Governance

Though widely debated, governance is generally a more inclusive concept than 'government', reflecting the negotiation between society and government in effectively implementing socially acceptable allocation and regulation by mediating behaviour through values, norms and laws (Jordan 2008; Treib et al. 2007; Mayntz 2004). Governance is broader than just the government, incorporating both state and non-state actors, both private and public. According to the UNDP, governance has been defined as 'the exercise of economic, political and administrative authority to manage a country's affairs at all levels. It comprises the mechanisms, processes and institutions through which citizens and groups articulate their interests, exercise their legal rights, meet their obligations and mediate their differences' (UNDP 1997).

This evolution from top down, centralised and hierarchical concepts of government to governance, represents a shift to a new form of governing society that is more inclusive and cooperative than the traditional ordered rule of government and concept

M. Hill, *Climate Change and Water Governance: Adaptive Capacity in Chile and Switzerland*, Advances in Global Change Research 54,
DOI 10.1007/978-94-007-5796-7_2, © Springer Science+Business Media Dordrecht 2013

of political steering; moving from placing state governments and political authorities at the centre of action to control socio-political processes to meet socio-economic goals (Mayntz 2006). While government refers to the autonomous authority of the state regime, governance relates to the network of private and public actors and structures, which interact to solve societal issues (Grote and Gbikpi 2002). Non-governmental actors are no longer seen as passive 'citizens' but as active 'stakeholders' (Grote and Gbikpi 2002), through their participation in public-private networks and interactions.

Governance thus allows us to conceptualise the complex arrangement of relationships and rules needed to manage and distribute resources in today's world, where traditional federal and top down structures of command and control may no longer suffice. However, since forms of governance still take place within the jurisdiction of nation states, scholars have acknowledged that higher jurisdictions, i.e. constitutionally superior states, are likely to not only steer networks but also unilaterally change the rules of the game (Rhodes 2007). Therefore, the role of government in the setting and application of legislation and regulation remains key. Academics have endeavoured to bring clarity to the conceptualisation of governance through the classification of different forms of governance, through the dimensions of politics, polity or policy (Treib et al. 2007). Politics represents the process of how (collective) actors translate different preferences into policy choices and different interests into unified action. Policy denotes the political steering and decisions made for and implemented in a society. Polity is the framework of formal and informal rules of the game (i.e. institutions) that direct the behaviour of actors within a society (Keman 2006; Héritier 2002). The institutionalist approach is linked to the polity mode, conceiving governance as a system of rules that shape actors' actions (Ostrom 2005).

Institutions and governance are interlinked and often synonymous concepts from a definitional perspective. For example, in the field of new institutionalism, North (1990) has described institutions as the rules that govern the behaviour of actors. In the same field, institutions are seen as including the governance structure and organisation, demoting the institutional arrangement (Saleth and Dinar 2004 in Herrfahrdt-Pähle 2010b). Ostrom (2007) defines institutions as laws, regulations, policies and property rights that define ownership, disposition and use rights to a natural resource, as well as the policies for protection and exploitation of a resource. Institutions can therefore be rules, or sets of rules (i.e. arrangements), that structure social interaction by shaping or constraining actor behaviour (Helmke and Levitsky 2004; North 1990). In a narrower sense, however, institutions are often synonymous with formal bodies and organisations (e.g., national ministries, sub-national agencies, multi-stakeholder management institutions, and planning departments; and the policies, plans, and other actions carried out by those organizations).

Institutions are also categorised as having formal or informal forms, differentiating the 'nature of processes of development, codification, communication and enforcement' (Pahl-Wostl 2009, p 356). Formal institutions tend to have their rules enforced by a state actor and are openly codified and officially accepted (e.g. legally binding documentation: regulation, constitutions, resource ministries, formal basin management organizations). Informal institutions convey norms of behaviour and socially shared rules that may be self-enforcing or enforced outside official channels, i.e. the unwritten rules of the game (e.g. traditions, social and cultural norms, organisational codes of behaviour, personal networks, community saving groups, black market) (Helmke and Levitsky 2004). Informal institutions can be as influential and shape behaviour as effectively as formal institutions, and thus often play an important a role in natural resource management at the local level (Berkes and Folke 2001; Helmke and Levitsky 2004; North 1990).

Traditions and customs encompassing what is right and wrong, or acceptable from a risk perspective (e.g. value structures) can shape formal institutional outcomes and support or undermine the trust in and effectiveness of governance outcomes. For example, Helmke and Levitsky (2004) cites an example from Chile's executive-legislative power-sharing mechanisms, for how an informal institution can create incentives for behaviour that alter the substantive effects of formal rules without directly defying them, thus reconciling actors' interest with existing formal institutions circumventing the process of formal reform. They present how the 'Leaders of the Democratic Concertación inherited an "exaggeratedly strong presidential system" and a majoritarian electoral system that ran counter to their goal of maintaining a broad multiparty coalition. Lacking the legislative strength to amend the 1980 Constitution, Concertación elites created informal mechanisms of inter-party and executive-legislative consultation aimed at counteracting its effects. These power-sharing arrangements "enhanced coalitional trust" in a formal constitutional setting that otherwise "provided very few incentives for cooperation." (Siavelis 1997 in Helmke and Levitsky 2004).

Both formal and informal institutions play an important role in water resources management in their potential to set rules and demarcate responsibilities between actors; co-ordinate mechanisms to minimize jurisdictional overlaps or deficiencies; bridge the gap between political and natural boundaries; match responsibilities, and serve as authorities and facilitators of action (GWP 2004, p 44). Young (2002) and Herrfahrdt-Pähle (2010a) elucidate the importance of ensuring a close fit (across and between different spatial, institutional or jurisdictional scales) between social and ecological systems through the existent institutional interface. Effective and fitting institutional arrangements (including compatibility between formal and informal institutions) to demarcate and coordinate rights and responsibilities are critical in areas or periods of abundance, but become even more crucial during periods of change (more extreme drought or flooding), scarcity, or generally more arid climates (Herrfahrdt-Pähle 2010a; Meinzen-Dick 2007; Ostrom 1990; Helmke and Levitsky 2004; Young 2002).

At this point, some theoretical clarity is called for, not only to elucidate how these terms will be used for the purposes of this book, but also to clarify the difference between the terms water governance and management, which are often used interchangeably. For the purposes of this piece of research, institutions are defined according to their broad and formal definition, i.e. more than just an organisation, but officially enforced or recognised (i.e. not informal social norms). Governance relates to the different processes of making and setting rules and institutions that takes into account the different actors and networks that negotiate acceptable positions in balancing trade-offs in policy and its instruments (Pahl-Wostl 2009). As the UNDP (1997) states, 'Governance is seen as encompassing institutions, as well as the broader laws, regulations, policies and actions with which natural resources are managed, as well as the networks of influence beyond just government, such as civil society, private sector actors, and non-governmental organisations'. Management on the other hand, is concerned with the application of these rules and operationalisation of the policy visions with the practical aspects of water allocation, protection, prevention of harm from extremes (Folke et al. 2005; Pahl-Wostl 2009).

The negotiation of different roles (state and non-state; formal and informal) in policy formulation and implementation in governance has been elucidated through the discourse on different types of governance (Rhodes 2007). The classification of different governance modes also has been defined by distinguishing between bureaucratic hierarchies, networks and markets (Thompson et al. 1991). The concentration of these modes in different national settings tends to be influenced by the political regime within the country depending on the diverse 'economic, cultural and political norms of a country and the behaviours or the legislature and legislators' (Rogers and Hall 2003, p 8), namely the informal institutional setting. Hierarchical governance refers to the traditional model of top down political system with highly centralised government and institutions. In many developed countries, this form of governance has been supplanted by the growing implementation of the concept of subsidiarity (the performance of functions at the lowest appropriate level).

The network mode is dominated by informal institutional arrangements and the participation of state and non-state actors and together with market based governance has received increasing attention over the past few decades for its flexibility and ability to provide access to new forms of knowledge (Kooiman 2003), but is vulnerable to challenges of accountability and legitimacy[1] if membership is not representative. Market based modes tend to be dominated by non-state actors across formal and informal institutions, and became a trend with increasing voracity from the 1970s in attempts to resolve issues previously assigned to traditional centralised command and control regulation, such as economic growth, social inequity and environmental pollution (Meinzen-Dick 2007; Freeman and Kolstad 2007). The concept of market led governance transfers resource allocation mechanisms to the private sector and the market, seen as more efficient than hierarchical forms of regulation. The laissez faire market model, prominently supported by Milton Freedman and the Chicago School, enjoyed considerable attention, particularly in the context of the neo-liberal governance approach taken in Chile during the Pinochet regime (Bauer 1997; Klein 2008; Valdes 1995). However, there is growing recognition that pure market modes are quite rare and that market based mechanisms require effective regulation to ensure social and environmental needs are met (Bakker 2003; Freeman and Kolstad 2007).

[1] For a definition of legitimacy see later discussion of good governance in Sect. 2.2.

As the notion of authority has scattered from the central state – to networks to market based – attention has turned increasingly to the multi-level and distributed forms of governance, initially through studies comparing federal and centralised systems (see Ammom et al. 1996 in Pahl-Wostl 2009) then more prominently through the study of the complex multi-level interactions in the European Union (EU) by Gary Marks (Hooghe and Marks 2003). Likewise, polycentric governance systems have long been discussed in the social sciences, but have recently been increasingly focussed on in relation to complex adaptive systems (Pahl-Wostl 2009), which shall be discussed in more detail later. Polycentric governance is determined to be 'a system of many centres of decision making which are formally independent of each other' (Ostrom et al. 1961; Huitema et al. 2009) and thus, like multi-level governance (MLG) 'implies the decision making authority is distributed in a nested hierarchy and does not reside at one single level', be that a central government, regional governments or municipalities, or indeed individuals or markets (Pahl-Wostl 2009, p 357).

Normative assumptions have been made about the linkages between different forms, modes and types of governance and their legitimacy, as well as their ability to adapt to a changing environment (Pahl-Wostl 2007). However, the discourse is seen to be gradually moving from ascribing one panacea as superior to another, rather to looking at issues of fit, interaction and compatibility (Meinzen-Dick 2007; Young 2002; Freeman and Kolstad 2007). In reality, however, neat conceptual constructs tend to be replaced by hybrid forms and thus many academics and organisations alike tend to encapsulate all three modes within their definitions of governance (Pahl-Wostl 2009; UNDP 1997). Distributed governance is one concept that has arisen to more effectively encompass the combination of formal and informal institutions (Kooiman 2000), representing a more dynamic relationship between different societal forces. It arises out of the recognition that neither the state nor the market can resolve social and environmental problems alone. This interpretation of governance is less 'Statist', more society orientated, and is primarily concerned with the manner in which governance systems provide a balance of power between different formal and informal state/society interactions, as well as the role of civil society and policy networks.

2.2 Good Governance

In the 1980s the concept of good governance was taken up from a more normative perspective, with the development of criteria of normatively 'good governance' (Pierre 2000; WB 2002). These criteria sought to guide the repair of the failures of the decreasingly legitimate top down governance structures, by focussing on alternative modes of actor constellations helping to resolve common issues from different perspectives. By the 1990s it was becoming used from a more analytical perspective in the social sciences as a mean of assessing public policy arrangements in empirical research (Kooiman 1993). The concept of good governance has become popular over recent decades, in response to the notion that 'more effective governance regimes or systems need to be designed/created to overcome government failure, market failure and system failure or a combination of these' (Rogers and Hall 2003, p 24).

Legitimacy is seen as a core concept of good governance (as an output), dependent on a number of inputs that are represented by a number of key components of governance identified as determinants of good governance. These indicators include participation, leadership, accountability and trustworthiness, effectiveness and transparency (see following section for more detail). The UNDP Regional Project on Local Governance in Latin America elucidates governance legitimacy is the proper functioning of institutions and their acceptance by the public, which is in part enabled by the efficacy of government and the achievement of consensus by democratic means as well as the ability of political, social and economic rules to solve conflicts between actors and adopt decisions. This highlights the role of actor networks and institutions in participative and effective governance processes for achieving good or legitimate governance, on which the following section shall expand.

A number of studies and institutes have defined diverse key components central to achieving good governance, which tend to encompass a range of normative values and public policy objectives which are seen as socially desirable (e.g. accountability, transparency, participation, justice, efficiency, rule of law and absence of corruption' UNDP 1997). The World Bank considered four key components as being central to achieving good governance: public sector management, accountability, legal framework for development, and transparency and information. In addition, the Asian Development Bank (ADB) also identifies four elements of key importance to governance; predictability, participation, transparency and accountability (Allan 2008). Generally, 'good' governance tends to relate to a 'regulatory system that shows qualities of accountability, transparency, legitimacy, public participation, justice, efficiency, the rule of law and absence of corruption' (Brugnach et al. 2008, p 423). These indicators tend to encompass a range of normative values and public policy objectives which are seen as socially desirable.

There are, however, practical challenges for these conceptual constructs of good governance inputs. For example, a central question that relates to participation is whether the involvement of non-governmental organisations and sub-national actors and authorities empirically leads to easier management of diverse interests, increased compliance with formal rules, or just further complicates the decision making process and decreases efficiency through resource intensive participative processes (Pahl-Wostl 2009). Rogers and Hall (2003) note that the proliferation of non-accountable NGOs that have often filled the governance vacuum from weak local governments results that often organisations calling for action have little responsibility for the actions which they propose. Furthermore, country specific contexts, in terms of contrasting political, cultural and geographical particularities, inevitably mean that a sweeping definition of good governance cannot be applied to all situations; rather different interpretations of what good governance means, depending on the national context, are available and governance reforms should be framed within the context of local conditions and existing public policy practices in order to be successful.

2.3 Water Governance: The Rise of New Standards

> Water governance refers to the range of political, social, economic and administrative systems that are in place to develop and manage water resources, and the delivery of water services, at different levels of society. (GWP 2000)

At its simplest, water governance systems not only decide who gets how much water, when and how but also protect resources from pollution, through the implementation of socially acceptable allocation and regulation of water resources and services (Rogers and Hall 2003). Water governance develops and sets the rules, roles and responsibilities of all involved stakeholders (local and national government, private sector, civil society) regarding ownership, administration and management of water resources (Rogers and Hall 2003). Hurlbert et al. (2008) describes this as an institutional process that defines the organisation and management of the interrelationships between society and water resources. It is a process that is constituted by many different stakeholders, each with their own interests, decision making processes and instruments including legal rulings, norms and acts. Since the water sector does not exist in isolation, but is intricately connected to broader political, economic and social developments, water governance is influenced by issues relating to the current governing regime and the wider concerns of civil society that may help or hinder the development of water governance arrangements (Rogers and Hall 2003). The external impacts of political power and competing priorities often define the relationships between different organisations and stakeholders (Hurlbert 2008).

Property and use rights are a central element in water governance due to the potential role different forms of ownership can play in the internalisation of externalities,[2] the realisation of efficiencies and the added security for long term investment (Thobani 1995; Demsetz 1967). Rights can be land-based or riparian, or use-based (including market-based or based on historical use) and tend to categorised into different forms of ownership (communal, private, state, open access) (Demsetz 1967). The link between water law and property rights is important, partly through the role of formal or informal institutions (state, courts or community) in enforcing, monitoring and protecting the relevant property or use rights. Furthermore, clear and suitable definition of water rights and appropriate accompanying legislative and regulatory frameworks are seen to assist in the reduction of negative hydrological effects on third parties when water is transferred to other economic activities (Thobani 1995). There are however, a number of outstanding questions and different versions of what the appropriate definition, valuation and measurement of water

---

[2] Demsetz 1967, p 348: *Externality is an ambiguous concept. For the purposes of this paper, the concept includes external costs, external benefits, and pecuniary as well as non-pecuniary externalities. No harmful or beneficial effect is external to the world. Some person or persons always suffer or enjoy these effects. That converts a harmful or beneficial effect into an externality is that the cost of bringing the effect to bear on the decisions of one or more of the interacting persons is too high to make it worthwhile, and this is what the term shall mean here. "Internalizing" such effects refers to a process, usually a change in property rights, that enables these effects to bear (in greater degree) on all interacting persons.*

rights (and natural resources in general) may be in the quest for efficiency and effectiveness, as well as the minimisation of social and environmental externalities. This is further complicated in the case of water rights because hydrological realities are not as fixed, regular or constant as land, building and other commodities.

A number of studies have drawn heavily on the empirical examples of water governance and have suggested there were certain levels and combinations of institutional organisation and regulation which allowed effective water governance to develop (Maas and Anderson 1978; Keohane and Ostrom 1995; Netting 1981). These studies signaled that no single model of effective governance could be set, since different systems should fit the social, political, cultural, economic and environmental contexts within which they must operate. However, certain principles and criteria have been considered essential elements in assessing water governance frameworks. Many organisations (World Bank, ADB, GWP, UNDP, and World Water Council Water Action Unit) and researchers have employed criteria established by a variety of actors in the international water industry, which draw on the aforementioned empirical evidence, to guide the assessment and reforms of water institutions globally. In the past decade, a number of different principles of good governance have been proposed 'to assist in the fair, effective and environmentally sensitive management of water' (Brooks 2002, p 4 in Hurlbert 2008).

Much of the discussion about water governance has moved towards debating the administrative and geographical levels at which it should be managed, the weak governance of public or private water utilities, the issue of private sector participation, the context specific nature of water governance (i.e. which laws/modes of governance work in which countries) as well as the importance and means of reducing water demand. In recent years, the intensity of debate in both academic and practitioner's fields has led to internationally agreed standards such as the Dublin Principles being adopted by the water community, as a means to set a common bar for water governance.

2.4 Integrated Water Resources Management

After the dominance of steady state resource management and the 'hydraulic mission' from the 1950s to early 1990s, a new approach of good governance and IWRM became more dominant in the face of the crisis of governance and continuing degradation of global water systems. In recent years, the intensity of debate in both academic and practitioner's fields has led to internationally agreed standards such as the Dublin Principles being adopted by the water community, as a means to set a common bar for water governance. The Dublin Statement on Water and Sustainable Development was adopted at the International Conference on Water and the Environment (ICWE) in Dublin, Ireland, in January 1992. The statement was adopted in response to what was seen as a growing threat posed to sustainable development through the misuse and growing scarcity of freshwater resources. It was then commended to world leaders at the UN Conference on Environment and Development in Rio de Janeiro in June, 1992. The conference statement and associated principles expressed a holistic and multi-disciplinary approach to global water issues that covered environmental, social, political and economic issues (Solanes and Gonzalez-Villareal 1999). In this context, the principles are seen to be nested in the concept of distributed governance (Rogers and Hall 2003, p 14), primarily through its focus on the participative role of civil society and non-governmental organisations.

The guiding principles define the need for concerted action to reverse present trends of overconsumption, pollution and threats from drought and flooding. In doing so the principles defined the need for a holistic approach to water management to effectively take account of the linkages across land and water uses over catchment areas; for a participatory approach that allows decisions to take place at the lowest appropriate level and with public consultation; that women should be granted a central role in water management and the protection of water resources and that this should be better reflected in institutional arrangements; and finally that as water has an economic value it should also be recognised as an economic good, as a means to achieve efficient and equitable use, while still recognising the basic right of all human beings to have access to clean water and sanitation at an affordable price.[3]

The Dublin Principles are one of many influencing sets of guidelines which are relevant to the management paradigm of IWRM, and both have risen to almost universal acceptance in the current environment of increasing pressures on water resources from the nexus of population pressures, consumer patterns, management issues, climate change, biodiversity loss, growing destruction and pollution of aquatic ecosystems and increasing cross-sectoral competition. IWRM is one response to the goal of managing these increasing pressures while balancing the need to protect and conserve water resources and water based ecosystems. The Global Water Partnership defines IWRM as 'a process which promotes the co-ordinated development and management of water, land and related resources in order to maximise the resultant economic and social welfare in an equitable manner without compromising the sustainability of vital ecosystems' (GWP 2000, p 22). In IWRM, greater emphasis is placed on collaborative governance of the multiple values of water, and also seen as one means of increasing the capacity of water management in the face of climate change.

IWRM has taken on significant currency as a means of ensuring 'equitable, economically sound and environmentally sustainable management of water resources and provision of water services' (Rogers and Hall 2003, p 4). As can be seen from this statement, as well as the conceptual criteria of IWRM proposed by the Global Water Partnership-Technical Advisory Committee (GWP-TEC) (GWP 2000), the three overriding criteria IWRM reflect social, economic and environmental conditions, namely, economic efficiency (in order to use increasingly scarce water resources with maximum efficiency), equity (basic right of access to adequate water quantities and quality) and environmental and ecological sustainability (ensuring sustainability of water resources and riparian ecosystems that support it for use by future generations).

[3] Refer to: http://www.gwpforum.org/servlet/PSP?iNodeID=1345

Water managers and scientists have progressively looked to IWRM to help mitigate not only governance failures of the past, but also increasing uncertainty in the future (Huntjens et al. 2011; Lach et al. 2006; Medema et al. 2008). For example the GWP has developed an IWRM toolbox with a range of instruments that may address governance failures (GWP 2000). The GWP prescribes three groups of substantive elements that should support the implementation of these criteria. The complementary components of an effective water resources management system are seen to include core elements of the governance system, including relevant management instruments (operational instruments for allocation, regulation, monitoring and assessment, informational and economic instruments), an enabling environment (general framework of policies, legislation, mechanisms for participation and cooperation) and clear institutional roles of different levels and stakeholders (levels of action, management boundaries and capacity building) (GWP 2000, p 30).

The GWP purports that the general consensus of the water community is that IWRM is the 'only viable way forward for sustainable water use and management' (Rogers and Hall 2003, p 30). Yet there is still considerable debate on how the paradigm is implementable in the governance realities that must apply them and whether the prescriptions of IWRM actually generate successful outcomes in practice (Engle et al. 2011; Ingram 2011; Medema et al. 2008; Meinzen-Dick 2007). Many of the principles for good water governance (framed within IWRM) from organisations such as GWP, WWC are also seen as providing important insights in establishing best practice criteria for developing adaptive policy in the face of climate change (Hurlbert 2009; IISD 2006). However, just as with previous trends, the greater range of conditions identified by researchers as being conducive to creating effective institutions, the more loss of nuance in broader policy application of such principles (Merrey et al. 2007).

References

Allan A (2008) Governance assessment methodology. Centre for Water Law, Policy and Science, University of Dundee, Dundee

Bakker K (2003) Good governance in restructuring water supply: a handbook. Federation of Canadian Municipalities (FCM) and the Program on Water Issues (POWI), Canada

Bauer CJ (1997) Bringing global markets down to earth: the political economy of water rights in Chile, 1976–95. World Dev 25(5):639–656

Berkes FC, Folke C (2001) Back to the future: ecosystem dynamics and local knowledge. In: Gunderson LH, Holling CS (eds) Panarchy: understanding transformations in human and natural systems. Island Press, Washington, DC

Brugnach M, Dewulf A, Pahl-Wostl C, Taillieu T (2008) Toward a relational concept of uncertainty: about knowing too little, knowing too differently, and accepting not to know. Ecol Soc 13(2): [online] URL: http://www.ecologyandsociety.org/vol13/iss32/art30/

Demsetz H (1967) Toward a theory of property rights. Am Econ Rev Pap & Proc 79th Ann Meet Am Econ Assoc 57(2):347–359

Engle NL, Johns OR, Lemos MC, Nelson DR (2011) Integrated and adaptive management of water resources: tensions, legacies, and the next best thing. Ecol Soc 16(1):19. [online] URL: http://www.ecologyandsociety.org/vol16/iss11/art19/

Folke C, Hahn T, Olsson P, Norberg J (2005) Adaptive governance of social-ecological systems. Annu Rev Environ Resour 30(1):441–473. doi:10.1146/annurev.energy.30.050504.144511

Freeman J, Kolstad CD (2007) Moving to markets in environmental regulation: lessons from twenty years of experience. Oxford University Press, New York

Grote JR, Gbikpi B (2002) Participatory governance: political and societal implications. Leske + Budrich, Opladen
GWP (2000) Integrated water resouces management TAC background paper no. 4. GWP. Global Water Partnership – Technical Advisory Committee (GWP-TAC), Stockholm, Sweden
GWP (2004) Integrated water resources management (IWRM) and water efficiency plans by 2005. TAC background papers no. 10. GWP. Global Water Partnership – Technical Advisory Committee (GWP-TAC), Stockholm, Sweden
Helmke G, Levitsky S (2004) Informal institutions and comparitive politics: a research agenda. Perspect Polit 2(4):740–775
Héritier A (2002) New modes of governance in Europe: policy-making without legislating? In: Héritier A (ed) Common goods: reinventing European and international governance. Rowman & Littlefield, Lanham, pp 185–206
Herrfahrdt-Pähle E (2010a) South African water governance between administrative and hydrological boundaries. Clim Dev 2:111–127
Herrfahrdt-Pähle E (2010b) The transformation towards adaptive water governance regimes in the context of climate change. Universität Osnabrück, Osnabrück
Hooghe L, Marks G (2003) Unravelling the central state. But how? Types of multi-level governance. Am Polit Sci Rev 97(2):233–243
Huitema D, Mostert E, Egas W, Moellenkamp S, Pahl-Egas W, Moellenkamp S, Pahl-Wostl C, Yalcin R (2009) Adaptive water governance: assessing the institutional prescriptions of adaptive (co-)management from a governance perspective and defining a research agenda. Ecol Soc 14(1):26. [online] URL: http://www.ecologyandsociety.org/vol14/iss21/art26/
Hurlbert M (2008) An analysis of trends related to the adaptation of water law to the challenge of climate change: experiences from Canada. Paper presented at the Climate 2008, Online, 3–7 Nov 2008
Hurlbert M (2009) The adaptation of water law to climate change. Int J Clim Change Strateg Manage 1(3):230–240
Hurlbert M, Diaz H, Corkal DR (2008) Government, water governance and adaptive capacity in the South saskatchewan river basin. Institutional Adaptation to Climate Change Project, Canada.
Huntjens P, Pahl-Wostl C, Rihoux B, Schlüter M, Flachner Z, Neto S, Koskova R, Dickens C, Nabide Kiti I (2011) Adaptive water management and policy learning in a changing climate: a formal comparative analysis of eight water management regimes in Europe, Africa and Asia. Environ Policy Gov 21:145–163
IISD (2006) Designing policies in a world of uncertainty, change and surprise: adaptive policy-making for agriculture and water resources in the face of climate change. International Developmcnt Research Centre & The Energy and Resources Institute, Winnipeg/Manitoba/New Delhi
Ingram H (2011) Beyond universal remedies for good water governance: a political and contextual approach. In: Garrido A, Ingram H (eds) Water, food and sustainability. Routledge, New York
Jordan A (2008) The governance of sustainable development: taking stock and looking forwards. Environ Plan C: Gov Policy 26:17–33
Keman H (2006) The comparative approach: theory and method. In: Pennings P, Keman H, Kleinnijenhuis J (eds) Doing research in political science: an introduction to comparative methods and statistics, 2nd edn. Sage Publications, London, pp 3–16
Keohane R, Ostrom E (1995) Local commons and global interdependence. Sage, London/Thousand Oaks
Klein N (2008) The shock doctrine: the rise of disaster capitalism. Penguin Press, London
Kooiman J (1993) Modern governance: new government-society interactions. Sage, London/ Newbury Park
Kooiman J (2000) Societal governance: levels, modesl and orders of social-political interaction. In: Pierre J (ed) Debating governance: authority, steering, and democracy. Oxford University Press, Oxford, pp 133–164
Kooiman J (2003) Governing as governance. Sage, London
Lach D, Ingram H, Rayner S (2006) You never miss the water till the well runs dry: crisis and creativity in California. In: Clumsy solutions for a complex world, Verweij M, Thompson M (eds) Palgrave Macmillan, New York, NY. pp 226–240

Maas A, Anderson RL (1978) The desert shall rejoice: conflict, growth and justice in arid environments. MIT Press, Cambridge, MA
Mayntz R (2004) Governance im modernen Staat. In: Benz A (ed) Governance – Regieren in komplexen Regelsystemen. Verlag für Sozialwissenschaften, Wiesbaden, pp 65–76
Mayntz R (2006) Governance Theory als fortentwickelte Steuerungstheorie? In: Schuppert GF (ed) Governance-Forschung: Vergewisserung über Stand und Entwicklungslinien. Nomos Verlag, Baden-Baden, pp 11–20
Medema W, McIntosh BS, Jeffrey PJ (2008) From premise to practice: a critical assessment of integrated water resources management. Ecol Soc 13(2):29. [online] URL: http://www.ecologyandsociety.org/vol13/iss22/art29/
Meinzen-Dick R (2007) Beyond panaceas in water institutions. Proc Natl Acad Sci USA 104(39):15200–15205
Merrey DJ, Meinzen-Dick RS, Mollinga PP, Karar E (2007) Water for food, water for life. In: Molden D (ed) A comprehensive assessment of water management in agriculture. Earthscan, London, pp 193–232
Netting RM (1981) Balancing on an Alp: ecological change and continuity in a Swiss Mountain Community. Cambridge University Press, Cambridge
North DC (1990) Institutions, institutional change, and economic performance. Cambridge University Press, New York
Ostrom E (1990) Governing the commons: the evolution of institutions for collective action. Cambridge University Press, Cambridge
Ostrom E (2005) Understanding institutional diversity. Princeton University Press, Princeton
Ostrom E (2007) A diagnostic approach for going beyond panaceas. Proc Natl Acad Sci USA 104(39):419–422
Ostrom V, Tiebout CH, Warren R (1961) The organization of government in metropolitan areas: a theoretical inquiry. Am Polit Sci Rev 55:831–842
Pahl-Wostl C (2007) Transitions towards adaptive management of water facing climate and global change. Water Resour Manage 21(1):49–62. doi:10.1007/s11269-006-9040-4
Pahl-Wostl C (2009) A conceptual framework for analysing adaptive capacity and multi-level learning processes in resource governance regimes. Global Environ Change Hum Policy Dimens 19(3):354–365. doi:10.1016/j.gloenvcha.2009.06.001
Pierre J (2000) Debating governance. Oxford University Press, Oxford
Rhodes RAW (2007) Understanding governance: ten years on. Org Stud 28:1243–1264
Rogers P, Hall AW (2003) Effective water governance. GWP TAC background papers no. 7. Global Water Partnership Technical Committee, http://www.gwpforum.org/servlet/PSP?iNodeID=215&itemId=197
Solanes M, Gonzalez-Villareal F (1999) The Dublin principles for water as reflected in a comparitive assessment of institutional and legal arrangement for integrated water resources management. Global Water Partnership: Technical Advisory Committee (TAC). http://info.worldbank.org/etools/docs/library/80638/IWRM4%5FTEC03%2DDublinPrinciples%2DSolanes%26Gonzales.pdf. Accessed 9 Dec 2008
Thobani M (1995) Tradable property rights to water: how to improve water use and resolve the water conflicts, Finance & private sector development note no. 34. World Bank, Washington, DC
Thompson G, Frnaces J, Levacic R, Mitchel J (1991) Markets, hierarchies and networks: the coordination of social life. Sage, London
Treib O, Bähr H, Falkner G (2007) Modes of governance: towards a conceptual clarification. J Eur Public Policy 14(1):1–20
UNDP (1997) Governance for sustainable human development. United Nations Development Programme [Online]. http://mirror.undp.org/magnet/policy/
Valdes JG (1995) Pinochet's economists: The Chicago School of Economics in Chile (Historical perspectives on modern economics). Cambridge University Press, Cambridge
WB (2002) Toward more operationally relevant indicators of governance. Prem notes no 49, Dec 2002
Young O (2002) The institutional dimensions of environmental change: fit, interplay, and scale. MIT Press, Cambridge, MA

Chapter 3
Adaptive Capacity, Adaptive Governance and Resilience

Abstract Different challenges arising from increasingly uncertain and unpredictable hydro-climatic conditions have been accompanied by a shifting focus of water governance solutions. More recently, the water resources research community has paid increasingly close attention to climate change adaptation and adaptive processes in relation to water governance, recognizing the need to better understand adaptive processes that seek to embrace, rather than control uncertainty. This chapter presents these issues and introduces the linked concepts of adaptive capacity, adaptive governance and resilience in social ecological systems. It provides a review of how these topics approach the challenges presented in previous chapters, and how scholars have sought to develop these frameworks to better take into account the need to foster and mobilise adaptive capacity within water governance structures.

Keywords Adaptive capacity of water governance • Resilience of social ecological systems • Managing increasing uncertainty • Climate resilient water management • Climate change challenges for water institutions • Adaptation to hydro-climatic changes

3.1 New Approaches for New Challenges: Integrating Uncertainty and Climate Change

The previous chapter discussed the internal and external pressures on governance systems, for which good governance and IWRM prescriptions have sought to provide solutions. External influences include physical climate factors, which pose novel risks on societies, even though many have long demonstrated adaptive practices. Climate change was discussed earlier as the 'great accelerator' of processes of change and 'threat multiplier' (Downing 2009) pushing systems past environmental thresholds, in terms of droughts, glacial retreat and heat waves. Some scholars have argued that the speed of such change is leaving governance systems increasingly

M. Hill, *Climate Change and Water Governance: Adaptive Capacity in Chile and Switzerland*, Advances in Global Change Research 54,
DOI 10.1007/978-94-007-5796-7_3, © Springer Science+Business Media Dordrecht 2013

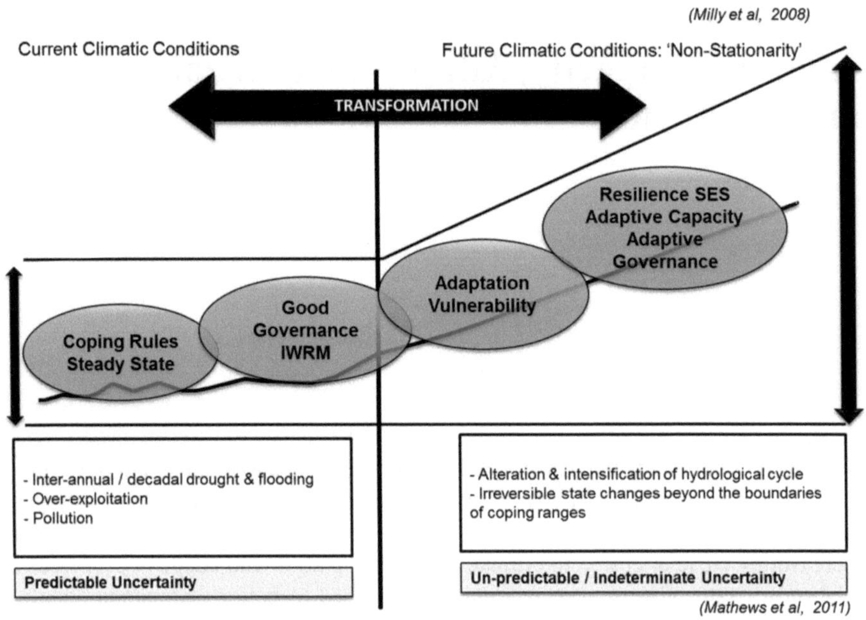

Fig. 3.1 The shifting focus of water governance solutions: The *left side* of the graph represents temperature ranges from the previous 150 years, while the *right hand side* represents future temperature projections according to a range of emission scenarios. The *black arrows* on the *left* and *right* depict the range of uncertainty associated with the two different ranges of climatic conditions, underlining the need for scholars to better understand adaptive processes that seek to embrace, rather than control uncertainty

unable to apply prior lessons to current problems (Ostrom et al. 1999), especially where these risks are outside of the range of both human and ecological frames of reference (e.g. impacts related to droughts, heat waves and glacier loss) (Parry et al. 2007).

Good governance systems are likely to have greater capacity to cope with climatic changes than poor governance systems, but both may struggle to develop and adapt existing coping mechanisms to more intense and frequent droughts or floods. For example, while governance systems may have had mechanisms to cope with issues such as seasonal or decadal drought or flooding, this might not imply the ability to adapt these coping techniques to more frequent or intense droughts or floods in relation to climate change impacts. Climate change can therefore be seen as attenuating threats that exacerbates other economic, developmental and environmental challenges relating to the governance of water resources.

This convergence of issues has therefore changed the questions that need to be asked and the frames of analysis by which we evaluate the ability of water governance arrangements to manage resources efficiently and effectively (Fig. 3.1), while simultaneously respecting the ecological constraints and social realities in which they must operate. While water managers, farmers, dam operators and other water resource stakeholders have long dealt with changes and stresses relating to climate variation, the projected speed and development of anthropogenic climate change exacerbates underlying variation, rendering future situations less manageable and less predictable (IISD 2006). Thus, as climatological and hydrological patterns shift from one set of parameters to a wider range of uncertainty and risks (as shown by the vertical arrows), the theoretical paradigms that inform water management are also shifting.

Water governance and management systems tend to have rules or tools to cope with normal ranges of uncertainties, or moderate deviations from the norm, such as wet years followed by dry years on an inter-annual or decadal timescale (Smit and Wandel 2006; Yohe and Tol 2002). For example, from a governance perspective, prioritisation rules may kick in when indicators suggest a dry year is underway. From a management perspective, reservoir storage could tie over water provision during dry years, or flood management strategies such as dykes and early warning systems might protect against high precipitation events (Herrfahrdt-Pähle 2010b; Huntjens et al. 2011; Smit and Wandel 2006).

However, climate change embodies a more unpredictable uncertainty or irreversible changes in state (reduced run off contribution from glacier and snow melt, shifts in seasonality, increasingly consecutive dry years) that may lie outside the coping ranges, or beyond the boundaries of past and present coping ranges of water management and governance regimes[1] (Smit and Wandel 2006; Yohe and Tol 2002). Stakhiva and Stewart (2010) note that the 'big and intractable difference is the much increased degree of uncertainty when dealing with climate change, uncertainty that requires implementing certain strategies that incorporate more redundancy into connected systems, thereby increasing reliability and robustness' (Stakhiva and Stewart 2010, p 107).

Traditional governance or management approaches have been criticised as being characterised by a command and control paradigm and fragmented regulatory and institutional landscapes (Gleick 2003; Pahl-Wostl 2009) that do not take the complex inter-linkages of the SESs for which they are responsible into account and seek to reduce uncertainty rather than attempt to manage and live with it. For example, ministries and regulation often are siloed along sectoral lines, while rulemaking on water resources does not take into account the needs, challenges and reality at the local level. Command and control is not limited to the governance of the social system, but also of the bio-physical system, where management approaches have favoured the control of the hydrological cycle (i.e. dam construction, dyke enforcement) in order to reduce natural threats and produce more predictable outcomes (Jewitt 2002). This form of approach is seen as reducing the natural range of variation, impacting riparian ecosystems and their services, and thus reducing the resilience of the system (Jewitt 2002).

[1] *Adaptive capacity has been analyzed in various ways, including via thresholds and "'coping ranges'", defined by the conditions that a system can deal with, accommodate, adapt to, and recover from (de Loe and Kreutzwiser, 2000; Jones, 2001; Smit et al. 2000; Smit and Pilifosova, 2001, 2003). Most communities and sectors can cope with (or adapt to) normal climatic conditions and moderate deviations from the norm, but exposures involving extreme events that may lie outside the coping range, or may exceed the adaptive capacity of the community* (Smit and Wandel 2006, p 287).

Furthermore, they have been criticised over their integration or management of uncertainty, in that approaches aim to disregard underlying uncertainties and shocks from climatic stimuli, or by viewing uncertainty as a challenge that needs to be reduced, rather than embracing it and taking it into account as a part of the system (Isendahl et al. 2009). Institutions are an important interface between the social and ecological system, since they embody and implement the rules of water use, protection and pollution (Herrfahrdt-Pähle 2010a; Young 2002). However, speeds and scales of current and potential change in the bio-physical system must also be reflected in the social and governance systems that frame their management (Herrfahrdt-Pähle 2010a).

Since the promises of successful water management outcomes through the IWRM panacea have often not materialised in the reality of highly variable water governance frameworks and physical settings (Engle et al. 2011; Medema et al. 2008; Meinzen-Dick 2007), theoretical attention has shifted to the better understanding of adaptive processes in water governance systems as a means of managing uncertainty and non-stationarity in future climatic conditions. Interrelated concepts such as vulnerability, adaptation, adaptive capacity, adaptive governance and management, resilience and sensitivity have more recently enjoyed broad application in the field of global change science (Smit and Wandel 2006). The following sections will discuss their pertinence to this study, and some of the issues with their application in reality.

3.2 Adaptation, Vulnerability and Adaptive Capacity

It is becoming increasingly clear that failure to mitigate climate change will require society to learn how best to respond to the challenges of living in a climatically altered world. Even if the political will is found to adopt the measures requisite to halt the rise of global emissions tomorrow, the time lag in the climate system is likely to mean an increase in global temperatures by at least 2°C by end of century (New et al. 2011; Rogelj et al. 2011). The associated shifts in climatological patterns will require us all, but water managers in particular, to adapt in a timely and effective manner, and in a way that is sustainable (i.e. balancing the different competing economic, social and environmental interests) and factors in the underlying drivers of degradation in the natural system. Climate impacts on hydrology are set to include alterations in seasonality, a rise in the frequency or intensity of extreme hydrological events (drought or flood), higher variability of precipitation patterns and increased glacial melt leading to increased or decreased run off.

The growing awareness and observation of climate change impacts and different forms of adaptation have led to a recent surge in the amount of research within the field of 'adaptation' (Adger et al. 2005; Dovers and Hezri 2010). Studies on adaptation have increasingly moved beyond a scenario based approach, to those that include assessments of current and future adaptations to climate, and adaptive capacity (Carter et al. 2007). Increasingly, adaptation in itself has been tackled from different disciplinary fields, either through vulnerability (exposure to threats or harm: Smit and Wandel 2006) or through resilience related fields (ability to persist in the face of change: Folke 2006; Engle 2010), such as adaptive governance or adaptive co-management. More on these alternative, but linked, perspectives shall be discussed later in this section.

Adaptation specifically refers to the 'process, action, or outcome in a system (household, community, group, sector, region, country) in order for the system to better cope with, manage, or adjust to some changing condition, stress, hazard, risk, or opportunity' (Smit and Wandel 2006, p 282). Often, adaptation choices to climate variability (as well as climate change) have taken the form of technical, 'hard' infrastructure projects. Anthropological studies of how past societies have or have not adapted to change (Brooks 2006; Orlove 2005; Diamond 2004) demonstrate that communities have always had to integrate some form of adaptation into their socio-political system. An IISD (2006) report provided the example of continuous adaptation in the global pastoral range lands where uncertainty has always needed to be managed due to perpetual disequilibrium, chaotic dynamics and where 'predictability and control are false hopes' (Scoones 2004, in IISD 2006, p 5). This ability to change and adapt to new threats or realities that have manifested, known as reactive or autonomous adaptation (Tompkins and Adger 2005), has in some respects become taken for granted.

However, as society moves closer towards increasingly large climate induced changes in environmental conditions that lie outside our range of experience (be that personal or community memory, or institutional memory), our ability to adapt declines. Social systems, in terms of individuals and communities, have exhibited differing levels of ability to respond and cope with climate variability, such as seasonal and decadal drought and floods (Adger and Brooks 2003). However, not all responses to climate variability may be sustainable (i.e. maladaptation) and larger scale changes in the planetary climate system have been shown to have major impacts on societies in the past (Tompkins and Adger 2004; Diamond 2004). This erosion in ability to adapt heightens exposure to the risk of climate change impacts, i.e. increasing vulnerability of the system.

Vulnerability is the extent to which a system (individuals or communities) is susceptible to, and unable to cope with, conditions that adversely affect their well-being, such as climate change, through climate variability and extremes (Plummer and Armitage 2007). Vulnerability comprises a number of components including exposure to impacts, sensitivity, and the capacity to adapt (Adger and Vincent 2005; IPCC 2001). There are a number of well-defined indicators of vulnerability to climate change: economic well-being and inequality, health and nutritional status, education, physical infrastructure, governance, geographic and demographic factors, agriculture, ecosystems, and technological capacity (Brooks et al. 2005). Scholars have suggested that the 'fundamental contribution of governance to reducing the vulnerabilities of people rests on its ability to anticipate problems and to manage risk and challenges in a way that balances social, economic, and natural interests'(Adger et al. 2007).

Often, resilience has been seen as the flip side to vulnerability. Instead of focussing on the exposure of a system to a threat, resilience refers to the ability of a socio-ecological system to absorb disturbances while retaining the same fundamental structure, function and identity, including the capacity to adapt to stress and change, through either recovery or reorganisation in a new context (Chapin et al. 2009). The underlying resilience or coping range of a system refers to the range of change and variability over short periods that a system can manage, or in which major disturbances and significant negative consequences are not observed (Yohe and Tol 2002).[2] Coping capacity is the means by which people or organisations use available resources and their abilities to face adverse consequences that could lead to a disaster. In general, this involves managing resources, both in normal times as well as during crises or adverse conditions. Coping is a related, yet distinguishably different concept to that of adaptation and adaptive capacity, most notably in that beyond the boundary of a coping range, the resilience of a system may reach its limit, degrading or tipping over into a less desirable state.

Adaptation may not always yield beneficial and positive outcomes, and thus adaptation that does not moderate vulnerability, but instead increases it, is considered to be maladaptive (Barnett and O'Neill 2010); likewise, adaptation may enhance resilience to change at one scale, may erode it at another. Promoting adaptation that reduces vulnerability and builds resilience avoiding maladaptation is a challenging and complicated process, primarily because adaptation takes place within the context of different temporal and spatial scales, but equally due to the competing cultural contexts and social goals that must be negotiated and balanced to achieve successful adaptation (Adger et al. 2005; Wilbanks and Kates 1999).

There has often been a disjuncture between the complex interactive nature of adaptation actions in reality and the levels at which the different adaptation foci tend to take place. Often, climate change policy decisions, including technical, 'hard' infrastructure adaptations, adaptation or vulnerability assessments are focused (but not exclusively) at the national levels, while the consequences of these decisions, adaptation implementation and climate impacts are experienced (again not exclusively) at regional scales or local community levels (Brunner 2010; Kane and Yohe 2000). Governance and institutional frameworks within which these policies and decisions are formulated, implemented and experienced need to take better account of these multi-scale dimensions. Researchers have suggested that greater inclusion, integration and focus on institutional and social learning are requisite for addressing these scale related challenges (Kane and Yohe 2000). Moreover, attempts to understand adaptation should take these different scales into account in their research.

Human actors not only are able reactively to adapt to changes, but are also seen to have the potential to proactively adapt by planning for anticipated outcomes or impact (Dovers and Hezri 2010; Engle 2010). This unique gift of foresight and the ability to imagine a potential future outcome, supplemented with the ability to learn from experience, can allow actors to develop anticipatory or proactive adaptation

[2] Also refer to Downing et al. (1997) and Pittock and Jones (2000).

plans and measures. Successful adaptation, both reactive and proactive, is reliant on the capacity to adapt, which becomes especially important to understand (along with associated governance and institutional frameworks that help or hinder its development) under increasingly uncertain conditions, where prior experiences may no longer be as indicative of the future.

Adaptive capacity is defined as the *capacity of actors (collectively or individually), to respond to, create, and shape variability and change in the state of the system* (Adger et al. 2005; Chapin et al. 2009; Walker et al. 2004). It can be characterised as the preconditions needed to enable adaptation, both proactive and reactive, including social and physical elements, and the ability to mobilise these elements (Nelson et al. 2007). It has also been described as relating to the 'ability to mobilise (scarce) resources to anticipate or respond to perceived or current stresses' (Engle 2010, p 33).

Adaptive capacity is also closely related to concepts of robustness, adaptability, flexibility, resilience, and coping ability (Smit and Wandel 2006). Rather than use the concept of adaptive capacity interchangeably with these concepts, it is clearer to characterise it as contributing to these aspects of a system, i.e. the presence of adaptive capacity leads to a better ability to cope with climate risks. In this regard, a system's coping range is seen as a feature of adaptive capacity. Yohe and Tol (2002) elucidate the ability for adaptation to expand the coping range of a community, region or country. Some authors use the term coping capacity to refer to a system's ability to just deal with or survive environmental and climate shocks (some also refer to this form of short term coping as adaptability[3]), while adaptive capacity represents the ability to develop longer term or more sustainable adjustments to changes (Bohle 1993 in Smit and Wandel 2006; Vogel 1998).

In the preceding decade, adaptive capacity has become significantly more mainstream a concept for those researchers who focus on the stresses and potential impacts from climate change and other sources of physical or external stresses (Yohe and Tol 2002, p 25). Through the adaptation and vulnerability literature it was termed as a function of a number of factors including; economic and physical resources; access to technology, information and skills; infrastructure; and institutions (Smit et al. 2000; Yohe and Tol 2002) and it was posited that by fostering or contributing to the presence of these factors in communities, adaptive capacity would be fostered, reducing vulnerability.

Of the different pillars of adaptive capacity, the institutional and governance pillar has received wide recognition as an important mechanism. Governance and institutional components include the law, policies, rights, formal & informal institutions, public policy, and have all been shown to be key for building adaptive capacity and resilience at local, regional and national levels (Adger et al. 2005; Agrawal 2008; Brooks et al. 2005; Engle and Lemos 2010; Gupta et al. 2010; Nelson et al. 2007; Smit et al. 2000; Yohe and Tol 2002). More recently, more nuanced evidence has been presented from the resilience community, suggesting that in the process of developing adaptive capacity in instituional contexts, flexible

[3] See Watts and Bohle (1993) and Vogel (1998).

approaches that embrace the concept of experimentation and ‘learning by doing’ have high importance (Brooks et al. 2005; Gunderson 1999; Olsson et al. 2004a; Pahl-Wostl et al. 2007a).

Within the water sector itself, there is a general call for all new water management measures to be climate resilient. Adaptation strategies therefore aim at reducing vulnerability, including the possibility of increasing adaptive capacity (UNECE 2009). To meet the challenges that water-related institutions face from climate change, the water resources and research community have in recent years focussed more heavily on better understanding adaptive processes (Pahl-Wostl et al. 2007b) as a crucial component to sustainable water resources management. Different approaches have been lauded as being relevant for building adaptive capacity at institutional levels in the water sector. These approaches have included IWRM (as detailed above) as well as adaptive management, and adaptive governance (as will be detailed in the following section) (Adger et al. 2005; Brooks et al. 2005; Eakin and Lemos 2006; Yohe and Tol 2002), so that institutions and governance frameworks can better cope with an uncertain climate future and its interaction with other, non-climate drivers of change.

In summary, enhancing adaptive capacity aims to broaden existing coping ranges, thus minimising vulnerability to increasing intensities or frequencies of hazards or shocks that could erode the resilience of the system (Yohe and Tol 2002). For conceptual clarity, an outline of some of the key terms discussed, features below:

- Coping capacity represents the range or boundary within which a system can deal with environmental/climate stresses through underlying resilience within the system (Smit and Wandel 2006; Yohe and Tol 2002).
- Adaptive capacity represents the ability of the system to respond to environmental and climate variability in order to enable and mobilise adaptation, both in anticipation and reaction to potential or current stresses (Nelson et al. 2007).
- Determinants of adaptive capacity are the social, economic, ecologic and political preconditions that influence the ability of the system to adapt (Smit and Wandel 2006; Wilbanks and Kates 1999).
- Adaptations are manifestations of adaptive capacity, represented by changes in the system to better deal with impacts from climate and environmental change to which that system is vulnerable (Smit and Wandel 2006).
- Reactive adaptation refers to autonomous reaction to events, mobilising adaptive capacity (Dovers and Hezri 2010; Engle 2010; Tompkins and Adger 2005).
- Proactive adaptation refers to planning for future climate change, developing adaptive capacity (Dovers and Hezri 2010; Engle 2010; Tompkins and Adger 2005).

3.3 Building Adaptive Capacity Through Adaptive Governance and Management Approaches

Adaptive governance is one facet of the theoretical movement beyond single policy concepts (as discussion in earlier sections in this chapter), that reflects the general call for water management solutions to be more nuanced and fitting to local social, economic and bio-physical settings (Meinzen-Dick 2007; Pahl-Wostl et al. 2007b). The concept of adaptive governance is closely related to that of adaptive capacity. It is often touted as the most appropriate approach for building adaptive capacity within a system, for it is seen as meeting the call for dealing with increased uncertainty and change, arising from the 'growing number of failures among current approaches and increasing vulnerability of social-ecological system' (Olsson et al. 2006, p 1). If adaptive capacity is an end goal, then adaptive governance can be seen as one means to that end. Adaptive governance has many different iterations and interpretations, but generally refers to the need to move from the conventional view of institutions as static, rule-based, formal and fixed organisations with clear boundaries to one that is more dynamic, adaptive and flexible for coping in future climatic conditions (Pahl-Wostl et al. 2007b).

Adaptive management is closely aligned with that of adaptive governance, and is also seen as a vital tool in building resilient SES. Adaptive management approaches see each management step as an opportunity for further adaptive learning, thereby embracing uncertainty through the navigation of changing circumstances, i.e. learning to manage by managing to learn (Pahl-Wostl and Sendzimir 2005). Adaptive management therefore focuses on methods such as learning by doing, scenario planning and social learning, with the aim of improving flexibility to re-address management approaches that might be perceived as inappropriate with hindsight (Pahl-Wostl 2007a). Allowing decision makers more flexibility for reflexive and evaluative planning processes (i.e. testing and refining responses) has been seen as highly appropriate in the context of climate change, where assumptions based on hydrological or ecological baselines may need to be revised more regularly (Tompkins and Adger 2005). The more iterative process of planning and management sees a more balanced approach to and rigour in the monitoring of policies, as for the initial formation of those policies (Garmestani and Benson 2010).

Adaptive co-management has in turn emerged from adaptive management itself, by combining its iterative learning dimension with collaborative management approaches as emphasised in IWRM, where rights, responsibilities and obligations are jointly shared (Huitema et al. 2009). The focus is on the creation of a community of institutional learning that takes place at the collective rather than just individual level (Berkes and Folke 2001), which aims to draw from and build memories and experience of an entire institution. This focus on participation and multi-level governance for incorporating different forms of knowledge and learning is also shared with IWRM approaches (Hurlbert 2010). Studies have suggested that the combination of collaborative management approaches with adaptive management builds more robust SES as it better accounts for cross-scale dynamics and linkages (with the process being emergent and self-organising, but facilitated by rules and incentives at higher levels), higher complexity, and focuses on the process of dynamic learning (Berkes and Folke 2001; Pahl-Wostl et al. 2007a; Huitema et al. 2009).

Both adaptive governance and management approaches focus heavily on the idea of 'learning by judicious doing' (Holling 1978), which represents a departure from the more traditional approach of rigid and irreversible planning and anticipatory management to a concept of policy experimentation. It also represents a blend of the IWRM focus on 'stakeholder participation and sectoral integration through systematic processes of experimentation, learning, collaboration, and monitoring that attempt to reduce uncertainty' (Folke et al. 2005; Nelson et al. 2007; Olsson et al. 2004b; Plummer and Armitage 2007; Huitema et al. 2009, in Engle et al. 2011). Constructive methods of dealing with and managing uncertainty are integrated into adaptive management approaches by adopting learning techniques (social and policy learning, scenario planning etc.), so that systems may respond to change and unknown conditions (Herrfahrdt-Pähle 2010b; Huntjens et al. 2011; Pahl-Wostl et al. 2007c). At the heart of policy learning approaches is the premise that institutions can use their past experiences and learning from those past experiences to guide their responses and actions to future challenges (Sabatier 1988; Bennett and Howlett 1992 in Huntjens et al. 2011).

Learning is highlighted in a number of studies as a vital component for building experience and flexibility to cope with uncertainty and change (Berkes et al. 2003; Folke et al. 2005; Nelson et al. 2007; Olsson et al. 2004a; Pahl-Wostl 2007b). Learning in the field of adaptive management and governance is seen as a vital component, or supplement, to 'knowledge generation', which in and of itself is not enough to build adaptive capacity in social-ecological systems, but needs to be complemented with the requisite institutional framework that fosters learning as a means to navigate a constantly changing environmental and social context (Folke et al. 2005).

Adaptive management also promotes the role of bridging organisations (networks, associations, cross-sectoral partnerships, political coalitions, social movements) and collective learning amongst others, as a core component in sharing learning experiences and findings and for promoting active continuous learning as a means of continuously improving and adapting management strategies (Pahl-Wostl et al. 2007a). This becomes particularly relevant in relation to climate impacts, where lack of information and high uncertainty about potential impacts is twinned with long decision making time-frames for adaptation (Keeney and McDaniels 2001). This raises the potential for actors to become locked into a set of responses that may no longer be suitable further down the line (Keeney and McDaniels 2001; Tompkins and Adger 2005). Keeney and McDaniels (2001) propose a combination of shorter timeframes in which to set policy objectives (less than 20 years) together with greater emphasis on testing and evaluation in order to overcome these challenges.

The learning related tools promoted within adaptive management provide a means to allow greater flexibility in the system to cope with the connectivity between processes and scales. Related to social learning and learning by doing, is the concept of institutional learning (Berkes and Folke 2001), which takes place at the institutional rather than individual level (Lee 1993). Concerning natural resources, institutional learning involves drawing from and carrying forward into the future memories and experience which provide the context for making modifications of resource-use rules. This experience may integrate local and traditional as well as scientific and formal forms of information and knowledge to develop strategies to respond to environmental change (Berkes and Folke 2001).

Not all learning is created equal though and categories of different levels of learning distinguish between the different intensities of learning and the resulting policy changes that they lead to (Argyris and Schön 1978; Herrfahrdt-Pähle 2010b; Huntjens et al. 2011; Pahl-Wostl et al. 2007b). Adaptive management studies elucidate three different forms of learning: single, double and triple loop learning.

The following summary of these stages of learning is taken from Huntjens et al. (2011). Single loop learning refers to the alteration of past policies and actions to generate better outcomes in the absence of changing underlying assumptions or considering alternatives forms of action. The example for single loop learning is a water manager increasing the height of dykes to improve flood protection. Double loop learning refers to a shift in the frame of reference and underlying assumptions that guide policy making and management actions. The example for double loop learning is the augmentation of boundaries for flood management and enhancing transboundary river basin collaboration. Triple loop learning is characterised by a transformation of the underlying assumptions and context that determine the frame of reference within which decisions are made, and this leads to a transition of the whole regime with new values and norms. An example of triple loop learning is the emergence of a major structural change in the regulatory framework for flooding or droughts.

While IWRM can be seen as a guiding principle for a sustainable water future that takes into account multiple water uses and services, adaptive management emphasises techniques to scope and plan interventions for learning about a system's behaviour (Pahl-Wostl and Sendzimir 2005) as a means of guiding more flexible management techniques for coping with change and uncertainty. Since its initial development in the 1970's at the UNESCO International Conference on Water (1977), IWRM has been broadly accepted a goal in the development of more sustainable water management practices (Medema et al. 2008). IWRM relies on a strong governance (legislative, policy, institutional and management instruments[4]) framework that enables the mismatch between ecological and administrative boundaries to be addressed, in the interests of better integrating and coordinating the management of land and water for more holistic and sustainable water management (GWP 2004; Koudstaal et al. 1992). It also requires the governance framework to address the challenges of sector integration (emphasising connections rather than integration between water-dependent ministries), to reduce challenges of implementing more sustainable management policies that take account of ecological and societal stakeholders that tend to be weaker than economic interests.

However, limitations to the IWRM approach in relation to climate change have been identified with regards to flexibility and uncertainty (Galaz 2007; Pahl-Wostl and Sendzimir 2005; Medema and Jeffrey 2005). The ability for a water governance and

[4] An enabling legislative and policy environment that sets up and empowers; an appropriate institutional framework composed of a mixture of central, local, river-basin-specific, and public–private organisations that provides the governance arrangements for administering; and a set of management instruments for gathering data and information, assessing resource levels and needs, and allocating resources for use.

management system to deal with uncertainty and surprise in relation to environmental rather than societal issues (e.g. surprise in climate related shocks rather than challenges of implementation) while ensuring equitable and efficient allocation and prioritisation of uses are essential requirements for sustainable water resources management in an age of enhanced rates of planetary climatic and environmental change. Therefore, integrative management goals should be supplemented with adaptive goals that are enabled through governance structures and management frameworks that not only addresses challenges related to transparency, participation and cooperation but also complexity, uncertainty and change (Pahl-Wostl 2007b; Engle et al. 2011).

Challenges in the implementation of both IWRM and AM approaches are both documented in the research literature (Biswas 2004; Engle et al. 2011; Jewitt 2002; Medema et al. 2008; Meinzen-Dick 2007). In IWRM, the challenges in achieving an integrationist agenda are recognised as being major due in part to the timeframes of policy and planning processes and limitations in institutional capacity at different levels of governance (White 1998). In adaptive management institutional and organisational technical barriers have been identified in relation to the resource and time intensive process of developing, implementing and monitoring policy experiments (Medema et al. 2008), as has the issue of scaling up results from case level projects to river basins (Levine 2004). Tensions in between the two approaches also have been identified, most notably in the balance between the search for flexibility and experimentation in AM and for legitimacy through deliberative, participatory and pluralistic forums in IWRM (Engle et al. 2011), which can take time to self-organise to face new challenges.

Despite the challenges in both approaches, there has been an increasing trend to combine aspects of the IWRM and AM approach, with the resultant AIM framework. Adaptive and integrated water resources management (AIWM) is an approach coined to propose a set of desirable characteristics for a system that encourages both a holistic and participative approach to water management as well as designing and learning from strategic interventions to address uncertainty and complexity in social-ecological systems. AIWM therefore is described as an approach that emphasises 'polycentric governance with a broadly based constituency, cross-sector analysis to support holistic understanding of system behaviour, transparent approaches to communication and knowledge sharing, and diversified funding through private and public sources' (Pahl-Wostl et al. 2007b, c). The NeWater project has recently emphasised the need for these blended frameworks, focussing on adaptive governance, adaptive co-management for addressing complexity and uncertainty in current and future water governance and management challenges (refer to newater.info and Pahl-Wostl and Sendzimir 2005).

3.3.1 The Role and Rule of Law in Adaptive Governance

Laws, regulations and other 'rules' associated with water resources are vitally important elements of any governance system. Climate change has significant ramifications for water law and governance (Tarlock 2009), yet, globally, there is strong evidence that legal regulations have failed to protect environments or promote sustainable development (Ebbesson 2010; Ostrom 2005). The level of degradation of ecosystems (including freshwater ecosystems) and loss of biodiversity (MEA 2005) is proof in point that law and environmental law have not been able to or set up to adequately balance the weighting of economic and environmental interests for different social groups (Cosens 2010). One challenge has been that while legal systems have tended to be fragmented, yet principle based and rigid, the ecological systems for which they are constructed to govern, tend to be interconnected, non-linear, complex and dynamic (Garmestani and Benson 2010). Another issue has been the challenge of reconciling the scientific quest for truth (in continual revision and questioning) within legal systems that are designed to provide finality (conflict resolution, legal codes). Cosens (2010) argues that while finality serves the interests of economic actors, it is science that serves environmental interests, implying a skewed balance in current frameworks.

This is not to suggest that the law has totally failed in areas of preservation, conservation and restoration of the environment. Law and regulation relating to natural resources management and territorial management has tended to be based on preservation and restoration paradigms, which have assumed that ecological change is both predictable and reversible (Craig 2009). In many countries (e.g. Europe and North America), the development of environmental law has provided for increased protection of the environment over the course of the twentieth century (refer to Table 6.2). However, as the twenty-first century dawned, some environmental scholars have argued that the tools used to address the pollution problems of the mid-twentieth century were no longer adequate for tackling the complex climate related challenges (Shellenberger and Nordhaus 2004). This argument suggests that the structure and rigidity of the preservation paradigm may be better suited to areas of pollution control regulation and conservation (Verschuuren 2007), than to the interpretation of legislative challenges that relate to complex and interacting social-ecological systems. The preservationist components of environmental law themselves are linked to assumptions of stationarity and uniformitarianism (Ruhl 1997), limiting the ability to confront emerging, cross-scale and cross-boundary challenges (Garmestani and Benson 2010). Therefore, while the form of regulation and control may be apt for the environmental quality challenges of the twentieth century for which much environmental legislation and regulation was designed, the interlinked, unpredictable and potentially irreversible impacts of global environmental change (with climate as a major component), the aptness of prior approaches may not be as relevant.

Legal scholars suggest that there is a burning need for the law itself to become better able to support more adaptive and flexible frameworks to meet the challenges posed by climate change. Craig (2009, p 23) suggests that 'both regulatory goals and the legal mechanisms for accomplishing them will have to be centred on the concept of change itself'. However, this increasing focus on enhancing flexibility is juxtaposed by the search for stability and certainty within legal frameworks (Barnes 2010; Craig 2009). The challenge in reconciling legal frameworks, that imply the rule of law and legal certainty, with the complex challenges posed in the governance and management of social-ecological systems is seen to focus on a set of resilience based requirements for coping with surprise and uncertainty: flexibility in social systems and institutions, openness of institutions, effective multi-level governance and the promotion of learning and adaptability (Ebbesson 2010; Folke 2006).

Craig (2009, p 24) argues that 'climate change means that regulatory objectives based on the pre-climate-change characteristics of particular places can and will become increasingly obsolete. Climate change adaptation law must be able to accommodate the transforming ecological realities of particular places and not attempt to freeze ecosystems and their components into some prior state of being'. While Craig (2009) focuses here on climate change adaptation law, the broader ramifications of his arguments extend to legislation and regulation of natural resources in general. He calls for the law to be able to meet the dichotomy of ecological dynamism and legal stationarity by embracing pervasive uncertainties and allowing for a 'a multiplicity of techniques to be brought to bear in crafting adaptation responses to particular local impacts while still promoting actions consistent with overall ecological and social goals' (p10). The law will therefore need to be sensitive to the complexity in the system it seeks to structure and embrace a multitude of techniques that can be employed to enable adaptation responses as they befit local needs.

Some jurists have thus started to turn their attention to means of balancing the search for stability and predictability in legal frameworks with the complexity in socio-ecological systems and the flexibility required within science based decision making (Cosens 2010; Craig 2009; Ebbesson 2010; Ruhl 1997). Scholars that have examined the place that flexibility has in a legal framework (defined by normative texts and fixed, predictable rules) purport that concepts such as 'principled flexibility' (Craig 2009) or 'measured stability' (Cosens 2010) could be useful in addressing the complex inter-relationship between predictability and dynamic systems and thus develop capacity to adjust to continual transformation.

Principled flexibility would see provisions that allowed the law and environmental management goals to reflect shifting baseline hydrological or ecological conditions (e.g. exemption clauses based on continuous informational inputs). This would imply potential amendments to administrative law (the rules that guide government agency rule making) to ensure that judicial review enabled a process of review and adaptability in the law (Craig 2009). Cosens' (2010) concept of 'measured stability' refers to the integration of adaptive processes through the utilisation of measured timeframes based on both the economic and ecological criteria, thus focussing more heavily on the sound scientific basis to develop the legal toolbox.

In a study of local Canadian water governance contexts, Hurlbert (2010) suggests that focusing on participative structures through the concept of 'living law' (of local communities) would increase adaptive capacity, instead of nurturing path dependency as the law is practiced by socio-technical experts (i.e. lawyers and government bureaucrats). This reinforces the important role that legislation and regulation at different scales of governance plays in adaptive and integrated resource management, particularly in the case of water resources governance (Ebbesson 2010). One of the aims of IWRM is in fact to retain flexibility in water management systems, by relegating different management mechanisms (monitoring, regulation etc.) to more 'dynamic parts of the legislative system' (GWP 2000, p 38), such as regional and local regulations and ordinances. Garmestani and Benson (2010) also point to the importance of matching institutions to the appropriate scales in the interest of balancing stability with flexibility.

Some jurists suggest that legal amendments and new laws are required to develop a regulatory context that shifts from assumptions of stationarity, to a 'paradigm of increasing resilience and adaptive capacity, based on assumptions of continuing, unpredictable, and non-linear change' (Craig 2009, p 31). Others note that the law in itself is in fact more dynamic and less rigid than many scholars would suggest (Hey 2010). Not only is the word of the procedural law less certain and inflexible that one might assume, but substantive law, in terms of property rights and individual rights, are also subject to renegotiation and development (Langlet 2010), as Table 6.2 shows in its presentation of the dynamic development of the legislative and regulatory framework in the Swiss and Chilean cases.

Therefore, there is a greater latent potential to shift the discourse and understanding of property rights and legislation to take account of social-ecological complexity and uncertainty than might be apparent at the outset. A further challenge, however, is to achieve this operationally (Langlet 2010), and according to shorter timeframes, since climate change impacts affect both the speed and magnitude of physical change. With regards to substantive law, Craig (2009) highlights the importance of giving meaningful weight to public rights and values in private property in order to address the challenges governments might face in addressing the impact of climate change on what the public perceives as 'absolute' private property rights. Langlet (2010) has also argued that the integration of ecologically subjective property rights would be an important step in overcoming the barriers that strong and fragmented property rights pose to integrated and adaptive governance.

Cosens (2010) also suggests that the notion of stability in legal frameworks is flawed precisely because of the imbalance between the finality of economic actors and the flexibility of environmental and scientific interest that has led to gridlock in and challenges to the current legal system, as dissatisfied actors struggle to have their interests served. Her concept of measured stability seeks to address this failure, since it represents a process that does not pit finality against science, but rather enables the measured integration of science for more effective conflict resolution.

Hey (2010) also suggests that legal certainty does not necessarily have to be non-adaptive, purporting that a blend between procedural certainty but changing substance might address the challenge of shifting baselines upon which rights and laws are based and the enhanced integration of science into environmental law. Some jurists have argued that less predictable uncertainty may in fact necessitate more stable and rigid legal structures, but with more flexible content, as well as flexible instruments which combine both rigidity and flexibility (Hey 2010). This refers to instruments that balance the regulatory and enabling function of the law (i.e. the law as goal oriented), with a process for reviewing and revising those goals once the specific baselines upon which they are set may become obsolete.

Focussing on the balances between these two elements might be a more productive framing of the problem than seeing social-ecological resilience as a black and white trade-off to legal certainty. This argument has also been used to suggest that the development of stable and predictable structures at higher levels (law, regulation, government institutions) might also allow for greater flexibility and experimentation at lower levels (Cosens 2010; Garmestani and Benson 2010). This would posit the role of law as one of stability within change, as opposed to stability versus change.

Legal rules and rights are a central and critical part of the social-ecological system, but perhaps not the defining influence as many lawyers would see them. For example, Hurlbert (2010) shows that, across three different case areas in Canada, processes of formal re-evaluation of the law are occurring whether or not they are statutorily required or not, suggesting that other factors are more influential than the legal framework in embedding adaptive processes in those case areas. Nevertheless, finding the right balance between structured and reflexive aspects of the legal framework will be a core element in enabling adaptive capacity within water governance regimes.

3.4 Navigating Change in Socio-ecological Systems

Resilience is the ability of a socio-ecological system to absorb disturbances while retaining the same fundamental structure, function and identity, including the capacity to adapt to stress and change, through either recovery or reorganisation in a new context (Chapin et al. 2009; Parry et al. 2007). While resilience theory emerged from the field of ecology (Holling 1973) and theoretical and mathematical modelling sciences (Gallopin 2006), the advent of the anthropocene has meant an increasing recognition that physical processes should no longer be examined in isolation from the human processes that are now becoming the dominant driver (Folke et al. 2005; Olsson et al. 2004a). The social system is intrinsically linked to the physical system, yet the institutions that manage them tend to be fragmented and constructed on borders unrelated to the ecological systems they manage (Cumming et al. 2006).

Today's mounting complex socio-ecological challenges require more integrated and adaptive approaches to resolving past, present and future resource management problems. Since few of the drivers of changes are purely physical, ecological or social, neither should be the framework within which we interpret these changes (Chapin et al. 2009). From the resilience perspective, the growing recognition of the importance of human-made influences on ecosystems, means that researchers need to pay close attention to both the human and environmental components of socio-ecological systems (SES) (Walker et al. 2006).

The concept of SES is key to understanding the complex interrelated changes that water institutions face. SES are comprised of interconnected socio-economic properties, or human components, (human-made infrastructure, institutions, governance system, economic systems, etc.) and ecological properties or environmental components (species, climate, biota, etc.) (Gallopin 2006). They represent the interrelated nature of the resources and ecosystem services upon which humanity relies, and the human activities which influence these ecological dynamics (Berkes et al. 2003). Social-ecological systems are typical examples of complex adaptive systems that are defined as having components that interact in a manner that drive the system to adapt and adjust in response to changing conditions (Chapin et al. 2009, p 14). Framing resource related issues in terms of how SES are impacted and adjust allows such problems to be analysed and remediated in a more integrated manner.

Within the specific context of river basins, scholars have pointed to the need to focus more attention on understanding and managing a transition from current management regimes to more adaptive regimes that 'take into account environmental, technological, economic, institutional and cultural characteristics of the basin' (Pahl-Wostl 2007b, p 49). Transformations or transitions to more adaptive governance approaches include shifts in social and governance features of SES in order to redirect attention and resources to "restoring, sustaining and developing the capacity of ecosystems to generate essential services" (Olsson et al. 2006, p 2). Three broad categories of outcomes are recognised in terms of directional changes that occur in the governance of SES; transformation, adaptation and passive (Herrfahrdt-Pähle 2010b; Chapin et al. 2009).

Transformation of SES into trajectories that sustain and enhance ecosystem services, societal development and human well-being (Folke et al. 2010) may occur through innovation and enhanced adaptive capacity, allowing the system to transition to a different, potentially more desirable state (Chapin et al. 2009). Transformability has also been described as the 'capacity to create a fundamentally new system when ecological, economic, or social structure makes the existing system untenable' (Walker et al. 2004), transitioning it onto a trajectory that enhances ecosystem services, societal development and human well-being (Folke et al. 2010). Transformations are particularly needed when an SES is locked into a highly resistant state, where adaptation no longer seems an option (Herrfahrdt-Pähle 2010b; Walker et al. 2006). Transformational outcomes are also associated with triple loop learning in adaptive management terminology (Herrfahrdt-Pähle 2010b; Pahl-Wostl 2009).

Outcomes that allow for the *'persistence of the fundamental properties of the current system through adaptation'* (Chapin et al. 2009, p 20) are termed 'adaptation'. Such outcomes can be adjustments in either the social or ecological systems of an SES in response to experienced or expected climatic stimuli (Smit and Wandel 2006), revealing an indication of mobilised adaptive capacity. These definitions of adaptation and persistence have been used interchangeably to imply sustainability of properties of the system (Chapin et al. 2009), but for the purposes of this piece of research, 'persistent adaptation' shall be used for this category. This outcome alludes to the capacity of actors within a system to influence the resilience of it, in terms of its adaptability (Walker et al. 2004) and has been related to double loop learning (Pahl-Wostl 2009; Herrfahrdt-Pähle 2010a). Folke et al. (2010) associates an adaptive outcome with the 'capacity of a SES to learn, combine experience and knowledge, adjust responses to changing external drivers and internal processes and continue developing within the current stability domain or basin of attraction'.

In addition to the above two categories of adaptive responses, a third category, passive, captures outcomes that contribute to the degradation of the system to a less favourable state, resulting from either a failure to transform or adapt (Chapin et al. 2009, p 20), or even maladaptation. This can include unintended or forced transformations, which are transformations of a SES that is not deliberately introduced by the actors (Folke et al. 2010). Single-loop learning is the category of multi-loop learning that is partially associated with this outcome classification (Herrfahrdt-Pähle 2010b; Pahl-Wostl 2009). However, it should be noted that single loop learning implies that some learning has taken place, and therefore can also be partially associated with persistent adaptation.

Institutional change theorists postulate that change is incremental due to institutional inertia and path dependency that limits the possibility for change within a linear framework (Lempert et al. 2004; North 1990; Pflieger et al. 2009; Pierson 2000). In contrast, resilience theory draws on dynamic process across multiple scales and integrated systems (i.e. SES). It recognises the legacy impact and path dependence of past events on the subsequent dynamics of future changes (Chapin et al. 2009, p 14, drawing on North 1990), acknowledging that events at one scale are likely to influences events at other scales in a linked system. However, it equally provides for non-linear processes of crisis and reorganisation (in the complex adaptive cycle) that allows for dynamic developments to be analysed (i.e. transformative outcomes; persistent adaptation; passive). The importance of both institutional change and complex adaptive cycles to this issue of adaptive outcomes within a river basin, reinforces the importance of cross-scale linkages in studying social-ecological systems and the value in studying them at multiple spatial and temporal scales (Berkes et al. 2003; Chapin et al. 2009, p 14).

References

Adger WN, Brooks N (2003) Does environmental change cause vulnerability to natural disasters? In: Pelling M (ed) Natural disasters and development in a globalising world. Routledge, London, pp 19–42

Adger WN, Vincent K (2005) Uncertainty in adaptive capacity. Comptes Rendus Geosci 337(4):399–410. doi:10.1016/jcrte.2004.11.004

Adger WN, Arnell NW, Tompkins E (2005) Successful adaptation to climate change across scales. Glob Environ Change 15(2):77–86

Adger WN, Agrawala S, Mirza MMQ, Conde C, O'Brien K, Pulhin J, Pulwarty R, Smit B, Takahashi K (2007) Assessment of adaptation practices, options, constraints and capacity. In: Parry ML, Canziani OF, Palutikof JP, van der Linden PJ, Hanson CE (eds) Climate change 2007: impacts, adaptation and vulnerability. Contribution of working group II to the fourth assessment report of the intergovernmental panel on climate change. Cambridge University Press, Cambridge

Agrawal A (2008) The role of local institutions in adaptation to climate change. Working paper. International Forestry Resources and Institutions Program # W081-3. University of Michigan, Ann Arbor

Argyris C, Schön D (1978) Organizational learning: a theory of action perspective. Addison Wesley, Reading

Barnes R (2010) The capacity of property rights to accommodate socio-ecological resilience (tranboundary issues). Paper presented at the Law for Social-Ecological Resilience Stockholm University, Stockholm, 17–19 Nov 2010

Barnett J, O'Neill S (2010) Maladaptation. Glob Environ Change 20(2):211–213

Berkes FC, Folke C (2001) Back to the future: ecosystem dynamics and local knowledge. In: Gunderson LH, Holling CS (eds) Panarchy: understanding transformations in human and natural systems. Island Press, Washington, DC

Berkes F, Colding J, Folke C (eds) (2003) Navigation social-ecological systems: building resilience for complexity and change. Cambridge University Press, Cambridge

Biswas AK (2004) Integrated water resources management: a reassessment. Water Int 29(2):248–256

Brooks N (2006) Cultural responses to aridity in the Middle Holocene and increased social complexity. Quat Int 151:29–49

Brooks N, Adger WN, Kelly PM (2005) The determinants of vulnerability and adaptive capacity at the national level and the implications for adaptation. Glob Environ Change 15(2):151–163. doi:10.1016/j.gloenvcha.2004.12.006

Brunner RD (2010) Climate adaptation and the drunkard's speech. Paper presented at the 29th annual Institute of the Society of Policy Scientists, Yale Law School, New Haven, CT, 16 Oct 2010

Carter TR, Jones RN, Lu X, Bhadwal S, Conde C, Mearns CG, O'Neill BC, Rounsevell MDA, Zurek MB (2007) New assessment methods and the characterisation of future conditions. Climate change 2007: impacts, adaptation and vulnerability. In: Parry ML, Canziani OF, Palutikof JP, van der Linden PJ, Hanson CE (eds) Climate change 2007: impacts, adaptation and vulnerability. Contribution of working group II to the fourth assessment report of the intergovernmental panel on climate change. Cambridge University Press, Cambridge, pp 133–171, http://www.ipcc.ch/pdf/assessment-report/ar134/wg132/ar134-wg132-chapter132.pdf

Chapin FS, Folke C, Kofinas GP (2009) A framework for understanding change. In: Chapin FS, Kofinas GP, Folke C (eds) Principles of ecosystem stewardship resilience-based natural resource management in a changing world. Springer, New York

Cosens B (2010) Resilience and administrative law in transboundary river governance. Paper presented at the Law for Social-ecological resilience conference, Stockholm Environmental Law and Policy Centre and Stockholm Resilience Centre, Stockholm, 17–19 Nov 2010

Craig RK (2009) 'Stationariy is dead' – Long life transformation: five principles for climate change adaptation law. Florida State University, Tallahassee

Cumming GS, Cumming DHM, Redman CL (2006) Scale mismatches in social-ecological systems: causes, consequences, and solutions. Ecol Soc 11(1):14. [online] URL: http://www.ecologyandsociety.org/vol11/iss11/art14/

Diamond JM (2004) Collapse: how societies choose to fail or survive. Viking Adult, New York

Dovers SR, Hezri AA (2010) Institutions and policy processes: the means to the ends of adaptation. Wiley Interdiscip Rev Clim Change 1(2):212–231

Downing T (2009) Experience and evidence: mainstreaming climate information into adaptation planning. Paper presented at the world climate conference 3, Geneva, Switzerland, 31 Aug–4 Sept 2009

Eakin H, Lemos MC (2006) Adaptation and the state: Latin America and the challenge of capacity-building under globalization. Glob Environ Change 16(1):7–18. doi:10.1016/j.gloenvcha.2005.10.004

Ebbesson J (2010) The rule of law in governance of complex socio-ecological changes. Glob Environ Change 20(3):414–422. doi:10.1016/j.gloenvcha.2009.10.009

Engle NL (2010) Adaptation to extreme droughts in Arizona, Georgia, and South Carolina: evaluating adaptive capacity and innovative planning and management approaches for states and their community water systems. University of Michigan, Ann Arbor

Engle NL, Lemos MC (2010) Unpacking governance: building adaptive capacity to climate change of river basins in Brazil. Glob Environ Change 20(1):4–13. doi:10.1016/j.gloenvcha.2009.07.001

Engle NL, Johns OR, Lemos MC, Nelson DR (2011) Integrated and adaptive management of water resources: tensions, legacies, and the next best thing. Ecol Soc 16(1):19. [online] URL: http://www.ecologyandsociety.org/vol16/iss11/art19/

Folke C (2006) Resilience: the emergence of a perspective for social-ecological systems analyses. Glob Environ Change 16(3):253–267

Folke C, Hahn T, Olsson P, Norberg J (2005) Adaptive governance of social-ecological systems. Annu Rev Environ Resour 30(1):441–473. doi:10.1146/annurev.energy.30.050504.144511

Folke C, Carpenter SR, Walker B, Scheffer M, Chapin T, Rockström J (2010) Resilience thinking: integrating resilience, adaptability and transformability. Ecol Soc 14(4):20. [online] URL: http://www.ecologyandsociety.org/vol15/iss24/art20/

Galaz V (2007) Water governance, resilience and global environmental change – a reassessment of integrated water resources management (IWRM). Water Sci Technol 56(4):1–9

Gallopin GC (2006) Linkages between vulnerability, resilience, and adaptive capacity. Glob Environ Change 16(3):293–303. doi:10.1016/j.gloenvcha.2006.02.004

Garmestani AS, Benson MH (2010) Actualizing panarchy within environmental policy: mechanisms for tweaking institutional hierarchies to mimic the social social-ecological systems they manage. Paper presented at the law for social-ecological resilience, Stockholm University, Stockholm, 17–19 Nov 2010

Gleick PH (2003) Global freshwater resources: soft-path solutions for the 21st century. Science 302(28 November):1524–1528

Gunderson L (1999) Resilience, flexibility and adaptive management: antidotes for spurious certitude? Conserv Ecol 3(1):7

Gupta J, Termeer C, Klostermann J, Meijerink S, Van den Brink M, Jong P, Nooteboom S, Bergsma E (2010) The adaptive capacity wheel: a method to assess the inherent characteristics of institutions to enable the adaptive capacity of society. Environ Sci Policy 13(6):459–471. doi:10.1016/j.envsci.2010.05.006

GWP (2000) Integrated water resouces management TAC background paper no. 4. GWP. Global Water Partnership – Technical Advisory Committee (GWP-TAC), Stockholm, Sweden

GWP (2004) Integrated water resources management (IWRM) and water efficiency plans by 2005. TAC background papers no. 10. GWP. Global Water Partnership – Technical Advisory Committee (GWP-TAC), Stockholm, Sweden

Herrfahrdt-Pähle E (2010a) South African water governance between administrative and hydrological boundaries. Clim Dev 2:111–127

Herrfahrdt-Pähle E (2010b) The transformation towards adaptive water governance regimes in the context of climate change. Universität Osnabrück, Osnabrück

Hey E (2010) Concluding panel debate and closure. Paper presented at the Law for Social-Ecological Resilence, Stockholm University, Stockholm, 17–19 Nov 2010

Holling CS (1973) Resilience and stability of ecological systems. Annu Rev Ecol Syst 4(1):1–23

Holling CS (1978) Adaptive environmental assessment and management. Wiley, London

Huitema D, Mostert E, Pahl-Egas W, Moellenkamp S, Pahl-Egas W, Moellenkamp S, Pahl-Wostl C, Yalcin R (2009) Adaptive water governance: assessing the institutional prescriptions of adaptive (co-)management from a governance perspective and defining a research agenda. Ecol Soc 14(1):26. [online] URL: http://www.ecologyandsociety.org/vol14/iss21/art26/

Huntjens P, Pahl-Wostl C, Rihoux B, Schlüter M, Flachner Z, Neto S, Koskova R, Dickens C, Nabide Kiti I (2011) Adaptive water management and policy learning in a changing climate: a formal comparative analysis of eight water management regimes in Europe, Africa and Asia. Environ Policy Gov 21:145–163

Hurlbert M (2010) Law, resilience and multi-level water governance in the Canadian Prairie Provinces. Paper presented at the Law for Social-Ecological Resilience, Stockholm, Sweden, 17–19 Oct 2010

IISD (2006) Designing policies in a world of uncertainty, change and surprise: adaptive policy-making for agriculture and water resources in the face of climate change. International Development Research Centre & The Energy and Resources Institute, Winnipeg/Manitoba/New Delhi

IPCC (2001) Climate change 2001: impacts, adaptation, and vulnerability technical summary. A report of working group II of the intergovernmental panel on climate change. WMO/UNEP, New York

Isendahl N, Dewulf A, Brugnach M, Francois G, Mollenkamp S, Pahl-Wostl C (2009) Assessing framing of uncertainties in water management practice. Water Resour Manage 23(15):3191–3205. doi:10.1007/s11269-009-9429-y

Jewitt G (2002) Can integrated water resources management sustain the provision of ecosystem goods and services? Phys Chem Earth 27(11–22):887–895

Kane SM, Yohe GW (2000) Societal adaptation to climate variability and change: an introduction. Clim Change 45(1):1–4

Keeney RL, McDaniels TL (2001) A framework to guide thinking and analysis regarding climate change policies. Risk Anal 21:989–1000

Koudstaal R, Rijsberman FR, Savenije H (1992) Water and sustainable development. Nat Resour Forum 16(4):277–290

Langlet D (2010) Concluding panel debate and closure. Paper presented at the law for social-ecological resilience, Stockholm University, Stockholm, 17–19 Nov 2010

Lee KN (1993) Compass and gyroscope: integrating science and politics for the environment. Island Press, Washington, DC

Lempert R, Nakicenovic N, Sarewitz D, Schlesinger M (2004) Characterizing climate-change uncertainties for decision-maker:. an editorial essay. Clim Change 65(1):1–9

Levine J (2004) Adaptive management in river restoration: theory vs. practice in western North America. University of California, Berkeley/Riverside

MEA (2005) Millennium Ecosystem Assessment. Millennium Ecosystem Assessment Board. Island Press, Washington, D.C.

Medema W, Jeffrey P (2005) IWRM and adaptive management: synergy or conflict? NeWater report series no. 7. Available at: www.usf.uni-osnabrueck.de/projects/newater/downloads/newater_rs07.pdf

Medema W, McIntosh BS, Jeffrey PJ (2008) From premise to practice: a critical assessment of integrated water resources management. Ecol Soc 13(2):29. [online] URL: http://www.ecologyandsociety.org/vol13/iss22/art29/

Meinzen-Dick R (2007) Beyond panaceas in water institutions. Proc Natl Acad Sci USA 104(39): 15200–15205

Nelson DR, Adger WN, Brown K (2007) Adaptation to environmental change: contributions of a resilience framework. Annu Rev Environ Resour 32:395–419

New M, Liverman D, Schroder H, Anderson K (2011) Four degrees and beyond: the potential for a global temperature increase of four degrees and its implications. Philos Transac R Soc A 369:6–19. doi:10.1098/rsta.2010.0303

North DC (1990) Institutions, institutional change, and economic performance. Cambridge University Press, New York

Olsson P, Folke C, Berkes F (2004a) Adaptive comanagement for building resilience in social-ecological systems. Environ Manage 34(1):75–90. doi:10.1007/s00267-003-0101-7

Olsson P, Folke C, Berkes F, Hahn T (2004b) Social-ecological transformation for ecosystem management: the development of adaptive co-management of a wetland landscape in southern Sweden. Ecol Soc 9(4):2. [online] URL: http://www.ecologyandsociety.org/vol9/iss4/art2/

Olsson P, Gunderson LH, Carpenter SR, Ryan P, Lebel L, Folke C, Holling CS (2006) Shooting the rapids: navigating transitions to adaptive governance of social-ecological systems. Ecol Soc 11(1):18. [online] URL: http://www.ecologyandsociety.org/vol11/iss11/art18/

Orlove B (2005) Human adaptation to climate change: a review of three historical cases and some general perspectives. Environ Sci Policy 8(6):589–600

Ostrom E (2005) Understanding institutional iversity. Princeton University Press, Princeton

Ostrom E, Burger J, Field CB, Norgaard RB, Policansky D (1999) Revisiting the commons: local lessons, global challenges. Science 284(9 April):278–282

Pahl-Wostl C (2007a) Requirements for adaptive water management. In: Pahl-Wostl C, Kabat P, Möltgen J (eds) Adaptive and integrated water management. Coping with complexity and uncertainty. Springer, Berlin/New York

Pahl-Wostl C (2007b) Transitions towards adaptive management of water facing climate and global change. Water Resour Manage 21(1):49–62. doi:10.1007/s11269-006-9040-4

Pahl-Wostl C (2009) A conceptual framework for analysing adaptive capacity and multi-level learning processes in resource governance regimes. Glob Environ Change Human Policy Dimens 19(3):354–365. doi:10.1016/j.gloenvcha.2009.06.001

Pahl-Wostl C, Sendzimir J (2005) The relationship between IWRM and adaptive water management. NeWater working paper no. 3

Pahl-Wostl C, Craps M, Dewulf A, Mostert E, Tabara D, Taillieu T (2007a) Social learning and water resources management. Ecol Soc 12(5):5. [online] URL: http://www.ecologyandsociety.org/vol12/iss12/art15/

Pahl-Wostl C, Kabat P, Möltgen J (eds) (2007b) Adaptive and integrated water management. Coping with complexity and uncertainty. Springer, Berlin/New York

Pahl-Wostl C, Sendzimir J, Jeffrey P, Aerts J, Berkamp G, Cross K (2007c) Managing change toward adaptive water management through social learning. Ecol Soc 12(2):30. [online] URL: http://www.ecologyandsociety.org/vol12/iss32/art30/

Parry ML, Canziani OF, Palutikof JP (2007) Technical summary. Climate change 2007: impacts, adaptation and vulnerability. Contribution of working group II to the fourth assessment report of the intergovernmental panel on climate change. In: Parry ML, Canziani OF, Palutikof JP, van der Linden PJ, Hanson CE (eds) Fourth assessment report of the intergovernmental panel on climate change. Cambridge University Press, Cambridge, pp 23–78

Pflieger G, Kaufmann V, Pattaroni L, Jemelin C (2009) How does urban public transport change cities? Correlations between past and present transport and urban planning policies. Urban Stud 46(7):1421–1437

Pierson P (2000) Increasing returns path dependence and the study of politics. Am Polit Sci Rev 94(2):251–267

Plummer R, Armitage D (2007) A resilience-based framework for evaluating adaptive co-management: linking ecology, economics and society in a complex world. Ecol Econ 61(1):62–74

Rogelj J, Hare W, Lowe J, van Vuuren DP, Riahi K, Matthews B, Hanaoka T, Jiang K, Meinshausen M (2011) Emission pathways consistent with a 2?°C global temperature limit. Nat Clim Change 1(8):413. doi:10.1038/nclimate1258

Ruhl JB (1997) Thinking of environmental law as a complex adaptive system: how to clean up the environment by making a mess of environmental law. Houst Law Rev 34(4)

Shellenberger M, Nordhaus T (2004) The death of environmentalism: global warming politics in a post-environmental world. http://thebreakthrough.org/PDF/Death_of_Environmentalism.pdf

Smit B, Wandel J (2006) Adaptation, adaptive capacity and vulnerability. Glob Environ Change 16(3):282–292

Smit B, Burton I, Klein R, Wandel J (2000) An anatomy of adaptation to climate change and variability. Clim Change 45:223–251

Stakhiva E, Stewart B (2010) Needs for climate information in support of decision-making in the water sector. Proc Environ Sci 1:102–119

Tarlock D (2009) Seminar. Paper presented at the UNESCO Centre for Water Law, Policy and Science, University of Dundee, Scotland, 16 July 2009

Tompkins EL, Adger WN (2004) Does adaptive management of natural resources enhance resilience to climate change? Ecol Soc 9(2):10. [online] URL: http://www.ecologyandsociety.org/vol19/iss12/art10

Tompkins EL, Adger WN (2005) Defining response capacity to enhance climate change policy. Environ Sci Policy 8:562–571

UNECE (2009) Guidance on water and adaptation to climate change. United Nations Economic Commission for Europe, Geneva

Verschuuren JM (2007) Adaptation to climate change: opportunities and barriers. Paper presented at the proceedings of the international colloquium on global warming, Rio de Janeiro, 21 May 2007

Vogel C (1998) Vulnerability and global environmental change. LUCC Newsl 3:15–19

Walker B, Holling CS, Carpenter SR, Kinzig A (2004) Resilience, adaptability and transformability in social–ecological systems. Ecol Soc 9(2):5. [online] URL: http://www.ecologyandsociety.org/vol9/iss2/art5/

Walker BH, Gunderson LH, Kinzig AP, Folke C, Carpenter SR, Schultz L (2006) A handful of heuristics and some propositions for understanding resilience in social-ecological systems. Ecol Soc 11(1):13. [online] URL: http://www.ecologyandsociety.org/vol11/iss11/art13/

White G (1998) Reflections on the 50-year international search for integrated water management. Water Policy 1(1):21–27

Wilbanks TJ, Kates RW (1999) Global change in local places: how scale matters. Clim Change 43(3):601–628

Yohe G, Tol RSJ (2002) Indicators for social and economic coping capacity – moving toward a working definition of adaptive capacity. Glob Environ Change 12(1):25–40

Young O (2002) The institutional dimensions of environmental change: fit, interplay, and scale. MIT Press, Cambridge, MA

Chapter 4
The Assessment of Adaptive Capacity

Abstract This chapter reviews the methods and challenges for the assessment of adaptive capacity. It presents and discusses the ranges of governance determinants of adaptive capacity as they have developed out of the different discourses such as good governance, adaptive governance, adaptive management, vulnerability and resilience. It concludes that the relative paucity of deep empirical examples exploring adaptive actions in periods that might be representative of a future warmer world remains a challenge in the operationalisation and characterisation of adaptive capacity as well as in the development in understanding how to mobilise it as climate change impacts take hold.

Keywords Adaptive capacity indicators • Governance determinants of adaptive capacity • Assessment of adaptive capacity • Analytical challenges to assessing adaptive capacity • Resilience based assessment of adaptation outcomes

4.1 Adaptive Capacity

In the preceding decade, adaptive capacity has become a more mainstream concept, yet significant challenges still remain in characterising and measuring it. To reiterate, climate change implies a speed and magnitude of change, which poses risks that are beyond the human experience and potentially at the boundaries of coping ranges (Adger et al. 2007). In order to better understand actions and means of expanding coping ranges, a growing body of literature has focused on identifying and developing determinants and indicators of adaptive capacity. Within this body of literature, indicators and determinants have tended to often be used without clear definition and sometimes interchangeably as can be seen in the discussion in the following section.

Determinants can be seen as a broad range of factors (technical, financial, institutional) that influence, affect or determine the outcome or nature of something. Indicators are seen as useful tools to interpret, monitor and provide information

M. Hill, *Climate Change and Water Governance: Adaptive Capacity in Chile and Switzerland*, Advances in Global Change Research 54,
DOI 10.1007/978-94-007-5796-7_4,

on the levels of presence or absence of factors that comprise determinants of a particular condition and are vital in the simplification, quantification and communication of complex processes (OECD 1997). Such indicators could be comprised of quantitatively or qualitatively measurable criteria that are indicative of the presence of the particular condition and can be useful in its assessment (Slocombe 1998). The following discussion of determinants and indicators of adaptive capacity reflect the discourse in the body of literature, which has tended to not always clarify between these different definitions. However, at the end of this section and in the following methodology chapter, the means in which these terms are used within this book shall be clarified.

Yohe and Tol (2002) suggested that determinants of adaptive capacity have a key role in defining the potential boundaries of coping ranges and the ability of SES's to effectively prepare for and respond to stresses. Early determinants of adaptive capacity were defined as including a variety of system, sector, and location specific characteristics (IPCC 2001):

- The range of available technological options for adaptation,
- The availability of resources and their distribution across the population,
- The structure of critical institutions, the derivative allocation of decision-making authority, and the decision criteria that would be employed,
- The stock of human capital including education and personal security,
- The stock of social capital including the definition of property rights,
- The system's access to risk spreading processes,
- The ability of decision-makers to manage information, the processes by which these decision-makers determine which information is credible, and the credibility of the decision-makers, themselves, and
- The public's perceived attribution of the source of stress and the significance of exposure to its local manifestations.

These determinants drew heavily on the vulnerability literature, and while they represented quite a broad brush stroke attempt at characterising the components of adaptive capacity, they were an important starting point from which gradually a more nuanced range of governance and institutional indicators of adaptive capacity could be developed (Engle and Lemos 2010). Since 2001, there has been a growing body of literature focusing in particular on institutional and governance determinants of adaptive capacity (Brooks et al. 2005; Bussey et al. 2010; Eakin and Lemos 2006; Engle 2011; Engle and Lemos 2010; Folke et al. 2005; Gupta et al. 2010; Medema et al. 2008; Olsson et al. 2004a; Pelling and High 2005; Wilby and Vaughan 2011; Yohe and Tol 2002).

Different disciplinary fields have developed alternate interpretations and characterisations of adaptive capacity (Engle 2010), ranging from a focus on cooperation, resources and incentives in geography and political economy (Adger 2003), to an emphasis on poverty reduction and climate injustice in development studies (Dow et al. 2006). There is however, still a long way to go, and comparatively little work on creating a robust framework to measure, characterise and foster components of adaptive capacity so that operationalised indicators could be transformed into meaningful and robust sets of choices for decsion makers. This is a crucial step towards more tangible and applicable methods for improving the adaptive capacity of water institutions and governance regimes.

Building adaptive capacity, by cultivating or contributing to the presence of its determinants in an SES, improves the ability of systems to be become resilient to surprises and longer term changes by shaping positive responses, even transformation or transition to a better state if this is required. The determinants of adaptive capacity listed above lay the foundations for a number of different features and principles, which are seen as useful indications of a systems' adaptive capacity. It is these indicators and principles that shall be discussed in this chapter. The following discussion builds on the body of research detailed earlier in this chapter, discussing the challenges in developing governance and institutional indicators to characterise and assess adaptive capacity, and thus presenting a synthesis of the current state of indicators and determinants of adaptive capacity.

The assessment of adaptive capacity is inextricably linked with that of adaptation. While the assessment of adaptation actions tend to be addressed within a framework of whether the outcome of such actions are equitable, effective and legitimate, there are also significant questions not just about *how we adapt*, but rather *whether we can adapt*. The concept of adaptive capacity is used as a point of departure to determine measurable indicators that 'could sustain comparable analyses of the relative vulnerabilities of different systems located across the globe and subject to a diverse set of stresses that lie beyond their control' (Yohe and Tol 2002, p 25). Such indices can be either qualitatively or quantitatively based, generated through formulaic or discursive data, but are critical for the management of risk in relation to climate change impacts.

Engle and Lemos (2010, p 3) note that 'decision makers are interested in identifying and nurturing specific system characteristics that will increase adaptive capacity and resilience'. The identification of determinants and indicators of adaptive capacity provide a broad suite of characteristics, among which governance and institutional processes are deemed particularly important for the development of adaptive capacity, reduction of vulnerability and prevention of overt and lasting damage from climate change (Brooks et al. 2005; Nelson et al. 2007). Previous studies of adaptation to climatic events have also highlighted the importance of institutional and governance aspects (Brooks et al. 2005; Engle and Lemos 2010; Hurlbert 2008).

As has been discussed in earlier sections of this chapter, a number of theoretical discourses have developed, such as adaptive management, adaptive co-management, and adaptive governance, in the quest for resilience in the face of uncertainty and climate change, and that take up the concept of adaptive capacity. Adaptive governance is seen to meet the call for dealing with increased uncertainty and change, arising from the 'growing number of failures among current approaches and increasing vulnerability of social-ecological systems' (Olsson et al. 2006, p 1). Along with the field of adaptive management, the concepts of learning by doing, social learning and scenario planning have become popular as a means of operationalising the need for flexibility and better integration of social and ecological factors. These approaches are seen as a response to the challenge of 'creating governance structures that are flexible and robust in the face of uncertainties and inevitable surprises' (Twin2Go 2010, p 3).

The adaptive co-management approach has also emerged from combining elements of adaptive management and collaborative management approaches, which also incorporate learning by doing and management flexibility, but emphasises collaboration and power-sharing within communities at the local level, as well as across regional and national levels (Resilience 2011). IWRM places more emphasis on collaborative governance and the recognition of the multiple values of water, and is seen as one means towards increasing capacity of water management in the face of climate change. Institutional capacity is also seen as a critical requirement in effective adaptation, particularly in the clarity of roles and responsibility of individual authorities, especially in extreme event situations (UNECE 2009). In the literature on good governance, and therefore in the governance assessment itself, adaptive capacity to climate change tends to be assumed if indicators of good governance are adequately met.

Tools and concepts used to measure the validity of outcomes of adaptive actions can also be employed to assess underlying states beneficial to the development of adaptive capacity. A number of determinants of adaptive capacity have been identified within the climate change impacts, adaptation and vulnerability literature. To recapitulate, common factors considered determinants can be categorised into the following groups; economic resources, technology, information and skills, infrastructure, institutions, equity, social capital, and collective action (Eakin and Lemos 2006; Engle and Lemos 2010; Yohe and Tol 2002). Yet, empirical verification of the merit of these norms for building adaptive capacity is sparse, particularly within the water sector (Engle and Lemos 2010; Wilbanks and Kates 1999).

4.2 Good Governance Determinants

4.2.1 Accountability, Participation, Transparency

As shall be discussed in the following chapter, the STRIVER governance assessment utilises three main indicators to assess good governance in the context of IWRM. These are accountability, participation, transparency (and IWRM is also employed). The indicators were not specifically designed to measure adaptive capacity, but were rather shaped in the context of good governance for IWRM. However, these indicators also play different roles in other adaptive capacity assessments (Engle and Lemos 2010; Hurlbert 2008; Iza and Stein 2009). Accountability, participation and transparency are often considered key principles in adaptive capacity. A recent IUCN report (Iza and Stein 2009) refers to different process principles in the discussion on reforming water governance, which are requisite to provide an enabling environment, including transparency, accountability and participation. Their definition of participation broadens out from more than just consultation in decision making to involvement in multi-stakeholder platforms and decision making at the lowest appropriate level. It is considered these elements of participation could effectively raise levels of awareness, co-management and citizen initiatives, all components deemed necessary for fostering effective water governance capacity as well as sources of resilience in social-ecological systems. Furthermore, the rule of law and legal certainty are seen as crucial for legitimacy of decision making and access to justice on environmental matters (Ebbesson 2010).

4.2.2 *IWRM & Integration*

IWRM is currently held up as the ideal framework for managing water in an integrated and sustainable way that would enhance the system's resilience to cope with the impacts of climate change on water resources. However, despite the concept's use in addressing the need for water governance processes to effectively and equitably manage the fair distribution and protection of the resource, it has weaknesses in terms of complexity, uncertainty and adaptive capacity (Timmerman et al. 2008). Timmerman et al. (2008) suggest that in addition to recognising multiple uses of water, that multiple sources of knowledge and information should also be integrated into management systems.

Olsson et al. (2006) explore the different features that contribute to the resilience of social-ecological systems in the face of change (in the context of adaptive co-management). Their criteria do not follow the neat normative categories of many of the other studies into adaptive capacity, but provide some useful insights into governance related criteria which can provide an enabling environment for enhanced resilience to environmental shocks and stresses. They suggest an 'enabling legislation that creates social space for ecosystem management' is requisite for the building of resilience. As vague as this may be, it deems that in order for resilience to be fostered, the institution of law should ensure that ecosystems and the environment are factored in as a relevant stakeholder. Not only should sectoral actors be integrated into legislation relating to resources (water in this case) but institutions also need to take account of ecosystem needs. This concept finds resonance with the element of integration and recognition for the non-economic uses of water within an IWRM context.

4.3 Resilience, Adaptive Governance and Adaptive Management Determinants

The following section reviews the common governance factors for adaptive capacity from the discourses relating to resilience and related concepts of adaptive governance and adaptive management.

4.3.1 *Leadership, Trust, Commitment*

Olsson et al. (2006) use the criteria of 'vision, leadership, and trust', which share some normative properties with accountability, in that an unaccountable system will not generate trust amongst its citizens. However, there is no reason to equate vision or leadership with the same norm, but both could be seen as requirements for the necessary political will requisite to foster proactive responses to climate change and develop relations across different networks and levels of decision making. Folke et al. (2005) also suggest that vision, trust and innovative leadership can provide key functions for adaptive governance, e.g. 'building trust, making sense, managing conflict, linking actors, initiating partnerships, compiling and generating knowledge, mobilizing broad support for change'. Other studies have reinforced the willingness to adjust to change from an individual (as well as societal) and this perspective is also seen as a key determinant in social ability to adapt to new pressures (Tompkins and Adger 2005).

The importance of these elements of leadership in building collaboration and resolving conflicts is underlined by their role as key components in bridging interests and stakeholders and to a certain extent driving realisation of other principles of adaptive governance. Leadership can be seen as an abstract concept, which can be highly subjective to personal opinion. Additionally, strong leadership may not always have a positive correlation with principles of adaptive governance, but it may be inferred that meeting the other principles of adaptive governance may not be as easily reached without the presence of leadership. Linkages may also exist with accountability, resources, networks, transparency and participation. Engle and Lemos (2010) also discuss the indicator 'commitment', which refers to the belief held by the different stakeholders that the institutional and governance structures in place are adequate for management of the resource as effectively and efficiently as possible.

4.3.2 Experience

Engle and Lemos (2010) note that more experience would correlate with a greater ability to deal with everyday events, as well as extremes, in an effective and efficient way. While experience can broadly be deemed as relevant, just as with the concept of leadership, precise measurement of this principle is very abstract. Yet, though an actor may have many years of experience, preconditioned ideals or values may subject his/her decisions to preconceived notions, which may or may not still be relevant for changing conditions. UNECE (2009) highlight the importance not just of career experience, but also fostering experience through training and simulation exercises on a regular basis.

4.3.3 Resources

Olsson et al. (2006) propose 'funds for responding to environmental change and for remedial action; capacity for monitoring and responding to environmental feedback' as indicators which both relate to the importance of human and financial resources for ensuring effective capacity for monitoring systems, enforcing laws and responding to extremes or feedbacks. The importance of information and knowledge sharing, not just in itself, but across different levels of stakeholder and decision makers is touched upon through criteria relating to information flow through social networks as well as the combination of various sources of information and knowledge. These criteria are also relevant for the creation of the appropriate level of public perception (Yohe and Tol 2002) for adaptation through sense making and collaborative learning (Olsson et al. 2006). Engle and Lemos (2010) also comment that levels of financial and human capital are critical for overall success of an organisation or governance structure. Yet, while more resources (financial and human) may increase the capacity of the system, it is how these resources are applied and organised that may be more important. Less could mean more. Therefore it is not just the presence of adequate resources, but perhaps the deployment of a suitable mix of financial and human resources across different scales that may be of relevance, emphasising the linkages with experience, networks, accountability, transparency and decentralisation.

4.3.4 *Networks & Connectivity*

Folke et al. (2005) explore the social elements of adaptive governance, which can enable adaptive ecosystem based management in the context of abrupt change. 'Connectivity across Networks' refers to connectivity across individuals, organisations, agencies and institutions through bridging organisations. Networks capture the various institutional levels and relationships involved with river basin management. Folke et al. (2005) also suggest that adaptive co-management requires more flexible social networks, which may be more innovative and responsive than bureaucracies in times of rapid change. Additionally, bridging and boundary organisations and networks (e.g. management councils, communities of practice, learning networks, associations, cross-sectoral partnerships, political coalitions and social movements) are seen as important central nodes of cross-scale interactions (Kofinas 2009). Challenges are, however, recognised in fostering adaptive learning between such bridging organisations and larger society as a whole (Kofinas 2009). It is assumed that the greater the networking and connectivity between groups and stakeholders involved in the management processes, the greater the adaptive capacity (Engle and Lemos 2010). While networks enable individuals to engage in the wider decision making environment, gain access to information and resources (technical or financial), the usefulness of such networks are determined by both social and institutional factors (Tompkins and Adger 2004). Hence, just as in the critique of participation, connectivity and networks alone may not imply a willingness to cooperate, which is requisite for systems to be adaptive (UNECE 2009).

4.3.5 *Predictability – Flexibility*

Flexibility is to be taken as the antithesis of irreversibility. This indicator is repeated across a number of the studies on adaptive capacity. The UNECE comments that 'the capacity to adapt requires flexibility. As a result, measures that are highly inflexible or where reversibility is difficult should be avoided' (UNECE 2009, p 78). In institutional terms, it refers to an ability to bend, but not break, and to learn iteratively, incorporating lessons learnt through experience efficiently and effectively (Engle and Lemos 2010). This concept of *iterative adaptive governance/learning by doing* is a key element of adaptive management and governance (Olsson et al. 2004b; Pahl-Wostl et al. 2007a). Tompkins and Adger (2004) also note that flexible management systems that incorporate learning-based processes (i.e. allow for modifications based on new information) are important for building resilience. Assumptions proposed are that the greater the flexibility of rules (legislation, institutions), the greater the adaptive capacity (Engle and Lemos 2010).

However, there is a struggle here between flexibility for adaptive management, and the need for certainty (Iza and Stein 2009; Tarlock 2009) or predictability (Hurlbert 2009; Engle et al. 2011) within the law, as emphasised in IWRM. Predictability suggests that all laws and regulations should be applied fairly and consistently. The assumption is that consistency in application of the law will enhance adaptive capacity. However, the discussion concerning the role and rule of law in adaptive governance (see Sect. 2.2.2) highlights the on-going challenge and discourse related to balancing predictability sought in the law, with flexibility requisite for adaptive behaviour. The IUCN (Iza and Stein 2009) use a similar concept in the process principle of 'certainty', rests upon the rule of law in terms of both predictability and enforceability. This would of course be dependent upon laws also reflecting principles of ecological integrity, equitable access for all and linkages between land and water resources. Otherwise, rigidity in the application of 'bad' laws and policies would diminish adaptive capacity.

4.3.6 Knowledge & Information

The UNECE (2009) cite the importance of supporting training and response systems with climate and hydrological information systems which are 'capable of delivering early warnings in a timely and efficient manner' (UNECE 2009, p 42). Folke et al. (2005) relate the idea of knowledge with the creation of an iterative learning environment. There are therefore important links with *flexibility* through the process of learning by doing. The goal here relates to an improved understanding of the dynamics of the whole system so that an understanding is established for how to manage periods of rapid change. The interpretation of knowledge is also highly linked with how to effectively deploy scientific information across different networks or levels of decision making for the management of resource issues in the context of change. Engle and Lemos (2010) also refer to the linkage of using scientific knowledge and information with the building of adaptive capacity, but add to the concept the importance of equality of decision making and knowledge use (in terms of power distribution among stakeholders and access to technical knowledge).

Nelson et al. (2007) also suggests that the ability to maintain a response capacity is predicated in part on the capacity for learning. Recent studies by Huntjens et al. (2011)

emphasise that in their study of eight different water governance regimes, positive correlations between knowledge indicators (information production, consideration of uncertainties, communication) and cooperation indicators (vertical, transboundary, joint/participation information) suggested that consensual knowledge is an important element in adaptive approaches when attempting to foster cooperation for managing uncertainty and change. This conclusion is also mirrored in studies by Tompkins and Adger (2004) and Olsson et al. (2006). Huntjens et al. (2011) go on to recognise the importance of socio-cognitive theory of information systems when recognising the interdependence of information management and social cooperation structures towards understanding the related challenges in developing adaptive water management regimes (Hemingway 1998, in Huntjens et al. 2011).

4.3.7 Decentralisation

Decentralisation and subsidiarity (Hurlbert 2008) refers to the delegation of responsibility and authority of water management to the lowest feasible level. Devolved decision making means that a system would be 'presumably, better able to recognize and respond to unforeseen circumstances' (IISD 2006, p 119). There is a theoretical link here to the IWRM component 'Basin/Watershed Approach', as well as to Olsson et al.'s (2004a) concept of enabling legislation that creates social space for ecosystem management. Yet, while a system may be highly decentralised, this does not imply that there are ecological based units of decision making. Nor does it always imply that sustainable solutions can be found in complex systems that contain multiple uses of water (i.e. river basins), where a measure of central top down control and guidance may provide some balance. Huntjens et al. (2011) concluded that in large scale complex systems, a centralised governance structure can help to facilitate participatory processes, set standards, build capacity and assist in building of cooperation across boundaries, conflict resolution and the provision of information not available to local level actors or institutions.

4.4 Analytical Challenges

This list of indicators captures the development in the analytical field of adaptation and vulnerability in the preceding decade. However, it is equally recognised that there has been fairly minimal empirical verification of the correlation between different principles and adaptive outcomes, particularly at local and regional scales, and more so within the water sector (Engle and Lemos 2010; Wilbanks and Kates 1999). There are a number of analytical challenges relating to the different principles and indicators of adaptive capacity listed above, five of which are discussed below. Firstly, much of the discussion around governance issues in adaptation and adaptive capacity has a strong normative edge. Normative principles such as accountability and participation tend to denote a stronger bias towards the researcher's analytical framework. More open indicators such as knowledge and levels of decision making are less prescriptive and therefore predisposed to be more iteratively developed through the research process, both theoretical and empirical exploration. While this distinction should be recognised, and normative bias to the analytical framework should be avoided where possible, it should not be seen as a major impediment to the development of more robust indicators.

Secondly, there is a difference between the process indicators as described in many of the studies, and the more outcome associated determinants in others. Requirements such as 'enabling legislation that creates social space for ecosystem management' (Olsson et al. 2004a) and institutional capacity (UNECE 2009), can be seen as requisite for both an enabling environment for adaptive capacity, but also as an outcome of sufficient adaptive capacity. A key issue is therefore how questions relating to enabling legislation and institutional capacity could be integrated into more open indicators. Or, are such concepts in fact outcomes of indicators such as 'levels of decision making' and 'networks', and therefore should not be separately tackled within the adaptive capacity assessment per se? More specifically regarding institutional capacity, one could perhaps infer that if indicators such as transparency, knowledge, networks, resources, decentralisation and subsidiarity as well as experience are met, then institutional capacity should be strengthened, and therefore it could be taken as an output.

Similarly, the issue of 'process vs. outcome' is pertinent to IWRM. While IWRM is not considered an indicator, its component parts could be seen as useful determinants of adaptive capacity. An indicator for 'integration' could encapsulate a key element of IWRM. Normative prescriptions could be avoided by not suggesting that an ideal level or type of integration pre-exists, but that different levels and types may enable adaptive capacity in varying sectors or geographies. Additionally, considering that numerous studies have shown that 'a substantial gap exists between promise and practice' (Ingram 2011, p 2) in IWRM, it would be make more sense to focus on how different types of integration rather than IWRM per se contribute adaptive capacity, rather than testing normative assumptions based on the criteria of IWRM.

The concept of environmental integrity or ecological system resilience (Nelson et al. 2007) appears regularly as a key determinant for adaptive capacity in the adaptive management discourse. Since the capacity of aquatic ecosystems to produce many of the goods and services on which societies depend is rapidly declining, the provision of water for nature or nature as a buffer can be seen as a key indicator of adaptive capacity in a system under stress. If the biological component of the system is already under stress, then adaption to more extreme conditions may be limited. Principles purported within the adaptive governance literature are linked with achieving these outputs, but again the question arises of how to define the relationship between ecological integrity and resilience with adaptive capacity.

Thirdly, preferences concerning the right mix of modes of governance (hierarchy/state, market/private and decentralisation/civil society) are rife within the literature on adaptation and vulnerability, despite the recognition by many that what matters is that prescriptions fit contexts (Ingram 2011). The focus on full participation and decentralisation in water management as desirable norms is reflected across a broad swath of the literature (Hurlbert 2009; Nelson et al. 2007; UNDP 1997; UNECE 2009;

WB 2002). However, other studies note the fact that decentralisation and participation per se are not *a priori* requirements for better management and enhanced resilience. Berkes in Nelson et al. (2007, p 409) suggests that 'the balance of evidence shows that neither purely local level management nor purely higher level management works well by itself', and Lemos and Agrawal (2006) highlight the development of emerging hybrid, multilevel and cross-sector forms of environmental governance.

Fourth, Ingram (2011, p 8) adds that 'participation is no panacea for water conflicts'. Other studies such as Iza and Stein (2009) elaborate that other factors such as coordination across levels, rather than pure participation and decentralisation hold significant importance. Thus, there is need to look beyond prescriptive norms such as participation and decentralisation and subsidiarity, to more exploratory indicators which allow examination of causal relationships between different indicators and adaptive capacity within different sectors as well as governance regimes.

Fifth and finally, in a number of studies the indicator of transparency is pinpointed as fundamental to good governance and adaptive capacity. However, drawing on studies and publications in the resilience framework and the wider climate dialogue, it might be worth broadening out from the normative prescription of transparency to a more thorough exploration of the contribution that different forms of knowledge and information play in enhancing resilience. By looking at knowledge as well, we therefore refer to not just scientific information and data (hydrological models, climate models, economic statistics etc.), but can also recognise the potential importance of local and indigenous knowledge. A recent report from Switzerland comments on the need to take into account and integrate traditional knowledge in climate data systems (Lugon 2010).

An awareness of the need for climate services also arose out of the 3rd World Climate Change Conference in Geneva (WCC-3 2009), which refers to the provision of climate information (both current climate variability and recent and future climate change) (Lugon 2010). It also calls for better management, communication and understanding of this information so that resource managers and the public alike can actually generate knowledge out of the wealth of data and information available. The HEID report comments that while today, people are likely to be inundated with information, often 'the hurdles are not the hard science but the communication' (Lugon 2010, p 64). It also notes that climate information per se is not enough; to be truly valuable it needs to be integrated with socio-economic and other environmental data. It is therefore important to investigate not just what kind of information decision makers are getting, but also how they use it, with whom do they share it and how relevant is it to the problem they need to resolve.

4.5 Developing the Approach

The understanding that past management approaches have led to a minimisation of choices through steady state resource management (Milly et al. 2008) and a focus on hard infrastructure and technical solutions (Gleick 2003), can be counter balanced by suggesting that future approaches should enable systems to have more choices to draw from in times of uncertainty and crisis. Drawing from the resilience literature, the inference is that higher adaptive capacity should correlate incrementally with an ability to transform or adapt to new challenges or states (refer to Sect. 2.4). Therefore one would expect positive fulfilment of the adaptive capacity indicators to correspond with more transformative and adaptive actions and management approaches, and negative fulfilment of adaptive capacity indicators to correspond with passive approaches. One may also then infer that the more transformative the approach, the better and larger the future choices should be.

To reiterate, *transformation* is seen as the transition of a system to a fundamentally different, potentially more desirable state (Chapin et al. 2009), onto a trajectory that sustains and enhances ecosystem services, societal development (including economic security) and human well-being (Folke et al. 2010). The concept of triple loop learning (Pahl-Wostl 2009) is associated with transformation. *Adaptation* refers to adjustments in response to actual or expected climate impacts, that allows the SES to persist within the current state or basin of attraction (Folke et al. 2010). This can be associated with elements of double loop learning and single-loop learning (Pahl-Wostl 2009). Passive change refers to the degradation of a system to a less favourable state resulting from a failure to adapt or transform (Folke et al. 2010). *Passive* change can be seen as the inverse of transformation, so while transformation is determined to be a positive transition to a more favourable state, passive change should be seen as transition to a more negative state (i.e. unintended transformation). Deeper operationalisation of these categories will be developed and discussed in Chap. 6.

Creating adaptive capacity in water governance regimes should be about creating options now and in the future, rather than limiting them and allowing a system to bend rather than break in the face of new challenges, ensuring that change is navigated in a way that leads to transformative and adaptive responses, rather than passive forced transformations with negative outcomes. Thus, for the purposes of this piece of research, adaptive capacity is conceptualised through *its role in the transformation potential of a system to a more sustainable state as a means to absorb future shocks and uncertainty, thereby creating not limiting future adaptation choices*.

Different forms of adaptive outcome can therefore be seen as manifestations of the presence or absence of adaptive capacity. Drawing on the literature and discussion on governance determinants and indicators of adaptive capacity above, a list of broad determinants was developed for the exploration of adaptive capacity across the case areas. These were Knowledge; Networks; Levels of Decision Making; Integration, Predictability-Flexibility; Experience; Resources; Leadership. Table 4.1 presents both the determinants and sub-criteria, which draw on current understanding and the different determinants and indicators (as often used interchangeably in the literature) in the discipline of adaptive capacity, adaptive governance and adaptive management, as well as the discourse on Integrated Water Resources Management. The more prescriptive and normative indicators employed within the STRIVER/BRAHMATWINN assessment

Table 4.1 Initial operationalisation of tentative determinants to explore adaptive capacity across the case areas

Governance determinants of adaptive capacity	
Tentative Indicators	Sub-criteria
Knowledge	Right to Information; Communication/Public Perception; Spatial Planning; Access to scientific/environmental information; Exchange of data & information; Integration of scientific expertise; Quality of Scientific Information; Use of traditional & local knowledge
Networks	Access to participation; Selection of non-state actors; Level of influence; Type of participation; Stage in the political process; Social Networks; Professions Networks; Willingness to Cooperate
Levels of decision making	Ecological based units of decision making; Institutional arrangements
Integration	Geographical integration; Sectoral/Uses integration; Political integration
Flexibility-predictability	Consistency in rule of the law; Rigidity of legal provisions; Iterative elements of law/institutions
Resources	Financial resources; Quantity/quality of human resources; Organisation of resources; Independence/impartiality of experts
Experience	Training & development; Years of experience
Leadership	Political Commitment; Facilitating role; Initiation of partnerships; Support mobilisation; Linking of actors; Trust amongst stakeholders

were replaced by more open determinants to better complement the iterative development of indicators within this research.

These governance and institutional related determinants are the platform from which adaptive capacity may be explored across the case areas. These determinants have been discussed as being important to the nature of adaptive capacity and to affecting the outcome of adaptive actions. While climate change risks have been well addressed in the academic literature, adaptation to climate change is often initially experienced through adjustments to variability and extremes (Tompkins and Adger 2004), but adaptation rarely takes place purely in relation to climate change alone (Parry et al. 2007). The potential inconsistency between using past extreme events as a proxy, when simultaneously enforcing the notion that the past may no longer be a prologue for the future, is fully recognised.

However, the focus on extremes specifically pinpoints situations that while currently recognised as an outlier event, may in the future become situated within the normal frame of management reference (e.g. 100-year floods recurring three times within the space of a decade). In this case past adaptations to climatic or hydrological stresses are likely to provide some useful insight into incremental step changes in the future hydro-climatic reality that are to be expected over the next 10–20 years. If in the coming decades (20–50 years) massive shocks do occur, where certain tipping points are crossed in the climate system, then one may insinuate that the learning generated through better understanding adaptive processes (rather than steady state resource management processes) should help decision makers better assess and develop responses to larger state changes. Therefore, concerning tensions and trade-offs across different scales, the assumption is that adaptations to current variability and experience of extremes should enable capacity to develop to longer term threats and challenges from climate change, but that inter-jurisdictional challenges and dynamics might hinder coherent adaptation.

Many studies have centered on theoretical development and in turn have been loaded with the assumption that these governance arrangements are desirable or key to increasing adaptive capacity. A common approach has been to define the key indicators and relevant policy or management prescriptions needed for adaptive capacity to be mobilised and then characterise how they are present within the system analysed (Adger et al. 2005; Brooks et al. 2005; Eakin and Lemos 2006; Smit and Wandel 2006; Yohe and Tol 2002; Pahl-Wostl et al. 2007c; Huntjens et al. 2010). It is a highly inductive approach that has partly led to a gap between theory and practice in establishing links between various water governance approaches and proven positive results in managing water resources in reality (Medema et al. 2008). Increasing the number of empirical studies in contrasting governance settings on the mobilisation and measurement of adaptive capacity can in part assist in addressing this gap. However, there are still few deep empirical examples exploring adaptive actions in periods that might be representative of a future warmer world, or even in attempting to measure the role of these approaches to support the theoretical assumptions. One aim of this book is to contribute to closing this gap.

4.6 Summary

The academic discourse on climate change adaptation in the water sector has seen a gradual realisation that hard path technical approaches (Gleick 2003) must be better balanced with soft path solutions, that also focus more on the enabling social infrastructure (governance, institutions, management) requisite for successful adaptive approaches (Pahl-Wostl 2007). Governance clearly plays a critical role in developing more adaptive and sustainable water management. Heightened vulnerability can erode resilience and so impede institutions from facilitating adaptation or resulting in maladaptation. Yet while the vulnerability, adaptation, and resilience frameworks are apt for defining the challenges that governance regimes face, their vulnerability in meeting those challenges, and the solutions to overcoming those challenges, they deal more with what those outcomes should look like than how they should be achieved; which is addressed by the concept of adaptive capacity.

While there are increasingly numerous calls for water governance and associated management institutions to be resilient and robust towards future uncertainty and climate change impacts, there is room for deeper discussion on what desirable outcomes would look like. As adaptation responses are shaped, it is important to question whether adaptation should lead to robust and resilient governance frameworks, or flexible and adaptive ones, or somewhere in between. Can an SES be both resilient and yet able to transform to be adaptable to new challenges and hydro-climatic realities? Where are the trade-offs implicit in the generation of institutional characteristics needed for climate resilient structures and adaptive elements. If we do presume that both robustness and transformative characteristics are desirable, then there is a need for cross case comparisons to show how these might be balanced and not mutually exclusive as well as to identify the means of negotiating and navigating these tensions within the governance framework.

There has been a set of incremental shifts in the focus on how to achieve better water management outcomes, from governance approaches that focus on the state, then the market, then decentralised role of user groups (Meinzen-Dick 2007). In the face of a number of converging disturbances in SESs, biodiversity loss, population growth and economic development, attention more recently turned to understanding governance approaches that fostered adaptability in water governance regimes. Generally, the bodies of research that have focussed on this issue have proposed that more flexible, participatory, collaborative, and learning-based designs and approaches will increase adaptive capacity and sustainability of water systems (Cromwell et al. 2007; Kallis et al. 2006; Pahl-Wostl et al. 2007b). Yet, scholars have also stressed the importance of acknowledging the difficulty in establishing links between concepts and management paradigms such as IWRM, adaptive management and adaptive governance with proven positive results in reality (Huitema et al. 2009; Medema et al. 2008).

In order to examine and define the underlying process that will enable governance regimes to respond to the challenges of the anthropocene, the concept of adaptive capacity has been used to refer to the latent conditions required for enabling successful and sustainable adaptation. The presence of adaptive capacity should allow a system to prepare for and adjust to the exposure of a stress, thereby reducing sensitivity and potentially embracing opportunities presented by that risk to not only adapt, but potentially transform to a new more sustainable pathway. In the field of resilience, adaptive capacity represents a more multi-faceted concept, both an ability to absorb shocks to maintain the system state, but also to facilitate transformations or transitions to a new, more desirable state.

For the purposes of this piece of research, adaptive capacity is conceptualised in relation to its role in the transformation potential of a system to a more stable and sustainable state as a means to absorb future shocks and uncertainty, thereby creating not limiting future adaptation choices. Thus, adaptive capacity should enable the system to prepare for, respond to and cope with challenges such as variability, uncertainty and surprise. The accommodation of uncertainty should enable the system to not constrain future options (creating choices), couching the understanding of adaptive capacity in the context of stationarity argument. Building adaptive capacity, by cultivating or contributing to the presence of its determinants in an SES, should therefore improve the ability of that SES to be resilient to surprises and larger scale changes, by proactive and reactively shaping positive responses, including transformations or transitions to a better state.

Conversely, the lack of adaptive capacity would lead to a narrowing of future choices (minimising choices), for example through a dominance of hard technical measures which are difficult to reverse when future hydrological or consumption patterns do not follow the decision maker's calculations. This lens of choice creation, posits adaptive capacity in the discourse on transformation and panarchy (Folke et al. 2010; Olsson et al. 2006; Walker et al. 2006; Schlüter and Herrfahrdt-Pähle 2011) and recognises the importance of ongoing dialogues within the policy sciences, such as path dependency, institutional inertia, and decision making under uncertainty (Lempert et al. 2004; North 1990).

Despite the growing body of evidence on adaptive capacity, governance and management, there is still significant scope for scientific validation and evaluation of many of the assumptions in the literature that has developed over the past decade, particularly in cases that cross both spatial and temporal scales (Chapin et al. 2009) rather than looking at single institutions in isolation (Meinzen-Dick 2007). Studies should therefore move beyond just assessing adaptation strategies and plans, to being able to investigating adaptive actions with a cross-scale lens. While a governance regime may not be a national plan or river basin plan for adaptation to climate change, local water users may already have techniques for coping with uncertainty that could provide valuable insights into the adaptive capacity of a particular sub-basin or even river basin system.

The current status of research into adaptive capacity and building of adaptive options is still in its infancy, despite an increase of interest in recent years (Engle 2011), and has only recently focussed more heavily on the practicalities of how to adapt (Dovers and Hezri 2010). Previous assessments and studies have focused on first showing that governance is important to adaptation and adaptive capacity, and then identifying certain approaches that are important in a system for being adaptable to change. The relative paucity of deep empirical examples exploring adaptive actions in periods that might be representative of a future warmer world remains a challenge in the operationalisation and characterisation of adaptive capacity as well as in the development in understanding how to mobilise it as climate change impacts take hold. The methodology employed for this research and described in the next chapter aims to address this gap, by drawing on the conceptualisation of adaptive capacity that draws from the multiple approaches described within this chapter.

References

Adger WN (2003) Social capital, collective action, and adaptation to climate change. Econ Geogr 79(4):387–404

Adger WN, Arnell NW, Tompkins E (2005) Successful adaptation to climate change across scales. Glob Environ Chang 15(2):77–86

Adger WN, Agrawala S, Mirza MMQ, Conde C, O'Brien K, Pulhin J, Pulwarty R, Smit B, Takahashi K (2007) Assessment of adaptation practices, options, constraints and capacity. In: Parry ML, Canziani OF, Palutikof JP, van der Linden PJ, Hanson CE (eds) Climate change 2007: impacts, adaptation and vulnerability (Contribution of working group II to the fourth assessment report of the intergovernmental panel on climate change). Cambridge University Press, Cambridge

Brooks N, Adger WN, Kelly PM (2005) The determinants of vulnerability and adaptive capacity at the national level and the implications for adaptation. Glob Environ Chang 15(2):151–163. doi:10.1016/j.gloenvcha.2004.12.006

Bussey M, Carter RW, Carter J, Mangoyana RB, Matthews J, Nash D, Oliver J, Richards R, Thomsen D, Sano M, Weber E, Smith TF (2010) Societal responses to significant change: an historical analysis of adaptive capacity. Report for the South East Queensland Climate Adaptation Research Initiative. Sustainability Research Centre, University of the Sunshine Coast, Sippy Downs, Qld, Australia

Chapin FS, Folke C, Kofinas GP (2009) A framework for understanding change. In: Chapin FS, Kofinas GP, Folke C (eds) Principles of ecosystem stewardship resilience-based natural resource management in a changing world. Springer, New York

Cromwell JE, Smith JB, Raucher RS (2007) Implications of climate change for urban water utilities. Association of Metropolitan Water Agencies, Washington, DC

Dovers SR, Hezri AA (2010) Institutions and policy processes: the means to the ends of adaptation. Wiley Interdiscip Rev: Clim Chang 1(2):212–231

Dow K, Kasperson R, Bohn M (2006) Exploring the social justice implications of adaptation and vulnerability. In: Adger N, Paavola J, Huq S, Mace MJ (eds) In fairness in adaptation to climatic change. MIT Press, Cambridge, MA

Eakin H, Lemos MC (2006) Adaptation and the state: Latin America and the challenge of capacity-building under globalization. Glob Environ Chang 16(1):7–18. doi:10.1016/j.gloenvcha.2005.10.004

Ebbesson J (2010) The rule of law in governance of complex socio-ecological changes. Glob Environ Chang 20(3):414–422. doi:10.1016/j.gloenvcha.2009.10.009

Engle NL (2010) Adaptation to extreme droughts in Arizona, Georgia, and South Carolina: evaluating adaptive capacity and innovative planning and management approaches for states and their community water systems. University of Michigan, Ann Arbor

Engle NL (2011) Adaptive capacity and its assessment. Glob Environ Chang 21(2):647–656. doi:10.1016/j.gloenvcha.2011.01.019

Engle NL, Lemos MC (2010) Unpacking governance: building adaptive capacity to climate change of river basins in Brazil. Glob Environ Chang 20(1):4–13. doi:10.1016/j.gloenvcha.2009.07.001

Engle NL, Johns OR, Lemos MC, Nelson DR (2011) Integrated and adaptive management of water resources: tensions, legacies, and the next best thing. Ecol Soc 16(1):19 [online]. URL: http://www.ecologyandsociety.org/vol16/iss11/art19/

Folke C, Hahn T, Olsson P, Norberg J (2005) Adaptive governance of social-ecological systems. Annu Rev Environ Resour 30(1):441–473. doi:10.1146/annurev.energy.30.050504.144511

Folke C, Carpenter SR, Walker B, Scheffer M, Chapin T, Rockström J (2010) Resilience thinking: integrating resilience, adaptability and transformability. Ecol Soc 14(4):20 [online]. URL: http://www.ecologyandsociety.org/vol15/iss24/art20/

Gleick PH (2003) Global freshwater resources: soft-path solutions for the 21st century. Science 302(28 November):1524–1528

Gupta J, Termeer C, Klostermann J, Meijerink S, Van den Brink M, Jong P, Nooteboom S, Bergsma E (2010) The adaptive capacity wheel: a method to assess the inherent characteristics of institutions to enable the adaptive capacity of society. Environ Sci Policy 13(6):459–471. doi:10.1016/j.envsci.2010.05.006

Huitema D, Mostert E, Egas W, Moellenkamp S, Pahl- Egas W, Moellenkamp S, Pahl-Wostl C, Yalcin R (2009) Adaptive water governance: assessing the institutional prescriptions of adaptive (co-)management from a governance perspective and defining a research agenda. Ecol Soc 14(1):26 [online]. URL: http://www.ecologyandsociety.org/vol14/iss21/art26/

Huntjens P, Pahl-Wostl C, Grin J (2010) Climate change adaptation in European river basins. Reg Environ Chang 10:263–284

Huntjens P, Pahl-Wostl C, Rihoux B, Schlüter M, Flachner Z, Neto S, Koskova R, Dickens C, Nabide Kiti I (2011) Adaptive water management and policy learning in a changing climate: a formal comparative analysis of eight water management regimes in Europe, Africa and Asia. Environ Policy Gov 21:145–163

Hurlbert M (2008) An analysis of trends related to the adaptation of water law to the challenge of climate change: experiences from Canada. Paper presented at the Climate 2008, Online, 3–7 November 2008

Hurlbert M (2009) The adaptation of water law to climate change. Int J Clim Chang Strateg Manag 1(3):230–240

IISD (2006) Designing policies in a world of uncertainty, change and surprise: adaptive policy-making for agriculture and water resources in the face of climate change. International Development Research Centre & The Energy and Resources Institute, Winnipeg, Manitoba & New Delhi, India.

Ingram H (2011) Beyond universal remedies for good water governance: a political and contextual approach. In: Garrido A, Ingram H (eds) Water, Food and Sustainability. Routledge, New York

IPCC (2001) Climate change 2001: impacts, adaptation, and vulnerability technical summary. A report of Working Group II of the Intergovernmental Panel on Climate Change. WMO/UNEP, New York

Iza A, Stein R (2009) RULE – reforming water governance. IUCN, Gland

Kallis G, Videira N, Antunes P, Pereira G, Spash CL, Coccossis H, Quintana SC, del Moral L, Hatzilacou D, Lobo G, Mexa A, Paneque P, Mateos BP, Santos R (2006) Participatory methods for water resources planning. Environ Plan C: Gov Policy 24(2):215–234

Kofinas GP (2009) Adaptive co-management in social-ecological systems. In: Chapin FS, Kofinas GP, Folke C (eds) Principles of ecosystem stewardship: resilience-based natural resource management in a changing world. Springer, New York

Lemos MC, Agrawal A (2006) Environmental governance. Annu Rev Environ Resour 31:297–325. doi:10.1146/annurev.energy.31.042605.135621

Lempert R, Nakicenovic N, Sarewitz D, Schlesinger M (2004) Characterizing climate-change uncertainties for decision-makers: an editorial essay. Clim Chang 65(1):1–9

Lugon R (2010) Climate information for decisionmaking: lessons learned from effective user-provider communication schemes. The Graduate Institute, Geneva

Medema W, McIntosh BS, Jeffrey PJ (2008) From premise to practice: a critical assessment of integrated water resources management. Ecol Soc 13(2):29 [online]. URL: http://www.ecologyandsociety.org/vol13/iss22/art29/

Meinzen-Dick R (2007) Beyond panaceas in water institutions. Proc Natl Acad Sci U S A 104(39):15200–15205

Milly PCD, Betancourt J, Falkenmark M, Hirsch RM, Kundzewicz ZW, Lettenmaier DP, Stouffer RJ (2008) Stationarity is dead: whither water management? Science 319(5863):573–574. doi:10.1126/science.1151915

Nelson DR, Adger WN, Brown K (2007) Adaptation to environmental change: contributions of a resilience framework. Annu Rev Environ Resour 32:395–419

North DC (1990) Institutions, institutional change, and economic performance. Cambridge University Press, New York

OECD (1997) Better understanding our cities: the role of urban indicators. Organisation for economic co-operation and development. OECD, Paris

Olsson P, Folke C, Berkes F (2004a) Adaptive comanagement for building resilience in social-ecological systems. Environ Manag 34(1):75–90

Olsson P, Folke C, Berkes F, Hahn T (2004b) Social-ecological transformation for ecosystem management: the development of adaptive co-management of a wetland landscape in southern Sweden. Ecol Soc 9(4):2 [online]. URL: http://www.ecologyandsociety.org/vol9/iss4/art2/. doi:Artn 2

Olsson P, Gunderson LH, Carpenter SR, Ryan P, Lebel L, Folke C, Holling CS (2006) Shooting the rapids: navigating transitions to adaptive governance of social-ecological systems. Ecol Soc 11(1):18 [online]. URL: http://www.ecologyandsociety.org/vol11/iss11/art18/

Pahl-Wostl C (2007) Transitions towards adaptive management of water facing climate and global change. Water Resour Manag 21(1):49–62. doi:10.1007/s11269-006-9040-4

Pahl-Wostl C (2009) A conceptual framework for analysing adaptive capacity and multi-level learning processes in resource governance regimes. Glob Environ Chang-Hum Policy Dimens 19(3):354–365. doi:10.1016/j.gloenvcha.2009.06.001

Pahl-Wostl C, Craps M, Dewulf A, Mostert E, Tabara D, Taillieu T (2007a) Social learning and water resources management. Ecol Soc 12(5):5 [online]. URL: http://www.ecologyandsociety.org/vol12/iss12/art15/

Pahl-Wostl C, Kabat P, Möltgen J (eds) (2007b) Adaptive and integrated water management: coping with complexity and uncertainty. Springer, New York

Pahl-Wostl C, Sendzimir J, Jeffrey P, Aerts J, Berkamp G, Cross K (2007c) Managing change toward adaptive water management through social learning. Ecol Soc 12(2):30 [online]. URL: http://www.ecologyandsociety.org/vol12/iss32/art30/

Parry ML, Canziani OF, Palutikof JP (2007) Technical summary. Climate change 2007: impacts, adaptation and vulnerability. Contribution of working group II to the fourth assessment report of the Intergovernmental Panel on Climate Change. In: Parry ML, Canziani OF, Palutikof JP, van der Linden PJ, Hanson CE (eds) Fourth assessment report of the Intergovernmental Panel on Climate Change. Cambridge University Press, Cambridge, pp 23–78

Pelling M, High C (2005) Understanding adaptation: what can social capital offer assessments of adaptive capacity? Glob Environ Chang-Hum Policy Dimens 15(4):308–319. doi:10.1016/j.gloenvcha.2005.02.001

Resilience (2011) Resilience alliance: adaptive co-management. http://www.resalliance.org/index.php/adaptive_comanagement. Accessed 8 Aug 2011

Schlüter M, Herrfahrdt-Pähle E (2011) Exploring resilience and transformability of a river basin in the face of socioeconomic and ecological crisis: an example from the Amudarya river basin, Central Asia. Ecol Soc 16(1):32 [online]. http://www.ecologyandsociety.org/vol16/iss31/art32/

Slocombe DS (1998) Defining goals and criteria for ecosystem-based management. Environ Manag 22(4):483–493

Smit B, Wandel J (2006) Adaptation, adaptive capacity and vulnerability. Glob Environ Chang 16(3):282–292

Tarlock D (2009) Seminar. Paper presented at the UNESCO Centre for Water Law, Policy and Science,, University of Dundee, Scotland., 16 July 2009

Timmerman JG, Pahl-Wostl C, Möltgen J (2008) The adaptiveness of IWRM; analysing European IWRM research. IWA Publishing, London

Tompkins EL, Adger WN (2004) Does adaptive management of natural resources enhance resilience to climate change? Ecol Soc 9(2):10 [online]. http://www.ecologyandsociety.org/vol19/iss12/art10

Tompkins EL, Adger WN (2005) Defining response capacity to enhance climate change policy. Environ Sci Policy 8:562–571

Twin2Go (2010) Synthesising research for adaptive water governance. Twin2Go Newsl Vol 1, January 2010

UNDP (1997) Governance for sustainable human development. United Nations Development Programme. http://mirror.undp.org/magnet/policy/

UNECE (2009) Guidance on water and adaptation to climate change. United Nations Economic Commission for Europe, Geneva

Walker BH, Gunderson LH, Kinzig AP, Folke C, Carpenter SR, Schultz L (2006) A handful of heuristics and some propositions for understanding resilience in social-ecological systems. Ecol Soc 11(1):13 [online]. URL: http://www.ecologyandsociety.org/vol11/iss11/art13/

WB (2002) Toward more operationally relevant indicators of governance. Prem Notes No49, December 2002

WCC-3 (2009) Better climate information for a better future. Conference Statement. http://www.wmo.int/wcc3/documents/WCC-3_Statement_Ex_Summary_04-09-095PM.pdf. Accessed May 2010

Wilbanks TJ, Kates RW (1999) Global change in local places: how scale matters. Clim Chang 43(3):601–628

Wilby RL, Vaughan K (2011) Hallmarks of organisations that are adapting to climate change. Water Environ J 25:271–281

Yohe G, Tol RSJ (2002) Indicators for social and economic coping capacity – moving toward a working definition of adaptive capacity. Glob Environ Chang 12(1):25–40

Chapter 5
Applying a Multi-pronged Approach to Assessing Adaptive Capacity

Abstract The chapter explains the methodological approach for the research presented in this book that aims to contribute to both theory and practice for developing climate adaptive water governance regimes. The chapter presents how suitable proxies and indicators for adaptive capacity were identified and developed through in depth empirical assessment of institutional adaptations and reactions to extreme events and stresses across the highly contrasting case areas. It details the multi-scale empirical approach taken in the context of recent extreme events to address some of the weaknesses from previous, often normatively driven, research on adaptive capacity that has often been focused either at national or local levels, but rarely across different jurisdictional, administrative and political levels.

Keywords Empirical assessment of adaptive capacity • Proxy extreme events • Assessment of adaptive actions • Governance determinants of adaptive capacity • Water governance assessment • Qualitative analysis of adaptive capacity

5.1 Introduction

The research presented in this book posed the following key questions:

- How do governance regimes (and mechanisms within these regimes) facilitate adaptive capacity in the water sector?
- What are the key tensions in building adaptive capacity that manifest across different contexts and scales, and how might these be navigated?

Focusing on adaptive capacity across multiple scales (Fig. 5.1) could redress the frequent disjuncture between the complex interactive nature of adaptive actions in reality and the levels at which the different adaptation foci tend to take place and at which research has primarily been targeted (Brunner 2010; Dovers and Hezri 2010).

M. Hill, *Climate Change and Water Governance: Adaptive Capacity in Chile and Switzerland*, Advances in Global Change Research 54,
DOI 10.1007/978-94-007-5796-7_5, © Springer Science+Business Media Dordrecht 2013

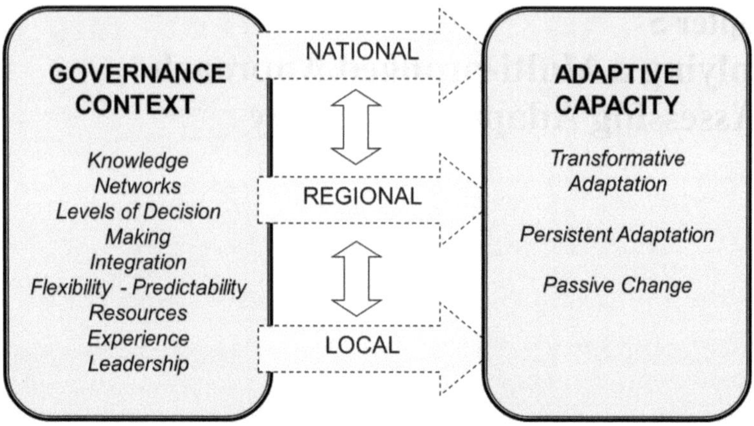

Fig. 5.1 Analytical framework for assessing relationship between the governance context at different scales (national, sub-national/regional and local) and adaptive capacity

Often, climate change policy and adaptation planning and assessments are made or focused (but not exclusively) at the national levels, while the consequences of those decisions – climate impacts themselves – are experienced (again not exclusively) at local community levels (Brunner 2010; Kane and Yohe 2000).

Independent variables relate to the governance context, operationalised through governance related determinants as discussed and set out in the previous chapters. The dependent variables relate to adaptive capacity, which has been theoretically defined in the preceding chapters and for measurement purposes has been operationalised through different categories of adaptive action (Transformation; Persistent Adaptation, Passive).

The social science field of climate change impact, adaptation and vulnerability is a still relatively young scientific discipline, and the literature on governance related indicators of adaptive capacity, specifically relating to the law, is particularly slim. Therefore, the broad aim of this research is to contribute to the growing understanding of adaptive capacity and resilience theory. Additionally, the case study and extreme event analysis aims to provide deeper empirical insights into the adaptive capacity of water governance in the specific contexts of the Valais in Switzerland and Aconcagua in Chile. Findings from this study could usefully inform policy makers and water managers on how to develop characteristics of the system, which can contribute to resilience in the face of uncertainty and future climatic change. The key objectives of the research are to:

- Contribute to the conceptualisation and operationalisation of adaptive capacity;
  - Identify the key components of adaptive capacity that can be empirically observed in the case areas in response to extreme events. In this context, adaptation strategies can be seen as outputs of the water management regime (Huntjens 2011);

- Better understand how the governance context and elements within those frameworks contribute to an enabling environment for adaptive capacity;
- Better understand the challenges in generating adaptive capacity across temporal and spatial scales.

Since this research took place within the context of the EU FP7 ACQWA project, there were certain parameters within which the research needed to take place, and a certain set of deliverables that needed to be met. Primarily, the case regions for the project were already defined as being the Rhône Basin, the Po Basin and the Aconcagua Basin, as well as a set of secondary case areas in Argentina and Kyrgyzstan that would inform the main cases. Additionally, as part of the main work package deliverables, a governance assessment in the context of IWRM was to be completed as a preliminary task for the research. This chapter on methodology will focus more heavily on the primary research on adaptive capacity across the disparate cases of the Chilean and Swiss basins, but will also detail the methods employed to complete the governance assessment (presented in Chaps. 7 and 8), as it was a piece of research that served to inform the findings, despite it not being a core part of the analytical framework developed for this research.

As detailed above, the case study selection was in part pre-defined by the parameters of the ACQWA project. While the case selection was driven primarily from the physical science context (in particular the fact that the Chilean region is facing *today* what might be the climate and glacio-nival environments of *tomorrow* in the Alps), as contrasting mountain zones with varying levels of climate data across which developments in hydrological and climatological models could be developed and transferred, the case areas also present a valuable opportunity for comparative analysis across highly contrasting governance contexts. Chapter 4 provides a deeper explanation of the case areas, which represent mountain watershed glacio-nival regimes where climate impacts on water resources (glaciers, snow pack, precipitation) have already been documented (Beniston et al. 2011; Pellicciotti et al. 2007). As the frame of analysis is interested in understanding multi-scalar process adaptive processes, the broad unit of analysis is river basin management (including a cross section of different uses across different scales), within which a set of sub-case areas were identified (see Chap. 6).

Research was divided into two complementary stages; an initial governance assessment and the main adaptive capacity assessment (Fig. 5.2). This section shall briefly detail the governance assessment methodology, but focus more heavily on the core methodological and analytical structure of the adaptive capacity elements that was the focus of the research. The governance assessment analysed the governance framework according to indicators of accountability, transparency and participation, within the wider context of IWRM. The next step aimed to more comprehensively address the adaptability of the governance system, by taking the impacts of climate change more specifically into account.

A set of extreme events were therefore identified within each case area to further test and refine a wider set of indicators in the context of past extreme events, such as drought and flooding events. These events served as a context in which to explore

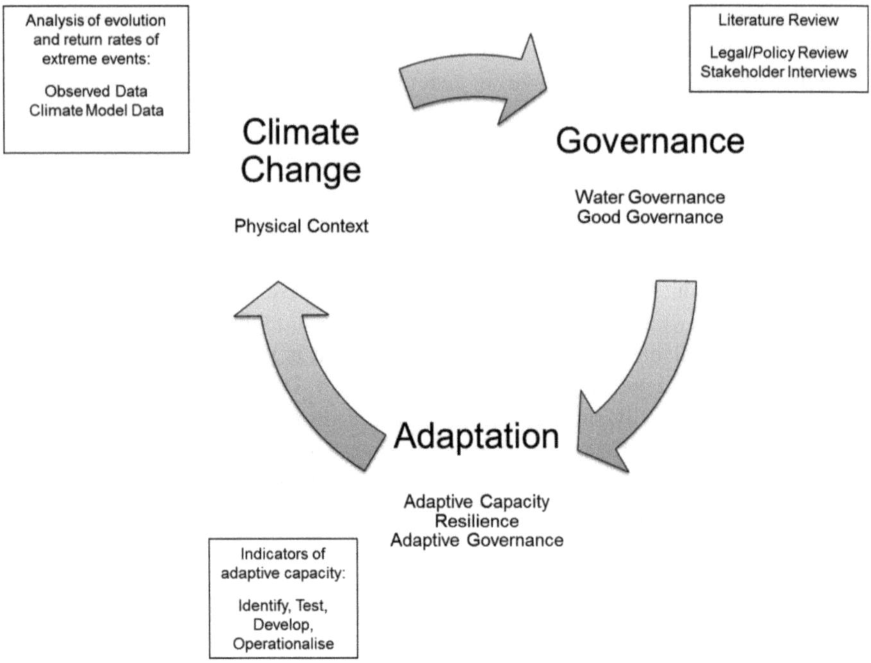

Fig. 5.2 Iterative methodological approach in relation to analytical background

the governance system's ability to cope with shocks; that is, using extremes as a proxy for assessing adaptive capacity to events that might become more recurrent or intense in a future, warmer world. While the research drew on climate data and information to contextualise the climate related challenges that the case areas face, the primary methodological approach chosen was a qualitative one.

5.2 Qualitative Research

Quantitative research presumes that the social world 'lends itself to an objective form of measurement, and that the social scientists can reveal the nature of that world by examining lawful relations between elements that, for the sake of accurate definition and measurement, have to be abstracted from their context' (Morgan and Smircich 1980, p 498). However, the more relaxed ontological assumption that human actors and the social world are a more open ended process that both shape and are shaped by external environments, requires more qualitative methodologies (Morgan and Smircich 1980). This is not to fully exclude the relevance of quantitative methodologies, or the fact that the appropriate quantitative techniques can be used to good effect in comparative qualitative analysis in social science (Ragin 1987), but instead to press the relevance of qualitative techniques in order to make sense of the complex nature of social phenomena that offer more positivist approaches to understanding a system's behaviour (Miles and Huberman 1994). Furthermore, it was decided not to employ quantitative techniques of data analysis on the qualitative data, to avoid adding more layers of interpretation to the data, and from that assigning quantitative meaning to biased data (Gibbs 2008).

The research questions posed and fields of study encapsulated in this book are focussed on the assessment and understanding of complex and multi-scale governance systems, which are not only characterised by a set of internal dynamics (e.g. actor networks, institutions at multiple scales) but also by the external dynamics (environmental change, climate change, socio-economic turbulence). Analysing these dynamics is challenging and calls for a more interdisciplinary approach to resource governance research and the underpinning conceptual frameworks employed, to be better able to take into account complex and context dependent dynamics of governance regimes (Ostrom 2008 in Pahl-Wostl 2009; Young 2007). Furthermore, social and biophysical scientists working on the urgent real world problems concerning environmental and climate related challenges are increasingly expected to develop knowledge that is context driven, problem focussed and interdisciplinary, by working collaboratively across stakeholder groups (Dovers 2005 in Aslin and Blackstock 2010). In this 'post disciplinary' world, scientists are being increasingly called upon to produce policy relevant (if not prescriptive) research (Aslin and Blackstock 2010), which in turn requires a broader mix of disciplines to match real world contexts and challenges.

Human-environment interactions have been characterised by a number of disciplines (including complex adaptive systems theory), by problems that are fundamentally systemic, multifaceted, complex, broad and 'wicked' (Rittel and Webber 1973) and thus require a mix of disciplines to better frame, understand, analyse and relate the problems under investigation, all this requiring input from a range of social and biophysical sciences (Aslin and Blackstock 2010). Inter- and trans-disciplinary approaches to problem oriented research questions, such as climate change, global environmental change and adaptation, allow the researcher to better account for and tackle complexity of science, fragmentation of knowledge and transcendence of different traditional disciplines engendered by the nature of the problem itself (Lawrence 2010).

Therefore, drawing on resilience-based literature, governance theory and climate change science, this book presents an interdisciplinary methodological approach, within the context of the ACQWA project, to better understand the problems and challenges in developing and mobilising adaptive capacity across highly contrasting governance regimes. Furthermore, a bottom up (inductive) and top down (deductive) approach has been combined to generate context sensitive analysis that is not specific and non-transferable (Pahl-Wostl 2009) in order to overcome the challenge of scaling up adaptation research (Smit and Wandel 2006) and develop both aggregate regional and national findings, as well as community specific findings. This approach takes advantage of and integrates different forms of hard and soft knowledge (local, scientific, specialised knowledge) identifies what can practically be achieved and what insights can be drawn across different scales and geographies, but inherently recognises the complexity of linked social and ecological systems, inter-connections across scales of governance and change, and the challenges of separating climate change from other scales (Lawrence 2010; Schoonfeldt 2010; Smit and Wandel 2006).

5.3 Governance Assessment

The initial governance assessment provides a basis from which the adaptive capacity assessment could take place in each case area, that focused more on how the systems dealt with climate related uncertainties and impacts from recent extreme events. The Striver governance assessment methodology has been developed by the Centre for Water Law, Policy and Science (CWLPS) at the University of Dundee, whose aim was to develop a systematic methodology to measure governance in the specific context of integrated water resource management. The legal analysis of water governance is a first step to its improvement at the local, regional and national levels. The methodology has already been applied and refined in a number of basins and sub-basins including the Sesan (Vietnam and Cambodia), the Tagus (Spain and Portugal) and the Lhasa (China/Tibet – upper Brahmaputra). Its application in the basins of the Rhône in the Valais, Switzerland and the Aconcagua, in Chile will aim to further refine and build on the methodology.

While there have been many attempts to assess the quality of governance, most relate to 'vague aspects of the broad context of governance but do not specifically address water and governance issues per se' (Rieu-Clarke et al. 2008). The indicator based approach to water governance is founded upon three core elements of governance; *accountability*, *transparency* and *participation*. Such an approach provides a mechanism not only to assess existing laws, policies and institutions, but also the extent to which such governance arrangements have been implemented in practice. For the purposes of this study, the governance context relates to: legislation; regulation; policies; formal organised institutions (ministries, government agencies); associations (self-organised groups); legal agreements amongst private actors; public-private agreements and judicial interpretation. Therefore within the STRIVER (www.striver.no) project, a comprehensive set of indictors for good governance, tailored directly to the IWRM context, were developed in the style of a questionnaire comprising 18 key questions, and 60 sub-questions. The resulting outputs are intended to provide a benchmark for identifying potential gaps and barriers to implementing IWRM (Rieu-Clarke et al. 2008).

The methodology comprises a series of key indicators which are allocated scores based on two key criteria: *Commitment*, which is the degree to which the governance context adheres to accepted standards of good governance; and *Process*, the extent to which this governance context is implemented in reality. Research was conducted initially through a desktop study to gather all relevant information on the laws, policies and institutions related to governance and IWRM, then followed by in-county interviews in order to ascertain the degree to which the law has been implemented in practice. The indicator approach to water governance is based on three core elements of good governance accountability, transparency and participation, and in addition, indicators of IWRM were also utilised:

- Accountability: Holding governments responsible for their actions; Contestability of political power
- Transparency: Right to information; Availability and clarity of Information
- Participation: Involvement of citizens in decision making
- IWRM: The integrated management of water resources in order to balance economic, social and environmental objectives

However, IWRM may not be seen so much as an indicator of good governance, but rather as a goal or aim, which may be useful in addressing the ability of water governance arrangements to effectively and equitably manage the fair distribution and protection of the resource. Results from the indicator assessment (assessment and desktop review) were then analysed according to the full list of indicator questions (available on request) and written up into a comprehensive report for the ACQWA deliverables (see acqwa.ch). Summarised versions of these reports can be found in Chaps. 7 and 8.

5.4 Adaptive Capacity

In developing the adaptive capacity component, the aim is to take better account of the dynamic interplay between the human (focussing on governance & institutions), hydrological and climate components of the system in order to understand the resilience of SESs in the face of future climatic uncertainty. Chapters 3 and 4 identified and presented the range of adaptive capacity determinants currently supported in the literature on adaptation to climate change from the adaptation, vulnerability and resilience literatures that are often focused at single temporal or spatial scales (Engle 2010). The indicators used for the governance assessment are representative of the broad brush indicators that are commonly found in the discourse on good governance and management approaches, but may not be as complete in the assessment of governance structures in the context of change and uncertainty.

Therefore a set of determinants was generated drawn from current theory (as presented in Chap. 4), in which to explore, measure, and develop a more nuanced characterisation of adaptive capacity. The complexity in assessing adaptive capacity, and the gap between the theory of adaptive and integrated approaches and the reality of positive outcomes can be seen as having two key aspects. The first concerns *what* it is that can be assigned as creating or increasing adaptive capacity? It is incredibly difficult to establish a causal link between a particular outcome and a specific governance approach or management tool, or even assigning a particular level of causation to one approach in comparison to another. The second issue relates to identifying *whether or not* adaptive capacity has actually been created. While it is possible to compare impacts between different places and times, it is not possible to compare the same event in the same place with a different set of variables.

5.4.1 Proxy Events

A number of studies have highlighted that there are significant challenges in assessing adaptive capacity (Engle and Lemos 2010; Smit et al. 2000) since 'adaptive capacity is latent in nature… it can only be actually measured after it has been realized or mobilized' (Engle and Lemos 2010, p 5). However, the empirical study of adaptation to climatic events, can to a certain degree overcome the pre-impact intangibility of adaptive capacity (Adger et al. 2007). In this case, the study of past extreme stresses (such as drought or flooding) can serve as a proxy for climate change impacts that are likely to manifest in a future, warmer, more uncertain world. Smit et al. (2000) show that by studying system responses to past climate variability (which tends to be experienced through the nature and frequencies of extremes) it is possible to identify attributes of the system which were key to either successful or failed responses. Likewise, the IPCC (Adger et al. 2007) also recognised that 'empirical knowledge from past experience in dealing with climate-related natural disasters such as droughts and floods… as well as longer term trends in mean conditions, can be particularly helpful in understanding the coping strategies and adaptive capacity' (p 138). Such studies of adaptation to extreme events have further highlighted the importance of institutions and governance mechanisms for the capacity or inability to deal with change (Brooks et al. 2005; Engle and Lemos 2010; Hurlbert 2009).

Given the challenges in investigating adaptive capacity, and in order to evaluate the validity of the assumptions in recent literature, it is therefore important to build and test more nuanced indicators in the context of recent past events (Adger et al. 2007). To achieve this, suitable proxies and indicators needed to be identified, and thus the case studies presented in this book have relied more heavily on empirical assessment of extreme stresses in order to test assumptions and build more nuanced indicators in the context of recent past events. This approach allows research not just to characterise adaptive capacity based on theorised and normatively driven determinants of adaptive capacity, but to attempt to understand it systematically through the presence or absence of adaptive behaviour (Engle 2011), as well as the form of adaptive behaviour.

Yohe and Tol (2002, p 26) determine that systems 'typically respond to variability and extreme events before they respond to gradual changes in the mean'. Therefore, this study used recent extreme events (drought and flooding) to serve as the primary context through which to explore the governance system's interaction with hydro-climatic events, eliciting information on planning and preparation for hydrological extremes, coping techniques and adaptation actions before, during and after the events.

However, when utilising extreme events as a means to analyse the governance framework, certain dynamics and limitations do need to be recognised. While looking at both heavy precipitation events and drought allows the study to explore the governance contexts of two contrasting hydrological events, the governance arrangements for responding to each event are very different. However, findings from the investigation of the individual governance mechanisms that shape responses to each form of extreme still are vital for comparative insights into the broader water governance framework within which they sit (Huntjens et al. 2011).

For the purposes of this study, adaptive mechanisms are termed as a response, institution or governance mechanism (law, regulation, policy, collective rule) that are undertaken at the national, regional or local level in order to prepare for or respond to different scales of environmental change (i.e. inter-annual variability, drought, floods, climate change impacts). This definition therefore takes into account both proactive and preparatory adaptation as well as reactive and autonomous adaptation. Adaptive actions and mechanisms analysed therefore represent legislation, or particular articles, regulation, policy frameworks or institutional actions (i.e. decisions or rules of user group associations) that provide guidance or mechanisms for drought or flood management, the prioritisation of users during particular peak periods (scarcity or high demand) and infrastructural adaptation to shifting hydrological patterns. While the Swiss case area covers adaptive mechanisms relating to both flooding and scarcity situations, the Chilean examples pertain only to drought and scarcity. The definition is deliberately broad and evades an exclusive linkage to climate change impacts since other studies have highlighted the difficulty in separating 'pressures exerted as a result of climate change from other economic, environmental or developmental pressures' (Tompkins and Adger 2004, p 564).

Beyond a certain tipping point of extremes, it is the response system that is invoked. However, the responses system is formulated through the governance system, born out of the institutional context and therefore this should not be seen as an impediment. The characteristics of the governance context are highly relevant to the effectiveness of the response system. Bearing this in mind, the study proposes that the governance framework can provide answers to managing flood situations that are not only pertinent to a response system. For instance, the rules that govern and frame land use zoning, natural flood plain management and river corrections are all elements of a water management system, independent to the response system, which would have an impact on the flexibility of a system to manage heavy precipitation events. This approach assumes that SESs will conduct some form of adaptation to such events, and that lessons can be drawn from them as a means of better understanding the attributes and variables of a system (governance and institutional mechanisms) that would foster sustainable adaptive actions to climate change impacts in the future.

The identification of extreme events within which to analyse adaptive processes is intended to improve the understanding of the current and future vulnerability to climate impacts on water resources, to document climate impacts on the water systems and to analyse the development of extreme events, using return period analysis according to different climate scenarios. Extreme events were identified through a mixture of expert interview, literature review and the analysis of meteorological data available (i.e. in the Swiss meteorological data from MétéoSuisse, while in Chile identification of the case events relied more heavily on expert interview though lack of data availability). In Switzerland, the development of the extremes of winter and summer temperatures, drought periods and heavy precipitation events have already been seen in incidents such as the 2003 European heat wave, warm winter spells of the mid 1990s (Beniston 2005; OcCC 2008) and the major flooding events such as 2005 in northern Switzerland and 2000 in the Valais.

In Chile, drought events are likely to be more dominant (Parry et al. 2007), exacerbated by the dwindling run-off from glacial melt and snow pack (Parry et al. 2007). However, while there is a strong influence of ENSO events on precipitation in central Chile, the potential development of ENSO events are still poorly modelled by GCMs, leaving a high level of uncertainty concerning the development of droughts and floods in central Chile (Garreaud et al. 2009; Kim and An 2011). During the period of study, the Aconcagua Basin in Chile was in the midst of one of their worst droughts for decades, which meant stakeholders were reviewing not only past actions but also relaying current events. In summary, in Switzerland the case events utilised as a focal point for exploration were the 2003 summer heat wave and the flooding events of 1993 and 2000. In Chile, the drought periods of 1996/1997 and 2010/2011 were used as the focal point for interview.

In addition to identifying these past extreme events, it is important to characterise the potential future frequency of such events, in order to judge what may or may not be a significant development in the intensity or frequency of the extreme events explored. To this end, the study also drew on return period analysis to understand the current development of such events (trend analysis) and to project the likelihood of such events happening under climate change scenarios (return period analysis). MétéoSuisse (CLIMAP) data was used in conjunction with results from the FP6 ENSEMBLES and FP7 ACQWA projects, in order to assess the current frequency and projected future frequency and intensity of the events.

5.4.2 *Data Collection*

Literature review (presented in Chaps. 1, 2, 3, 4), in combination with initial results from the Swiss governance assessment enabled a preliminary set of adaptive capacity determinants to be developed (as presented in Chaps. 3 and 4). These determinants provide the framework for exploring adaptive capacity through a series of interviews and archival data analysis. With the initial list of determinants, semi-structured qualitative interviews (Patton 1990) were carried out in the context of the extreme events, with national, regional and local level water governance stakeholders and experts. An open ended interview technique was chosen as being the most suitable to deploy across the two cases, for providing the most meaningful and deep information for the purposes of the research question.

Table 5.1 provides an overview of the experts and stakeholders interviewed across the different cases areas. Qualitative data consisted of: interviews n = 60; presentations, workshops n = 15. Water management stakeholders were identified at different levels from different sectors (both private and public as well as different water intensive economic sectors) as being most suited to providing insights into the connections between these different levels of water governance. This allowed insights to be gathered from both a top down and bottom up perspective, mainly focusing on the sub basin level, but also on the connections between sub basin level and other political, administrative and geographical boundaries. Interviewees were chosen to represent the operational level, as well as the political level and the challenges of balancing the reality of these two institutional settings.

The semi-structured interviews consist of questions relating to the different indicators, lasting between 50 and 90 min with a range of individuals with expertise and experience in water and natural hazards policy and management at the local and cantonal level. A list of interviewees (Table 5.1) and an example of the interview sheet can be found below. Three sub case areas were identified in the Valais, consisting of six communes that were representative of the different sector interests as well as the different micro-climates in the Valais. The qualitative data from these interviews serve to provide greater insight into the indicators and will be used to operationalise the criteria of the indicators, as well as assess its mobilisation in response to the events.

In order to ensure against bias, the mixture of snow ball technique and expert selection of interviewees aimed to ensure an equal and fair representation across the different stakeholder groups including government representatives (ministry Level, commissions, legal experts), NGOs (environmental and human rights groups), private actors (water users, utilities, hydropower, mining companies) as well as academic and scientific experts. In summary, the data used for both the governance and adaptive capacity portions of the research were:

- ~60 semi-structured qualitative interviews
- Swiss-German, French, Spanish & English
- ~15 workshop presentations
- Archival Data
- Legislation & Regulation
- Policy Documents
- Articles & Grey Literature

5.4.2.1 Interviewees

Table 5.1 List of interviewees, showing sector focus, governance scale and case area

#	Location	Scale	Sector	Responsibility	Language	Country
1	Bagnes	Local	Water Manager	Water supply	French	Switzerland
2	Sion	Regional	Hydropower	Hydropower	French	Switzerland
3	Visp	Local	Commune Administration	Flooding/water provision	Swiss-German	Switzerland
4	Brig	Regional	Cantonal Administration	Flooding	Swiss-German	Switzerland
5	Sion	Regional	Cantonal Administration	Groundwater protection	Swiss-German	Switzerland
6	Sion	Regional	Cantonal Administration	Water protection	French	Switzerland
7	Sion	Regional	Cantonal Administration	Agriculture	Swiss-German	Switzerland
8	Visp	Sub Regional	Agriculture	Training	Swiss-German	Switzerland
9	Martigny	Local	Water Manager	Water supply	French	Switzerland
10	Zermatt	Sub Regional	Hydropower	Facility operation	Swiss-German	Switzerland
11	Zermatt	Local	Hydropower	Facility operation	Swiss-German	Switzerland
12	Zermatt	Local	Water Manager	Water supply	Swiss-German	Switzerland
13	Visp	Local	Industry (Chemicals)	Facility operation	Swiss-German	Switzerland
14	Visp	Local	Water Manager	Water supply	Swiss-German	Switzerland
16	Stalden	Sub Regional	Hydropower	Facility operation	Swiss-German	Switzerland
17	Sion	Sub Regional	Cantonal Administration	Flooding (TRC)	French	Switzerland
18	Bern	Federal	Federal Administration	Water resources management	Swiss-German	Switzerland
19	Zürich	National	Research	Water resources management	Swiss-German	Switzerland
20	Lausanne	National	Research	Water resources management	French	Switzerland
21	Zürich	National	Research	Hydropower	Swiss-German	Switzerland
22	Zürich	National	Research	Environmental law	Swiss-German	Switzerland
23	Zürich	National	NGO	Water resources management	Swiss-German	Switzerland
24	Sion	Regional	Cantonal Administration	Natural hazards	Swiss-German	Switzerland
25	Sion	Regional	Cantonal Administration	Flooding	French	Switzerland

26	Sion	Regional	Cantonal Administration	Hydropower	Swiss-German	Switzerland
27	Brig	Regional	Cantonal Administration	Flooding	Swiss-German	Switzerland
28	Visp	Regional	Agriculture	Training	Swiss-German	Switzerland
29	Santiago	National	Government Administration	Environment/water resources management	English	Chile
30	Santiago	National	Government Administration	Water rights	English	Chile
31	Santiago	National	Research	Water resources management	Spanish	Chile
32	Santiago	National	Government Administration	Water rights	Spanish	Chile
33	Santiago	National	Research	Human rights	English	Chile
34	Santiago	National	Government Administration	Environment/water resources management	English	Chile
35	Santiago	National	Water Law	Water rights	English	Chile
36	Santiago	National	Government Administration	Water infrastructure	English	Chile
37	Santiago	National	Research	Water resources management	French	Chile
38	Santiago	National	Research	Water resources management	English	Chile
39	Santiago	National	Water Law	Water rights	English	Chile
40	Santiago	National	Water Law	Water rights	Spanish	Chile
41	Santiago	National	Water Law	Water rights	English	Chile
42	Quillota	Regional	Government Administration	Water rights	Spanish	Chile
43	Quillota	Local	Private User Group	Irrigator	Spanish	Chile
44	Quillota	Regional	Government Administration	Water infrastructure	Spanish	Chile
45	Santiago	National	Research	Water resources management	English	Chile
46	Santiago	National	Government Administration	Water rights	Spanish	Chile
47	Santiago	National	Research	Water resources management	English	Chile
48	Santiago	National	NGO	Environment/water rights	English	Chile
49	Santiago	Regional	Hydrologist	Water resources management	Spanish	Chile
50	Santiago	National	Government Administration	Water infrastructure	Spanish	Chile
51	Los Andes	Local	Private User Group	Irrigator	Spanish	Chile

(continued)

Table 5.1 (continued)

#	Location	Scale	Sector	Responsibility	Language	Country
52	Los Andes	Local	Private User Group	Irrigator	Spanish	Chile
53	Quillota	Local	Government Administration	Water rights	Spanish	Chile
54	Quillota	Local	Private User Group	Irrigator	English	Chile
55	Santiago	National	Government Administration	Agriculture	English	Chile
56	Santiago	Regional	Mesa Técnica Aconcagua	Water resources management	Spanish	Chile
57	Quillota	Regional	Water Manager	Water supply & sanitation	Spanish	Chile
58	Quillota	Local	Canal Association	Irrigator	English	Chile
59	Limache	Local	Canal Association	Irrigator	Spanish	Chile
60	San Felipe	Local	Private User Group	Irrigator	Spanish	Chile
61	Santiago	National	Government Administration	Irrigation	Spanish	Chile
62	Santiago	National	Research	Water resources management	English	Chile
63	Santiago	National	Government Administration	Water supply & sanitation	English	Chile

5.4.2.2 Interview Guideline

General Introduction:
Thank you very much for your willingness to be interviewed, and for taking the time to be speak with me. I am interviewing you to better understand how certain extreme periods (e.g. XXXX) of drought or water related stress (or flooding if it is relevant) has been managed in the Aconcagua region and how prepared the water management system may be to dealing with such events in the future, considering the potential impacts of climate change.

The interview should last between forty five minutes and an hour, and if it is ok with you, I would like to record your responses so that afterwards I can translate parts of the interview into English. Your responses are confidential and I will not publish any quotes without first seeking your consent. This interview will only be used for my own research, and when writing up my results, I will not refer to you by name or identifiable information.
Section 1: General Questions:

1. Could we start with a short description of your role and institution?
2. How long you have worked here?
3. How was your group/sector affected by these events?
4. Was there a change in the amount of water you were allocated during the event?
   (a) Were certain uses prioritised outside of normal rights/legal agreements?
5. Were water reserves affected during this event?

Section 2: Indicator Questions:

Experience

6. Had there been any previous experience of water stress (and flooding)?
7. What was learnt from these experiences?
8. Was there any training or preparation for such an event? (Workshops/Information at municipal/regional/national level)
9. Had you had previous experience of managing water supply issues in similar events?

Networks

10. Within your role, did you have regular involvement (coordination) do you have with other water managers/water stakeholders? (Actors & Political Levels – Municipalities; Regions; State actors)
    (a) If so, can you give examples of how you engage with them?
11. Did any of these groups/individuals particularly block or drive progress?
12. How do you transfer/share information across these different stakeholder groups?

Resources

13. How was the event managed: Adequate financial capacity/human resources for managing the event? Preparation – Relief, Quality/Quantity?

Levels of Decision Making

14. Who was involved in decision making about water supply during the event?
    (a) Were you?
    (b) Who else was?
15. Who/What institution/level was the primary decision maker?

Integration

16. Were there any concerns for environment/ecosystems taken into account in the management of the event?
17. Is climate change integrated into the planning process within your sector or within any committees you are involved in at the local/regional level?

Knowledge

18. What kind of technical or scientific information used to manage water supply in your region? (Examples)
    (a) Environmental Impact Studies
    (b) Weather forecasts
    (c) Climate Models
    (d) Hydrological Models
    (e) Water Quality Information
    (f) Monitoring stations
    (g) Traditional knowledge
    (h) Local observation of change
    (i) Disaster risk maps
19. How do you access this information? (Examples)
20. Can you give examples of how these technology/information/data was used to manage the extreme situation & how the information was integrated into decision making?
21. Was this information shared across different groups? How?

Flexibility-Predictability

22. Do legal provisions/guidelines exist for the management of water supply during periods of high demand/stress for water been managed in the area? Or for flooding events?
23. Was there any adjustment/change in the system following on from any of the events?
    (a) Lessons Learnt? Incorporated into the system?

Closing General

24. How able is the water management/governance system to respond to these stress periods?
25. What would you see as the main impediments (legal, policy, political, social) for the system to cope with stress periods? How could this be remedied?
26. Do you have any final thoughts about how climate change will impact (or already is) water resources, and the system's ability to cope with these impacts in terms of minimising damage or taking advantage of opportunities it may present?

5.5 Qualitative Data Analysis

Interviews were recorded, transcribed and translated from Spanish, French and Swiss-German into English. This time-consuming initial process of translation and transcription provided a valuable initial interpretation of key themes emerging from the qualitative data, as well as pointing to issues that needed to be covered in subsequent interviews. The transcripts were then coded according to the analytical framework (Fig. 5.1) and then analysed using MaxQDA software, a qualitative data analysis software tool. Archival data (legal, policy and presentation documents) were given the same treatment.

An initial round of coding was applied to interview data according to the set of determinants that had been used in the semi-structured interviews (Table 5.2), as well as a number of *free* codes (Miles and Huberman 1994) for the external drivers, perceptions of bridges and barriers to adaptive capacity and that arose out of the preliminary coding exercise. Coded extracts were then downloaded from MaxQDA and re-analysed, categorised and then the amended coding structure was uploaded into a new MaxQDA project. The aim of conducting three waves of coding was to increase the reliability of the interpretation of the data. Coding schemes should not be a 'catalogue of disjointed descriptors or a set of logically related units and sub units, but rather a conceptual web, including larger meanings and their constitutive parts' (Miles and Huberman 1994, p 63). Moreover, a 'conceptual structure must, whether pre-specified or evolving, must underlie the definitions' (Miles and Huberman 1994, p 63).

Both descriptive and analytical codes were given to the data. Initial analytical codes related to the determinants listed in Table 5.2, as well as to the categories of change, were used to analyse the adaptive responses (as a means of defining outcomes and establishing linkages and causation – transformation, persistent adaptation, and passive). Descriptive codes were used to identify relevant information to construct the case, i.e. information on impacts of the extreme events and the broader water management challenges in the different case areas. Miles and Huberman (1994, p 65) state that the 'ultimate power of field research lies in the researcher's emerging map of what is happening and why'. The practice of coding data allows the researcher to work through iterative cycles of inductive and deductive analysis to better map out these causal relationships. Through the different waves of analysis, codes may change, develop and emerge (Miles and Huberman 1994).

Table 5.2 Exploratory attempt at identifying initial determinants of adaptive capacity from the existing body of literature, and from which questions could be generated to explore their efficacy in interview

Tentative determinants	Related questions
Knowledge/information	Which information sources were used to deal with the event?
	How was this information accessed/integrated into decision making?
	Was there any exchange of data with other groups/regions?
	What information was needed but not available/accessible?
	Public perception? Public understanding of risk/climate change etc?
Networks	Which actors were involved/excluded?
	How do the actors/networks involved contribute to dealing with the situation?
	At which point in the process – consultation, decision making etc?
	At which administrative levels was the situation resolved/ negotiated? i.e. federal, cantonal, regional, communal?
	Are there issues in cooperation across/amongst networks?
Levels of decision making	Enabling legislation that creates social space for ecosystem management
	At which levels/within which boundaries of the watershed/region were decisions taken to deal with the case event?
	Which/what kind of institutions/groups were involved in managing the situation?
Integration	Most of these questions have been answered through the IWRM indicators previously
	How sectoral are departments – how are units of decision making organised?
Predictability – flexibility	Where is there consistency in the application of policies/laws across the different events?
	How differently were the situations handled?
	Framework (legislation & actor system)
	Did the legal framework need to be reinterpreted to accommodate the case event?
	Which uses were prioritised?
	Were trade-offs made?
	What mechanisms are in place for the management of stress/ extreme situations?
	Did the legal rules indicate how to handle the situation? If so, strictly/flexibly?
	Did the event lead to changes in rules?
	Did lessons learnt get incorporated into institutional/legal framework to inform future situations?
Resources	Level of resource involved in managing the case event? (Education, Training, Number of people, Type of financial assistance)
	Who provided scientific/technical information?
	Funds for responding to environmental change and for remedial action?
	Capacity for monitoring and responding to environmental feedback?

(continued)

Table 5.2 (continued)

Tentative determinants	Related questions
Experience	What training for such a situation had they received? What preparation for such an event had taken place? i.e. what planning for such a situation was already in effect? Who provided scientific/expert information?
Leadership	Vision, leadership, and trust Ascertain if there are particular individuals who directed the management of the event, helped to bring stakeholders together to deal with it Trust amongst the different stakeholders/institutions

In the first round of coding, the initial codes tended to be a priori/deductive/concept driven (relating to interview questions built from theory) as can be seen in Table 5.2. From this initial list of codes, a number of new codes surfaced as they emerged from the data collection (empirically grounded), many of which were more descriptive. Table 5.2 shows the starting list for the coding exercise, which were linked to the research questions and to the interview questionnaire and to a list of sub-indicators (bins of conceptual variables).

From this initial round of coding and analysis, the emerging themes helped refine the analytical structure. Out of this initial round of analysis the most significant emerging theme was that of the core tensions and trade-offs in mobilising adaptive capacity at different scales, of which the structural tension between flexibility and predictability was the most prevalent and significant element. In the second round of coding, coded segments were extracted and codes refined to remove redundancy and improve reliability. Additionally, a coding check was performed with another researcher to ensure the reliability of the coding methods employed, and accuracy of the definitions given to the codes. There was also an attempt to refine the code set to move from descriptive to more analytic codes.

Codes were refined and regrouped into the following categories to remove redundancy and streamline the analytical framework: Legal Framework/Regime (what actors can / cannot do); Knowledge (what actors know – is there information exchange, what kinds of knowledge /information are integrated); Networks (How actors interact – how information/data is exchanged, how it is integrated); Flexibility – Predictability. Beneath these first level codes, there were a number of second and third level codes that had been drawn from the original list of determinants. The descriptive and inferential codes also emerged in the first round of coding and needed to be sorted from the analytical codes and refined in order to remove redundancy. These included bridges and barriers (Illustrative), impacts (descriptive), water resources management issues (descriptive).

As mentioned earlier, in order to improve the robustness of the analytical process, a reliability check was taken in the second round of coding. According to Perreault and Leigh (1989), analytical reliability is seen to increase in correlation with the use of multiple coders. Coding checks therefore can aid definitional clarity and act as an important reliability check (Miles and Huberman 1994) when there is just one researcher working on a case study (as in this book), as opposed to multiple coders working in a research team.

However, there are studies that have shown that while multiple coders can add reliability; this may not be true with 'data that require coders to evaluate the importance of the information in the context of a larger response' (Crittenden and Hill 1971, in Engle 2010), such as complex processes across multiple scales (e.g. governance mechanisms for adaptive capacity). In such cases, multiple rounds of coding may provide better results for improving reliability (Engle 2010; Perreault and Leigh 1989). Therefore, an internal coding check was performed through the multiple rounds of coding (coding and recoding the same texts), but an inter-coder check with a researcher familiar with the research topic was also performed.

The inter-coder reliability check took place with another researcher, also studying water resources management. Due to the number of codes and the depth of codes, it was considered inappropriate to conduct a traditional form of inter-coder reliability test, as it would have taken too long for the researcher to become familiar with the codes and their detailed descriptions in the memos. Therefore, the researcher read through a section of coded passages and assessed the relevance and appropriateness of the codes assigned to those sections, and asked for clarifications where needed.

The result was positive with a fairly high agreement and also provided useful insights into where some of the codes may have been duplicated or needed to be re-organised within the coding hierarchy. The final issue in coding methods is to know when enough is enough. For the purposes of this project, the advice of Miles and Huberman (1994, p 62) was followed, who suggest that this stage of analysis is complete when the 'analysis appears to have run its course, incidents readily classified and categories are saturated, and regularities emerge'.

Having refined the core codes and variables that arose out of the interview process, the next process was to complete the investigation of the linkages between governance context across the different cases and scales (independent variables) and adaptive capacity (dependent variables). The analytical process relied mostly on qualitative data gathered through the interview process, but also drew from archival data to construct the case and attempt to answer the key questions at the start of this chapter. In order to construct the case, there were four main steps:

1. Characterising adaptive responses and outcomes according to different forms of adaptation (Transformation, Persistent Adaptation, Passive)
2. Correlating categorised adaptation actions (Transformation, Persistent Adaptation, Passive) with different governance mechanisms and contexts
3. Identifying bridges and barriers to mobilising adaptive capacity in each case area and with regards to governance mechanisms across different scales of governance
4. Deeper characterisation of governance indicators according to analytical steps above

The different steps comprised the multi-pronged approach to understanding and assessing adaptive capacity as detailed in Fig. 5.3. The starting set of governance determinants of adaptive capacity were used to frame the exploration of adaptive

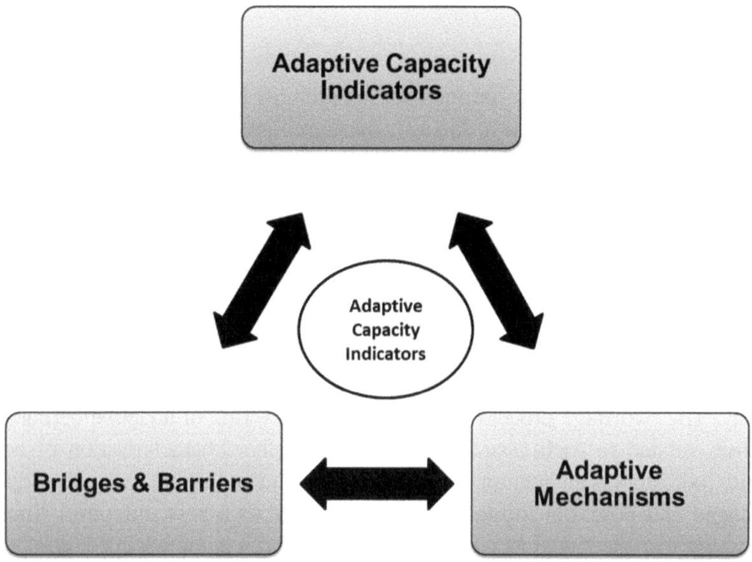

Fig. 5.3 Taking a multi-pronged approach to understanding and assessing adaptive capacity in order to triangulate towards a more nuanced and empirically based set of adaptive capacity indicators

behaviour within the context of extreme events. Interview and archival data were then used to categorise adaptive mechanisms, correlate them with associated governance mechanisms and identify set of bridges and barriers. These analytical steps then allowed for the operationalisation and measurement of more nuanced and empirically grounded indicators of adaptive capacity according to three broad categories (*Regime, Knowledge, Networks*) that are presented and discussed in Part III. Criteria (or sub-indicators) were qualitatively operationalised to allow for factors that could be monitored and assessed as being indicative of positive manifestations of adaptive capacity. Intersections of codes were extracted from MaxQDA to identify combinations of input variables (independent) and their associated output variables (dependent) to establish trends in correlations between governance mechanisms and adaptive outcomes. Different relationships were explored across the study cases, variables and scales or sectors. Intersections were extracted between coded segments relating to both adaptive outcomes and associated governance related determinants in order to identify the governance mechanisms associated with the different categories of adaptation. Descriptive codes (event impacts and water resource management issues) were extracted and analysed in order to better characterise the case events and the broader challenges within which adaptive capacity is mobilised. Finally, drawing on these initial analytical steps, the adaptive capacity determinants (according to the refined categories of *Regime, Knowledge, Networks*) were more deeply operationalised within the context of the emergent codes, i.e. core tensions in adaptive capacity, that of balancing flexibility with predictability.

5.6 Summary

While the previous chapter outlined the challenges relating to the assessment of adaptive capacity in relation to latency, measurement, characterisation and operationalisation, this chapter has focussed on the approach taken to attempt to address these limitations. The methodology detailed represents an inductive and iterative approach to the measurement and characterisation of adaptive capacity, in order to be able to not just fortify current theory, but contribute and advance it as well. This approach also allows researchers to combine the strengths of both inductive and deductive approaches, in order to minimise weaknesses and blind spots which are equally implicit in both (Bohensky et al. 2010).

The final outcome seeks to develop a more nuanced characterisation of adaptive capacity. The two stage process of categorising the forms of adaptive outcomes and responses related to each case area before the further characterisation of adaptive capacity indicators is an attempt to build, advance and contribute to theory. The major challenges and potential limitations of being able to assess outcomes (including governance response) against climatic events have been acknowledged, accepted and in part dealt with through a broader definition of response that evades an exclusive link to climate change impacts and pressures but recognises their interconnection with other environmental, social or economic pressures (Tompkins and Adger 2004).

To recap, the study draws primarily on qualitative data, using expert and stakeholder interviews, combined with archival research, to develop, explore, test and operationalise a set of governance related indicators of adaptive capacity within each water governance regime. Recent extreme events (drought and flooding) served as the primary context through which to explore the governance system's interaction with hydro-climatic events, eliciting information on planning and preparation for hydrological extremes, coping techniques and adaptation actions before, during and after the events. The exploration of past experiences in relation to climate related extreme events therefore acted as a means to understand and assess the institutional mechanisms that enhanced or hindered adaptive capacity across different scales in each context.

Stakeholder perceptions to climate variability and climate change were sought both in relation to the events, as well as more generally. While it is recognised that the governance arrangements for responding to drought and flooding events are very different, studies have shown that the investigation of individual governance mechanisms that shape responses to each form of extreme still are vital for comparative insights into the broader water governance framework within which they sit.

Beyond issues relating to the specifics of measuring and assessing adaptive capacity, there were also challenges connected to the nature of case study research, across different cultures and languages. Interviews were conducted in four different languages (English, Swiss-German, French and Spanish), therefore analysis relied on translation by a number of different parties and only a handful of interviews were conducted in the researcher's mother tongue. Fieldwork in Chile required the use of a translator and interpreter, while time and financial constraints meant that a greater amount of time and flexibility was allowed for the Swiss case than the Chilean. While most stakeholders contacted in Chile were open and welcoming towards being interviewed, it proved impossible to secure interviews with mining and hydropower stakeholders during the course of the field trip. Finally, the connection of the research to the ACQWA Project served as both a help and hindrance. It provided a productive framework and network within which to conduct this research, but limited the freedom in the set of case areas chosen.

References

Adger WN, Agrawala S, Mirza MMQ, Conde C, O'Brien K, Pulhin J, Pulwarty R, Smit B, Takahashi K (2007) Assessment of adaptation practices, options, constraints and capacity. In: Parry ML, Canziani OF, Palutikof JP, van der Linden PJ, Hanson CE (eds) Climate change 2007: impacts, adaptation and vulnerability. Contribution of working group II to the fourth assessment report of the intergovernmental panel on climate change. Cambridge University Press, Cambridge

Aslin HJ, Blackstock KL (2010) Now I'm not an expert in anything': challenges in undertaking transdisciplinary inquiries across the social and biophysical science. In: Brown VA, Harris JA, Russell JY (eds) Tackling wicked problems: through the transdisciplinary imagination. Earthscan, London/Washington, DC, pp 117–129

Beniston M (2005) Warm winter spells in the Swiss Alps: strong heat waves in a cold season? Geophys Res Lett 32:L01812

Beniston M, Stoffel M, Hill M (2011) Impacts of climatic change on water and natural hazards in the Alps: can current water governance cope with future challenges? Examples from the European "ACQWA" project. Environ Sci Policy 14:734–743

Bohensky E, Stone-Jovicich S, Larson S, Marshall N (2010) Adaptive capacity in theory and reality: implications for governance in the Great Barrier Reef Region. In: Armitage D, Plummer R (eds) Adaptive capacity and environmental governance. Springer, Heidelberg, pp 23–41

Brooks N, Adger WN, Kelly PM (2005) The determinants of vulnerability and adaptive capacity at the national level and the implications for adaptation. Glob Environ Chang 15(2):151–163. doi:10.1016/j.gloenvcha.2004.12.006

Brunner RD (2010) Climate adaptation and the drunkard's speech. Paper presented at the 29th Annual Institute of the Society of Policy Scientists, Yale Law School, New Haven, CT, 16 Oct 2010

Dovers SR, Hezri AA (2010) Institutions and policy processes: the means to the ends of adaptation. Wiley Interdiscip Rev Clim Chang 1(2):212–231

Engle NL (2010) Adaptation to extreme droughts in Arizona, Georgia, and South Carolina: evaluating adaptive capacity and innovative planning and management approaches for states and their community water systems. University of Michigan, Ann Arbor

Engle NL (2011) Adaptive capacity and its assessment. Glob Environ Chang 21(2):647–656. doi:10.1016/j.gloenvcha.2011.01.019

Engle NL, Lemos MC (2010) Unpacking governance: building adaptive capacity to climate change of river basins in Brazil. Glob Environ Chang 20(1):4–13. doi:10.1016/j.gloenvcha.2009.07.001

Garreaud RD, Vuille M, Compagnucci R, Marengo J (2009) Present-day South American climate. Palaeogeogr Palaeoclim Palaeoecol 281:180–195

Gibbs G (2008) Analysing qualitative data. Sage, Los Angeles/London

Huntjens P (2011) Beyond boundaries: learning to adapt. Nat Clim Chang 1(5):274

Huntjens P, Pahl-Wostl C, Rihoux B, Schlüter M, Flachner Z, Neto S, Koskova R, Dickens C, Nabide Kiti I (2011) Adaptive water management and policy learning in a changing climate: a

formal comparative analysis of eight water management regimes in Europe, Africa and Asia. Environ Policy Gov 21:145–163
Hurlbert M (2009) The adaptation of water law to climate change. Int J Clim Change Strateg Manage 1(3):230–240
Kane SM, Yohe GW (2000) Societal adaptation to climate variability and change: an introduction. Clim Chang 45(1):1–4
Kim BM, An SI (2011) Understanding ENSO regime behavior upon an increase in the warm-pool temperature using a simple ENSO model. J Clim 24:1438–1450. doi:10.1175/2010JCLI3635.1
Lawrence RJ (2010) Beyond disciplinary confinement to imaginative transdisciplinarity. In: Brown VA, Harris JA, Russell JY (eds) Tackling wicked problems: through the transdisciplinary imagination. Earthscan, London/Washington, DC, pp 16–30
Miles MB, Huberman AM (1994) Qualitative data analysis: an expanded sourcebook. Sage, Beverly Hills
Morgan G, Smircich L (1980) The case for qualitative research. Acad Manage Rev 5(4):491–500
OcCC (2008) Das Klima ändert – was nun? Der neue UN-Klimabericht (IPCC 2007) und die wichtigsten Ergebnisse aus Sicht der Schweiz. Organe consultatif sur les changements climatiques, Bern
Pahl-Wostl C (2009) A conceptual framework for analysing adaptive capacity and multi-level learning processes in resource governance regimes. Glob Environ Chang Hum Policy Dimens 19(3):354–365. doi:10.1016/j.gloenvcha.2009.06.001
Parry ML, Canziani OF, Palutikof JP, van der Linden PJ, Hanson CE (2007) Summary for policymakers. In: Climate change 2007: impacts, adaptation and vulnerability. Contribution of working group II to the fourth assessment report of the Intergovernmental Panel on Climate Change. Cambridge University Press, Cambridge, pp 7–22
Patton MQ (1990) Qualitative evaluation and research methods. Sage, Newbury Park
Pellicciotti F, Burlando P, Van Vliet K (2007) Recent trends in precipitation and streamflow in the Aconcagua River basin, central Chile glacier mass balance changes and meltwater discharge. Selected papers from sessions at the IAHS Assembly in Foz do Iguaçu, Brazil, 2005, IAHS Publication 318
Perreault WD, Leigh LE (1989) Reliability of nominal data based on qualitative judgments. J Market Res 26(2):135–148
Ragin C (1987) The comparative method: moving beyond qualitative and quantitative strategies. University of California Press, Berkeley
Rieu-Clarke A, Allan A, Magsig BO (2008) Assessing governance in the context of IWRM. Striver Policy Brief: strategy and methodology for improved IWRM – an integrated interdisciplinary assessment in four twinning river basins Issue No. PB6.2. http://www.dundee.ac.uk/water/projects/striver/
Rittel H, Webber M (1973) Dilemmas in a general theory of planning. Policy Sci 4:155–169
Schoonfeldt J (2010) Applying specialized knowledge. In: Brown VA, Harris JA, Russell JY (eds) Through the transdisciplinary imagination. Earthscan, London/Washington, DC, pp 139–147
Smit B, Wandel J (2006) Adaptation, adaptive capacity and vulnerability. Glob Environ Chang 16(3):282–292
Smit B, Burton I, Klein R, Wandel J (2000) An anatomy of adaptation to climate change and variability. Clim Chang 45:223–251
Tompkins EL, Adger WN (2004) Does adaptive management of natural resources enhance resilience to climate change? Ecol Soc 9(2):10. [online]. URL: http://www.ecologyandsociety.org/vol19/iss12/art10
Yohe G, Tol RSJ (2002) Indicators for social and economic coping capacity - moving toward a working definition of adaptive capacity. Global Environmental Change 12 (1):25–40. doi:10.1016/S0959-3780(01)00026-7
Young O (2007) Designing environmental governance systems: the diagnostic method. IDGEC synthesis conference, Bali summary published in IHDP Newsletter (1.2007):9–11

Part II
The Case Areas in Chile and Switzerland

Chapter 6
Introducing the Case Study Areas: Hydro-climatic and Governance Contexts

Abstract This chapter provides a comparative overview of the two case areas in Chile and Switzerland. It details the political backdrop of the country as well as introduces the river basin case areas of focus, the Rhône in Switzerland and the Aconcagua in Chile. An introduction to the climatic situation is given, along with an overview of the impacts of climate related extreme events on mountain watersheds. While both case areas are representative of mountain watershed nivo-glacial regimes, they are contrasted by highly differing water governance regimes. Despite these differences, both regions are contending with a wide range of pressure on water resources, including increasing observation of climate change impacts on water, snow and ice.

Keywords Comparative case analysis; Rhône • Canton Valais, Switzerland • Aconcagua, Region V, Chile • Climate change impacts on water management • Socio-economic pressures • Private market based governance • Decentralised governance

6.1 General Overview

Table 6.1, provides a comparative overview of the two case areas. Both regions represent mountain watershed nivo-glacial regimes, in which observed impacts of climate change on glacial melt and elevation of the snow line have been documented. Both regions have diverse demands on their water resources, for productive, consumptive and ecological services. However, the two regions are distinct in that they operate under highly different governance modes. The case areas cover a similar land area, but while the Chilean basin is fully encapsulated within the case area, the Swiss case area represents only the upper basin of the Rhône River. The Rhône is a major European trans-boundary river which traverses two more cantons in Switzerland before passing through France before it flows into the Mediterranean Sea.

M. Hill, *Climate Change and Water Governance: Adaptive Capacity in Chile and Switzerland*, Advances in Global Change Research 54,
DOI 10.1007/978-94-007-5796-7_6,

Table 6.1 Introducing the case areas

Parameters	Rhône, Canton Valais, Switzerland	Aconcagua, Region V, Chile
Water basin	Rhône Basin (source in Alps)	Aconcagua Basin (source in Andes)
Population	303,241 Average growth since 1996 has been 0.84%, but has increased since 2000	485,614/1,539,852 (basin/region) Growing population, particularly in urban centres (Reyes Carbajal 2007) around 3%/annum
Economic sectors	Irrigated Agriculture & Viticulture; Industry (chemicals); Tourism; Hydropower Area with a large concentration of Swiss hydropower production, viticulture, as well as ski and tourism area	Irrigated Agriculture; Industry (mining & cement); Hydropower; Tourism in the coastal urban zone Important agricultrual region (avocado and other table fruits)
Geography	Surface areas of 5,375 km2 (53.8% unproductive land) The nivo-glacial regime is characterised by low discharge in winter and high discharge in summer. The importance of glaciers within the basin is high, since in over 50% of the basin, precipitation falls in the form of snow	7,340 km^2 (45% of Valparaiso Region - 16,396.1 km^2). Basin-average annual runoff at the coast side is over 50 mm/year and in the central part less than 20 mm/year. Higher values occur in summer and spring as a result of glacier and snow melt, lower flows occur during autumn and winter especially in the upper part above 1,000 m altitude where the flow regime is nival. At lower elevations the regime becomes mixed; nivo-pluvial
Climate	Mediterranean climate Nivo-Glacial/Pluvial regime The Alpine interior, especially the canton of Valais, has a relatively dry continental climate – less than 600 mm of water per year on the plain	Semi-arid Mediterranean climate Nivo-Pluvial regime Two climate types are observed in the Aconcagua basin: warm Mediterranean and cold climate in the high Andes.
Climate impacts	Increased glacial melt & elevation of snow line Changes in the seasonality of river flows Increase in summer run off evident in most heavily glaciated regions	Increased glacial melt & elevation of snow line Changes in the seasonality of river flows Decrease in summer run off already evident
Extreme events	Flooding events (major events in 1993 & 2000), water stress periods and heat waves (2003 heat wave), and seasonal periods of peak demand due in part to its large tourism sector	Droughts in 1996–1998, 2003, 2008, and 2010 Decreases in summer run off in the upper basin as a result of a reduction in glacial coverage and snow depth in the drainage watershed; glaciers are now in a phase of diminishing contribution

(continued)

Table 6.1 (continued)

Parameters	Rhône, Canton Valais, Switzerland	Aconcagua, Region V, Chile
	Encircled by the Bernese and Valaisanne Alps; creating a rain shadow and extremely low precipitation Recession of glaciers, changes in seasonality, melting of permafrost, and changes in vegetation and precipitation. Periods of low precipitation have however so far been offset with increased glacier meltwater	
Local authorities	1 region, 12 districts, 143 communes Federal & Cantonal Water legislation	1 region, 7 provinces
Water governance	Mixture of private & public water rights Complex framework of use & protection regulation	Water rights market implemented in 1981 Water code Neo-liberal regulation Centralised institutions
Governance mode	Decentralised, federal system of both regulatory framework (with Federal and Cantonal Laws) and decentralised governance system known as 'Subsidiarity of Implementation' Mixture of public, private & common property rights for water resources Federal policy setting and legal framework informs watercourse management and flood provisions at canton and local levels	Centralised governance for to regulatory aspects, but 'laissez faire' approach to water management – leaving it primarily in the hands of private actors Neo-liberal agenda with water as a private commodity in the Water Market, regulated by the 1981 Water Code, as well as competing legal provision in laws governing Energy and Mining

Source: Mauch et al. (2000), Carrasco et al. (2005), Kundzewicz et al. (2007), Reyes Carbajal (2007), Pellicciotti et al. (2008), Valais (2009), Beniston et al. (2011)

While the two cases have a number of differences, the independent variable that defines the distinction between the two case studies is the different governance modes under which they operate. The governance models in Chile and Switzerland are founded upon from different paradigms but both will need to adapt in the face of future uncertainties and challenges from climate change. Chile is now an OECD country, therefore its status as an emerging economy could be challenged, despite the poverty and hardship that does exist for a large cross-section of the country. However, it is important to note that Chile does have a more solid institutional foundation than many other emerging economies, and therefore can be seen as a useful non-developed country upon which to conduct a comparative analysis. In Chile, the principle driver of water management is the definition of water rights as a market commodity. Even though water is defined in the country's national constitution as a public good, the perception of it as an economic good means that the private market based governance model dominates. There is a strong national centralised approach to water management, with most of the power relegated to private interests and private water rights.

By contrast, in Switzerland, the water governance framework is characterised by the 'principle of subsidiarity', which defines the decentralised implementation of public policy and law at the cantonal and municipal levels. Sovereignty over water often resides at the commune level, as in the Valais; thereby devolving a far greater share of power to the cantonal and communal level. The focus of Swiss water legislation has been guided by the development of three general responsibilities; the protection against water hazards; water exploitation (e.g. for hydropower); protection of water (Clivaz and Reynard 2008; Varone et al. 2002). The water governance regime is dominated by a government agency based model, but with influences of the user based (especially in certain areas in the Valais region) as well as private model.

6.2 Rhône Basin, Canton Valais, Switzerland

Within Switzerland there is a well-established decentralised framework for decision-making and planning. Movements towards the Swiss Confederation began in the thirteenth Century and were completed in 1848. In embryonic Switzerland, communes (essentially villages) and the cantons sought to protect and maintain their autonomy against the centralizing forces of the noble communities. This led to the formation of a confederation of sovereign states and a nation founded on essentially horizontal coalitions (Wiegandt 1977). The resulting alliances were held together by a federal structure, forming a highly decentralised state with a high level of freedom for the communes and cantons.

In the Swiss Confederation, these democratic traditions have led to a high number of public administrative units and a complex decision making structure, where a far greater share of power is still devolved to the cantonal and communal level. This is known as the Principle of Subsidiarity and is vital to Switzerland's highly decentralised system, allowing each administrative task to be carried out at the lowest level possible. Traditionally, political affairs took place predominantly at the cantonal level, but as the Confederation has progressed, tasks have been increasingly consolidated at the federal level. However, implementation of public policy (including environmental laws) has mostly remained within the cantons, often with considerable room for manoeuvre, known as the federalism of implementation 'Vollzugsföderalismus' (Mauch et al. 2000).

The research presented in this book focuses on the Canton Valais (VS), a bilingual canton in the south-western part of Switzerland (see Maps 6.1 and 6.2), and the third largest of the country's 26 cantons. Each canton has its own constitution, parliament and tribunal, as well as its own competency for education, health, transport, police and fiscal departments. Communal authority and competencies vary greatly between cantons. The Valais is situated in the south western part of Switzerland, bordering both France and Italy. The canton comprises of 143 communes divided between 12 districts (Valais 2009).

Map 6.1 Map of Switzerland, showing Canton Valais highlighted in the *darker colour*

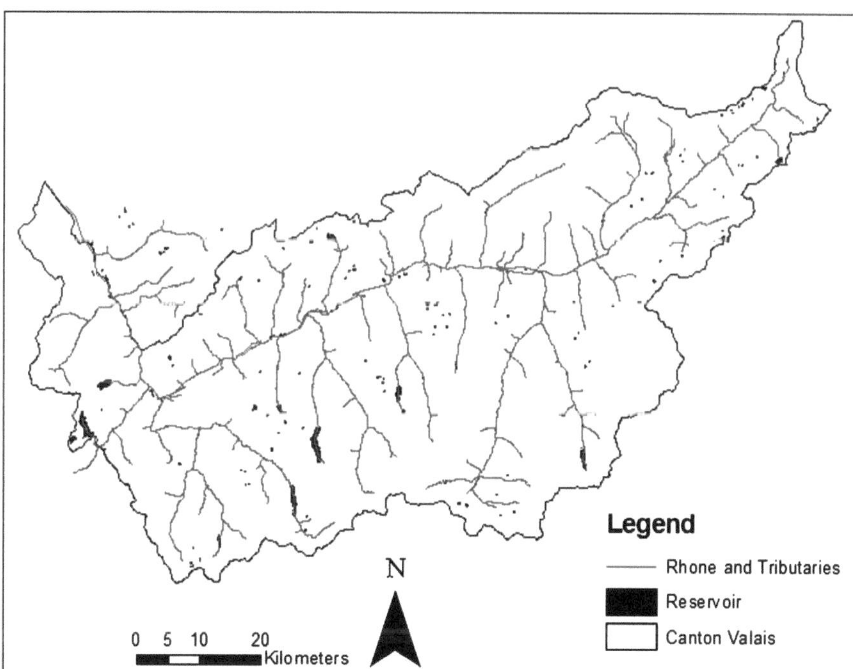

Map 6.2 Map of the Rhône basin in the Canton Valais, showing the main tributaries (*thin lines*) to the Rhône River (*thick line*). The *circles* highlight the sub-case areas of particular focus for interviews

The communes are organised into a legislative council, an orientation body, and an administrative council for a population of just under 300,000 (Valais 2009). The canton contains a number of different economic sectors (hydropower, tourism, agriculture as well as chemical, biotechnology, IT & communications, aluminium, electronics & micro-technology, metal industry), which requires a coordination of water allocation for these different uses. It is one of the poorer cantons in Switzerland, and is therefore reliant on federal subsidies for the implementation of projects. Management of the canton's waterways is delegated across these 143 communes, which in the Valais are significantly independent (Clivaz and Reynard 2008) for a surface area of 5,224 km^2 (53.8% is unproductive land) (Valais 2009).

Geographically, the Valais is characterised by an extensive valley corridor, which separates the Northern and Southern Alps. The river Rhône flows through this valley, eventually emptying into the Lake of Geneva (Lac Leman) at La Bretagne. The topography is defined by the altitudinal contrast of the valley and river floor, and the high mountain peaks, which influence the large precipitation differential between the different areas of the Valais. As the clouds rise over the peaks, it forces the water vapour to condense and precipitate. Since the valley is surrounded on both sides by mountains that often exceed 4,000 m in elevation, this leads to very low precipitation (some of the lowest in Switzerland) in the valley, and high precipitation in the mountain areas. For example in some areas in the Vispertal (the area leading from Visp up to Stalden), average annual precipitation is about 475 mm. Annual demand however for the maintenance of the cultural and agricultural landscape is in the range of 700–900 mm/year. In order to bridge this gap, 'Suonen/Les Bises' (water conducts) have been built in many areas of the Valais to transport water from higher altitudes to the more water stressed valley.[1]

6.2.1 *Climatic Detail*

Climatic influences on mountain water resources are especially since climatic changes are taking place within a broader context of rapid socio-economic transformation (Wiegandt 2008). Recent reports have indicated that Switzerland is becoming more and more urbanised, agriculture is intensifying and mobility is increasing, resulting in growing pressures on the environment (in particular bio-diversity) from the intensity of consumptive patterns despite any gains from eco-efficiency and environmental protection (FOEN 2007). Additionally, tourism has had a major effect on the alpine environment, contributing to increased water usage (increased

[1] Netting (1981) and Ostrom (1990) both have written at length about cooperation in the commons in the spotty environment of the Valais. While rainfall and snowfall is high in one section, but precipitation is low in another section, where it is very dry. Netting presented a detailed description of the rules that were used in Törbel (just above Stalden) that farmers used to co-manage resources through times or areas of plenty and of scarcity.

population, energy, artificial snow) as well as having an impact on water quality and related ecosystems. Furthermore, hydroelectric production has also impacted river quantity, quality and water related ecosystems not only through decreased river flows, but also through hydropeaking (Von Arx 2009; Bonzi 2009). Hydro-production is set to grow in coming years, especially from micro-hydropower plants due to Federal Office for Energy's promotion of renewable energy through the subsidy system *Kostendeckende Einspeisevergütung* (KEV) (BFE 2009).

As an alpine country, Switzerland is particularly threatened by climate change, since the most pronounced effects of global warming are projected to be over land, in the northern hemisphere in the winter months (Bürki et al. 2005). The resulting glacial retreat, melting of permafrost as well as the changes in vegetation and precipitation, are already raising considerable challenges across the country. In summary, for a double CO^2 simulation, higher winter temperatures and a more marked increase in summer temperatures are noted. Precipitation will also be higher and more intense during winter, but much reduced in summer months (Häberli and Beniston 1998); seasonal shifts in the timing of extreme precipitation events and their intensity will also result in impacts on the mountain environment and infrastructure different to what has been experienced in recent decades (Beniston 2006). Between the mid-nineteenth century to the mid-1970s alpine glaciers lost on average a third of their length and half their volume in response to temperature increases that in many parts of the Alps have exceeded 1–1.5 °C since 1900, i.e., about three times the global-average temperature rise (Beniston 2004). Since then, a further 20–30% of ice volume has melted. In the heat wave of 2003, glaciers lost 5–10% of their 2,000 total volume (Häberli et al. 2005). Reports have shown the impacts of increased warming on glacier retreat and their consequences for the energy and economic system of the Swiss Alps (Horton et al. 2005).

Increased glacial melt also is leading to an increase in flood risks and other natural hazard events (OcCC 2008). Increased flooding and extreme precipitation events are compounded by an increase in risk exposure due to infrastructure/housing development in vulnerable areas which are currently seen as 'safe' due to technical interventions. Temperature increases at alpine elevations also raise demand for water uses such as artificial snow making and summer cooling/drinking water leading to complex management shifts, compounded by changes in seasonality.

6.3 Aconcagua Basin, Chile

The case of Chile represents a very different political paradigm, one that is informed by colonial elitism, neo-liberal market ideology and dictatorship. It was the Pinochet military dictatorship (1973–1990) that pioneered the neo-liberal development strategy, which is still in place today. Neo-liberal ideology holds that social functions are best managed through the free markets, and that economic development should be undertaken by the private sector, with the state playing a facilitating regulatory role (Budds 2004). Prior to the military coup, Chile's 140 years of democracy was defined by a hierarchical style of politics imposed on a centralised state, with political institutions developing a top-down character, keeping civil society in check (Carruthers 2001).

The Pinochet regime narrowed the scope for political agency by legitimising repression of civic associations, opposition parties and labour unions in order to refashion class and labour relations in the mould of neoliberal politics. Carruthers (2001, p 346) notes that 'contemporary Chilean politics is characterized by three dominant, related trends: a decline in popular participation, the reconsolidation of elitism, and embedded neoliberalism'. The legacy of history, dictatorship and transition has led to a situation where the values and solidarity of grassroots movements are seen as anachronistic and idealistic. Social movements are marginal and political parties are seen as elitist (Carruthers 2001), creating a difficult context for environmental policy and activism to take place. Awareness of environmental issues is gradually reawakening, as an increasing number of environmentally destructive projects are inciting civic movements (see for example the Patagonia Sin Repressa campaign).

The preceding 30 years have therefore been a period of both turbulence and dynamic growth for Chile. Following on from its initiation by the Pinochet Regime in 1973, the 1980 Constitution came into effect, with Article 19.24 that formalises water rights as private property, changing water into a commodity like any other. The handover of power from Pinochet to the *Concertacion* came with the requirement that the Chilean Constitution would not be altered, and some stakeholders still allude to the latent threat from the powerful military and business leaders if this agreement would be reneged. The neo-liberal economic policies implemented during the dictatorship have since defined Chilean politics and driven its export oriented economy, subsidiary government role, as well as the privatisation of vital services. It has led some to suggest that the country, its watershed, rivers and ecosystems have been handed over to the forces of the market and the private sector. In the absence of state regulation that protects the environmental and social health of the country, citizen bodies and international NGOs are now attempting to raise awareness and fill in the protective, long term focussed, role that a government regulatory body should in theory provide.

Water governance similarly had to fall in line with the commitment to development based on an export-oriented economy (Rogers and Hall 2003), which led to a lack of transparency, participation and concern for ecosystems. Water rights in Chile are a marketable commodity, and the country's approach to water management is deemed unique in the world since it took the water use rights market as the basis for its water governance system (Bauer 2004). While other governance systems have utilised a water market as a means to improve effectiveness and efficiency of their governance regime (e.g. California, USA and Australia), the Chilean governance system is led by the market based focus on private property rights, with minimal environmental regulation and no sectoral prioritisation (Bauer 2004; Corkal and Hurlbert 2008; Solanes and Gonzalez-Villareal 1999).

The new regime sought to utilise water resources for expanded irrigation and economic development primarily from the growth of export goods through agribusiness (Table 6.2). In 1981, when the Water Code was established, initial market movements resulted in cases of hoarding and speculation of unused water rights (Corkal and Hurlbert 2008), propagating the sense of the Code's injustice towards the poor and the evidence that it limited the state's capacity to regulate water efficiently and equitably. Water management in La Ligua and Petorca valleys has been classified by Budds (2004, p 324), as highly politicised, due to the manipulation or ignoring of neutral laws and policies by elite political influence and the connections to large scale farmers and other forms of industry. Before 1990s a significant and accountable environmental regime did not exist (Table 6.2). Today an environmental law framework exists (Corkal and Hurlbert 2008), but is still seen as being highly flawed.

In Chile, the socio-economic situation places further stress on its water system from the growing need for energy (hydro-power), urban development and the neoliberal economic growth model. Growing industrialisation, increasing population and highly intensive agricultural production in the lower watershed are all playing a part. The Aconcagua basin is experiencing increasing pressure on its water resources from both climatic and non-climatic pressures, leading to heightened competition among different users for water allocation (Pellicciotti et al. 2007). Chile is one of the most urbanised countries in South America, with the overwhelming majority of the people living in central Chile (Central Valley) and approximately 88% population in urban centres. A third of these are in Santiago, the capital.

The Aconcagua Basin is situated in the southern part of the Valparaíso region, 50 km north of Santiago. The region is divided into seven provinces; Petorca, Los Andes, San Felipe de Aconcagua, Quillota, Valparaíso, San Antonio, Isla de Pascua and has an area of 16,396.1 km^2. The capital of the region, Valparaíso, is a legislative hub of power and an important commercial port. Total population of the Aconcagua basin is 485,614 with the highest density in the urban centres around the course of the main river (Reyes Carbajal 2007). Population is currently expanding, placing an increasing demand on water supply for main activities such as agriculture, mining and industry.

The mining industry not only impacts water resources by its demand in production, but also by its destructive impact on the glaciers on which it mines (Brenning 2007). In the past 15 years, the overwhelming growth of mines in rock glacier areas (e.g. Los Bronces & Division Mina Andina) has produced a strong geomorphologic impact. In the context of tremendous growth of the Chilean mining industry, and the great potential for future exploitation of reserves, further cases of future degradation and destruction of rock glaciers and glaciers have to be expected in Chile. The region is also important for agricultural production, wine production and industrial activity (copper mining, cement production). The largest oil refinery is situated on the mouth of the Aconcagua River, in Con-Con. Chemical and gas storage is another important industry for the region. In the agricultural corridor of the interior valleys, the export agribusiness is focused on avocados, cherimoyas and flowers. Recently, drip feed irrigation has been applied to the hill sides, to cultivate otherwise dry and unproductive land.

6.3.1 Climatic Detail

Again, a more detailed discussion of climate change impacts on the Aconcagua basin will be discussed in Chap. 9, but for now, I shall highlight some of the key characteristics of the basin as detailed by Reyes Carbajal (2007). The total area of the watershed (see Map 6.3) is 7,340 km^2 (45% of Valparaiso Region), which covers a highly variable region where the Andes rise to more than 5,000 m.a.s.l. within 200 km of the coastline (Vicuña et al. 2011). Two climate types are observed in the Aconcagua basin: warm Mediterranean and cold climate in the high Andes. Mean annual temperature is 5.2 °C but in summer it increases above 27 °C. Mean annual monthly precipitation at the coast is approximately 395 mm/year, in the central part of the basin it is 261 mm/year, presenting dryer areas and less amount of precipitation because of its relief. At higher elevations precipitation increases to 467 mm/year. Basin-average annual runoff at the coast side is over 50 mm/year and in the central part less than 20 mm/year. Higher values occur in summer and spring as a result of glacier and snow melt, lower flows occur during autumn and winter especially in the upper part above 1,000 m altitude where the flow regime is nival. At lower elevations the regime becomes mixed; nivo-pluvial.

In the Andes, glacial melt water supports river flow and water supply for tens of millions of people during the long dry season, and precipitation can be close to zero during summer (Pellicciotti et al. 2008). Hydrological impact studies have shown that warming leads to changes in the seasonality of river flows in areas where a high percentage of winter precipitation currently falls as snow (Kundzewicz et al. 2007). The IPCC (Kundzewicz et al. 2007) have shown that higher temperatures will generate increased glacier melt. Studies in the Aconcagua basin have suggested that the increase in the melt water production from the glaciers in the upper section of the Aconcagua basin have already taken place and that the glaciers are now in a phase of diminishing contribution to the basin stream flow (Pellicciotti et al. 2007).

Studies into temperature trend analysis have shown that in the Aconcagua region, and specifically in the upper Aconcagua River Basin a statistically significant increase in both summer and winter air temperature in the last decades has been observed (Pellicciotti et al. 2008). Carrasco et al. (2005) have observed an elevation in the snow line in central Chile by 127 m in the last quarter of the twentieth century. In another study by (Reyes Carbajal 2007), analysing hydro-climatic trends in the basin, decreases in summer run off were observed mainly in the upper part of the basin, as a result of a reduction in glacial coverage and snow depth in the drainage watershed.

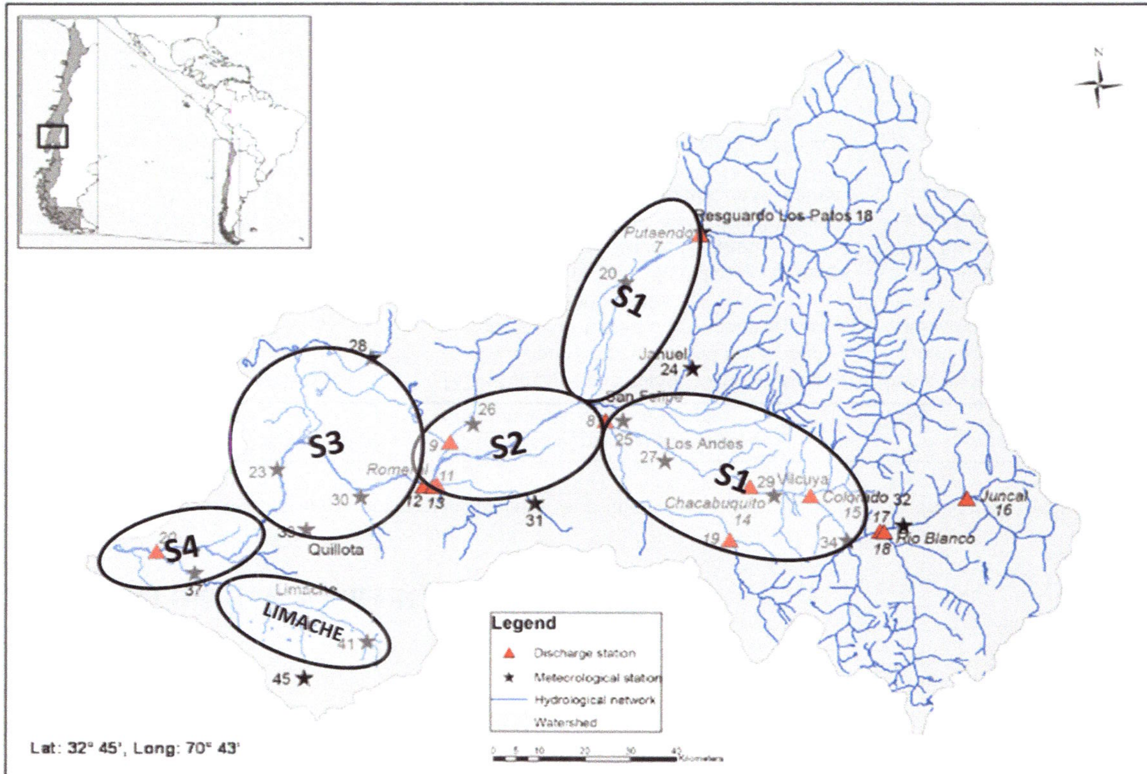

Map 6.3 Aconcagua Basin, Valparaíso, Chile showing the divisions of the Aconcagua Basin. Map indicators the main irrigation sectors of the Rio Aconcagua (Sections 1–4 and the Limache section). Each sector of the basin has its own Junta de Vigilancia that presides over the different associations of canal users (DGA 1996, in Ribbe and Gaese 2002)

6.4 Development of the Governance Regimes

Table 6.2 Timeline showing the developments in environmental laws, institutions and policy in relation to the corresponding political and economic landscape

Period	Country	Political & economic developments	Environmental/internal governance Environmental law & principles	Trend
1870–1940	Switzerland	After the end of WWII, economic expansion rapidly progressed Deforestation in mountain areas (especially the Alps) was considered the main culprit behind the damage caused by a series of catastrophic floods in the nineteenth century (Mauch et al. 2000). Growth in demand of energy = new competitive use of water	1874: Federal Constitution – Art 24 (assigns sovereignty over water and the superintendence of forests in mountain regions to the Confederation) 1876: Federal Law on the Policing of Forests 1877: Federal Law on the Hydraulic Engineering Policy: regulating flood and erosion protection. Regulation is centred at the federal level 1877: Federal Law on the Policing of Waters 1888: Law on Fishing with a paragraph prohibiting the dumping of wastewater and other substances by industry into water bodies if this represented a threat to fish and shellfish stocks	Simple regime, focussed on technical (and responsive) solutions against natural disasters and extreme events Gradual realisation of the economic worth in exploiting water power 1874 – Introduction of the confederation as a state actor – transfer of jurisdiction from the cantons (Mauch et al. 2000)

1908: 24bis FC regulated the assignment of concessions for the production of hydroelectric power, the transport of electricity, the charges to be raised by the Cantons, and limited the use of water bodies. Hydroelectricity production is founded on the system of 'water concessions', whereby the private surface water owners at the communal level may sell a 'concession' (right to exploit the power of the water) for a given time period (generally 80 years) for an annual hydraulic fee (Clivaz and Reynard 2008)

1912: water property rights defined in the Swiss Civil Code

1916: Federal Law on the Use of Water Power: improve electricity production and supply – Federation now has the superintendence over the use of private and public water bodies for hydroelectric power. Cantons retain responsibility to allocate concessions and ability to profit from the charges (within the framework of the federal legislation)

Wahlen Plan: during WW2, strongly promoted drainage measures, as a result of which substantial areas were made available for agricultural use

(continued)

Table 6.2 (continued)

Period	Country	Political & economic developments	Environmental/internal governance Environmental law & principles	Trend
	Chile	1891: Chilean Civil War Period following ushers in democratic tradition, though tightly controlled from above.	Civil Code: declared water as a national good for public use. Water management defined in the constitution. Settler politics usurped cultural water management of indigenous communities.	Water as a public good.
1950	Switzerland	Failure in the implementation (at canton level) of previous federal laws on water pollution leads to a growing discontent.	1953: Ratification and adoption of article 24quater FC (81% yes vote). Corresponding legal basis simultaneously developed leading to: 1955: Law on Water Protection (LWater). Introduction of the polluter pays principle, where polluters are now liable for damage	Environmental Protection becomes an issue at the federal level. Polluter Pays Principle enters legislative framework
	Chile		First Water Code: institutionalised water rights, recognised as private property and possible to register with pre-defined uses (Bauer 1997)	Water as a private property
1960	Switzerland		1966: Swiss law on participation and access to environmental information developed independently of EU Law, and in fact precedes it. The right of organisations to Appeal against Rulings on Installations was first introduced in the Federal Act on the Protection of Nature and Cultural Heritage (PNLA). It was hotly debated, but further strengthened in the 1983 EPA	Increased transparency and provisions in the law open up access to environmental information Movement of environmental legislative competence from the regional to the federal level. Implementation remains at local & regional level

	Chile	1964: Presidential election of Christian Democrat Eduardo Frei Montalva initiated a period of major reform 1967 -: Major economic depression, peaking in 1970	Chile's liberal governments experimented with reallocation of rights to water to incorporate new technology and planning criteria to promote transformative social justice aims First Agrarian Reform Law (Law No. 15,020) passed by Jorge Allessandri. Creation of CORA (Corporacion de la Reforma Agraria) Agrarian Reform: Revised Agrarian Reform Law (Law No. 16, 640). Significant land expropriation & redistribution start. Water law revised to reinstate state control over water resources, cancelling former concessions	Reform & Redistribution
1970	Switzerland	Development of a highly industrialised consumer based industry placed additional demands on the State Economic expansion becomes the dominant principle. Increases in consumption and demand for goods	Total revision of state policy and competencies (e.g. energy & Umweltbelastungen) 1971: More comprehensive Water Protection Act, to meet the consequences of growing pollution Federal Law against Pollution: Since the 1950s, a large number of technical regulations have been developed in order to protect against pollution, which the cantons are responsible for implementing. 1971: Federal Office of the Environment established	Population pressures met through more comprehensive legal framework regarding water Technical Approach to protection from water maintained

(continued)

Table 6.2 (continued)

Period	Country	Political & economic developments	Environmental/internal governance Environmental law & principles	Trend
	Chile	Economy in crisis, inflation out of control 1973: Military Coup Pinochet brings in military dictatorship which lasted until 1988. Highly violent regime Reversal of the state led reforms of the previous decade, leaving aspirations for social justice thwarted. Public administration was reshaped to foster neo-liberal development, with neoliberal policies institutionalised into the 1980 Constitution	Country forcibly neo-liberalised. Since then the Chilean government has remained most responsive to the demands of corporations and non-indigenous land owners that control the country's development agenda Aggressive economic expansion into export goods such as non-indigenous forestry, mining, aquaculture, agriculture as well as hydroelectric development	Neo-liberalisation of the economy Economic projects are given priority over social and environmental elements Economic regime
1980	Switzerland	Continued economic, population and urban growth	EPA 1983: Embodies all the fundamental environmental principles. Introduced a broader scope to environmental protection. Since enactment of EPA, environmental law is almost entirely federal law	Increased federal control over environmental protection. Federal philosophy to water governance shifts to a more integrated approach, but not at canton or local levels, where technocratic implementation remains

	Greater awareness of the fact that as an increasingly important source of water pollution, agriculture had hitherto been more or less excluded from the water protection legislation. Drying up of an increasing number of mountain creeks and small watercourses being forced underground (Mauch et al. 2000)	Throughout the 1980s, there was a long and hard fought negotiation around the revision of the water protection law (namely around agriculture and hydropower)	Quantitative aspect of water protection, which was understood as part of a more comprehensive protection of «aquatic environments» and biotopes, gained in significance (Mauch et al. 2000)
Chile	1980: Pinochet moves to being President. Regime gradually permitted greater freedom of assembly, speech and political activity Vigorous pursuit of free market economic policies Chilean economy moved away from state involvement towards a largely free market economy, which saw a large increase in domestic and foreign investment	1981: Water Code (revised). The previous water code was reformed in line with the new political-economic framework (Budds 2004). A neo-liberal economic water market was established (by basing the law on private water use rights), designed to promote economic efficiency and agricultural development. Rights could be freely traded with few restrictions, no prioritisations and little state regulation	Free-market approach to water law and economics and water resources management 'Chilean Model': water rights treated not just as private property, but also as a fully marketable commodity

(continued)

Table 6.2 (continued)

Period	Country	Political & economic developments	Environmental/internal governance Environmental law & principles	Trend
1990	Switzerland	Further development & strengthening of federal law & legal enforcement in environmental law	1991: Total revision of the LWater brought into law from the Referendum (Volksinitiative) 'Zur Rettung unsere Gewässer', which focussed on better waste water management at the legal level. The opportunity was then taken to implement further environmental regulations. There was strong opposition from hydropower interest groups to the introduction of quantitative protection of water, but a referendum opposing the new law (on quantitative measures) was rejected in a popular vote in 1992 1997: Further development of the PPP, through the vote on 'Finzierung einer nachhaltigen Abwasserbeseitigung un Abfallentsorgung' (BB1 1996 IV 1219). Transferred the burden of pollution clean up from the state to private companies/ polluters	Conflict with nature protection organisations (see examples of the Thur and Sihl in Zaugg 2002) lead to a growing awareness of the global character of environmental problems & bargaining processes and solutions which integrated the different uses of water WPA & EPA amended to strengthen the implementation of the implementation of the polluter pays principle (PPP) More integrated regime
	Chile	1989: Pinochet not re-elected as president. 1990–1994: Transitional Period New democratic centre-left coalition government came to power, still broadly committed to the neoliberal model	1992: Initial reforms to soften certain neoliberal elements to the 1981 Water Code were proposed (Bauer 1997), but strongly opposed by economic interest groups (e.g. agricultural groups) and neoliberal economics and rights wing politicians (Budds 2004) 1993: Indian Law: Reference Article 20, 23	Conflicts between consumptive and non-consumptive uses Hording and environmental issues not adequately addressed in the system

			1996: Second coalition government revised the proposal for modifications to the Water Code. Fees for non-use (absence of infrastructure for potential use instead of forfeiture (Bauer 1997). Minimal impact was guaranteed since the new proposals would only impact new rights, and omitted river basin associations (mainly due to the conflicts with the hydroelectric sector in southern Chile). Significant opposition was raised from the agricultural, hydropower and mining sectors, reforms are still waiting approval by the Senate. 1997: Environmental impact studies (EIA) and declarations become obligatory for most industrial and mining activities,	
2000	Switzerland	Continuing debate between business groups and fishing community & ecological groups on environmental legislation and provisions for protection	2007: Federal People's Initiative was initially proposed, called 'Lebendiges Wasser (Renaturierungs-Initiative)'. The initiative was started up by the community of fishermen in order to further strengthen the provision for residual flows in the law. The initiative seeks to make amendments to the Federal Constitution (18 April 1999) by adding a new clause (Art. 76a Renaturierung von Gewässern – Renaturalisation of water ways). This has been supported by various ecological groups, and in 2011 was passed in an amended form by parliament (See Appendix 9.6)	Increasing appreciation of integrative nature of the system, moving to an integrated approach Added complexity to the legal framework, in order to try to meet both ecological and economic demands on water

(continued)

Table 6.2 (continued)

Period	Country	Political & economic developments	Environmental/internal governance Environmental law & principles	Trend
	Chile	Political-legislative framework in Chile (i.e. neoliberal constitution) favours large scale economic interests Increase in advocacy and environmental groups	Some reforms proposed: But unrealised in implementation Advocacy community makes clear connections between indigenous rights, environmental protection and sustainable development, and is supporting indigenous contestation of water control. In 2005 a minor set of reforms was passed, relating to payment for non-use and sustainable use volumes, but were limited in their effect due to only being required for new water use rights	Strong national centralized approach to water management, through the Water Code, with most of the power relegated to private interests and private water rights
2010	Switzerland	Increased cooperation with the European Community	2009: Continued debate on Lebendiges Wasser in parliament	Awareness of Ecological Water (Residual Flows)

	Increase unregulated water use for the development of artificial snow making	Wasser Agenda 21: is a recent collaboration of the most important water management actors in Switzerland, who have come together to establish a platform for dialogue and a think tank for the generation of optional solutions to arising issues. It aims to support and enhance current collaboration and coordination across sectors, orientate the industry towards sustainable development principles, pre-emptively identify issues and develop an integrated approach to the management of water resources and water waterways. However, interview data has shown that WA21's remit may not be so welcome in the more independent cantons, where any intrusion from national or federal actors tends to be unwelcome	Further movement towards a more integrated and cross-sector watershed based approach in policy and law Implementation gap
Chile	Continued economic (expert oriented) and hydropower development	The manipulation or ignoring of 'neutral' laws and policies has been effected by the political influence and connections of large scale farmers and other forms of industry	Water rights are a highly politicised issue Chilean water market

Source: Bauer (2004, 1997), Budds (2004), Carruthers (2001), Corkal and Hurlbert (2008), Mauch and Reynard (2002), Petitpierre (1999), Varone et al. (2002), WA21 (2008)

The majority of the extreme events that will be used as the context in which to study the adaptive capacity of the governance regime have taken place in the final rows (1990–2010) of this table

6.5 Summary

Both case have undergone dynamic developments in their governance regimes over the preceding decades and as mountainous areas will face increasing challenges from altering bio-physical parameters from a changing climate (Viviroli et al. 2011). However, these shifts in seasonality and alterations in glacier melt are likely to take on particular significance in the Andean region where dependence on glacier and snow melt run off is high for water availability during the dry summer months (Pellicciotti et al. 2007; Souvignet et al. 2008). In these mountain watershed nivo-glacial regimes climate change (as experienced through glacier melt and snow pack changes) will correspond with changes in the seasonality of river flows. In both areas impacts of climate change have already been observed on glacial melt elevation of the snow line with associated impacts on the timing and amount of run off (Häberli and Beniston 1998; Pellicciotti et al. 2007) projected to increase (Christensen et al. 2007).

Global climate models show that warming and drying trends have already been observed and can be projected to intensify for the Andean region (Christensen et al. 2007) while temperature increases in the Alps have exceeded 1–1.5 °C since 1900 (about three times the global-average temperature rise), with corresponding implications for increased glacial melt and changes in snow pack (OcCC 2008; Solomon et al. 2007). Furthermore, in combination with the strong ENSO event currently occurring, the central-northern regions of Chile have been experiencing one of the worst drought periods in memory (DGA 2010).

The convergence of climate change impacts with the complex political and economic issues poses significant challenges across the two case areas that will need to be navigated through effective water governance frameworks. Table 6.3 provides an overview of the impacts on different goods and services (Mauch et al. 2000) for the different extreme events in mountain zones of interest to this study. Winter warm spells (strong positive temperature exceedances in winter) were included in this table, as events lead to high temperature anomalies and are associated with lower than average winter precipitation (Beniston 2005b).

Table 6.3 Detail of impacts on separate categories of water goods and services per each extreme event

Case event: geomorphological impacts	Impacts on water goods & services				
	Production	Energy	Environmental flows	Transport & absorption of effluents	Domestic use: water supply & sanitation
Heat waves/drought Permafrost degradation & reduction in the cohesion of slope material, acceleration of glacier retreat, reduction in snow pack	Heat stress, enhanced evaporation, droughts, soil-moisture depletion, acceleration of glacier retreat (could lead to decreased or increased run off depending on the state of glacial retreat)	Changes in hydro-power supply because of seasonal shifts in the filling of dams in the Swiss case and river flow in the Chilean case	Damage to eco-systems through excessive heat & drought Drying up of river beds, streams and estuaries	Reduced flow increasing concentration of chemicals in water ways	Despite reduced flows, drinking water is prioritised during drought situations in both areas (in Chile only once an official Presidential Declaration of drought)
Winter warm spell/drought Enhanced avalanche risk Without the buffering effect of snow, a heavy downpour can become a catastrophic event, low precipitation	Decreased snow availability/ seasonal snow cover Early run off into alpine river basins	Changes in hydro-power supply because of seasonal shifts in the filling of dams in Swiss area	Early start of the vegetation cycle		Reduction of mid-winter snow pack negatively impacts water reserves for the recharging springs with melt
Extreme precipitation Slope instability events in mountain regions, increased frequency and severity of floods	Sediments deposited in large quantities on agricultural lands, irrigation canals and streams – reductions in agricultural production	Erosion, discharge & sedimentation rates – damage hydropower infrastructure		Increased flows causing overflow of waste waters into water ways	Excess debris can negatively impact water treatment plants and canalisation infrastructure

Source: Beniston (2005a, b), Beniston et al. (2007), Kundzewicz et al. (2007)

References

Bauer C (2004) Siren Song: Chilean Water law as a model for international reform. Resources for the Future, Washington, DC

Beniston M (2004) Climatic change and its impacts. An overview focusing on Switzerland. Kluwer Academic, Dordrecht/Boston

Beniston M (2005a) Mountain climate and climatic change: an overview of processes focussing on the European Alps. Pure Appl Geophys 162:1587–1606

Beniston M (2005b) Warm winter spells in the Swiss Alps: strong heat waves in a cold season? Geophys Res Lett 32:L01812

Beniston M (2006) The August 2005 intense rainfall event in Switzerland: not necessarily an analog for strong convective events in a greenhouse climate. Geophys Res Lett 33. doi:10.1029/2005GL025573

Beniston M, Stephenson DB, Christensen OB, Ferro CAT, Frei C, Goyette S, Halsnaes K, Holt T, Jylha K, Koffi B, Palutikof J, Scholl R, Semmler T, Woth K (2007) Future extreme events in European climate: an exploration of regional climate model projections. Clim Chang 81:71–95. doi:10.1007/s10584-006-9226-z

Beniston M, Stoffel M, Hill M (2011) Impacts of climatic change on water and natural hazards in the Alps: can current water governance cope with future challenges? Examples from the European "ACQWA" project. Environ Sci Policy 14:734–743

BFE (2009) Kostendeckende Einspeisevergütung (KEV). Bundesamt für Energie. http://www.bfe.admin.ch/themen/00612/02073/index.html. Accessed 9 June 2009

Bonzi C (2009) Gewässerschutz darf nicht weiter verwässert werden. Pronatura Magazin, Feb 2009, pp 4–5

Brenning A (2007) The impact of mining on rock glaciers and glaciers: examples from Central Chile. In: Orlove B, Wiegandt E, Luckman BH (eds) Darkening peaks: glacier retreat, science, and society. University of California Press, Berkeley/Los Angeles/London

Budds J (2004) Power, nature and neoliberalism: the political ecology of water in Chile Singapore. J Trop Geogr 25(3):322–342

Bürki R, Elsasser H, Abegg B (2005) Climate change and tourism in the Swiss Alps. In: Hall M, Higham J (eds) Tourism, recreation and climate change. Channel View Press, Clevedon, pp 155–163

Carrasco J, Casassa G, Quintana J (2005) Changes of the isotherm and the equilibrium line in altitude in central Chile during the last quarter of the 20th century. Hydrol Sci J 50(6):933–948

Carruthers D (2001) Environmental politics in Chile: legacies of dictatorship and democracy. Third World Q 22(3):343–358

Christensen JH, Hewitson B, Busuioc A, Chen A, Gao X, Held I, Jones R, Kolli RK, Kwon RT, Laprise R, Magaña V, Mearns CG, Menendez CG, Raisanen J, Rinde A, Sarr A, Whetton P (2007) Regional climate projections. In: Climate change 2007: the physical science basis. Contribution of working group I to the fourth assessment report of the Intergovernmental Panel on Climate Change. Cambridge University Press, New York

Clivaz C, Reynard E (2008) Crans-Montana: water resources management in an Alpine tourist resort. In: Wiegandt E (ed) Advances in global change research 31: mountains: sources of water, sources of knowledge. Springer, Dordrecht, pp 103–119

Corkal DR, Hurlbert M (2008) Governance institutions and community vulnerabilities to climate -induced water stress – case studies in Canada and Chile. Paper presented at the World Water Congress, Montpellier, France, 1 Sept 2008–4 Sept 2008

DGA (2010) MOP declara zona de escasez hídrica en regiones de Coquimbo y Valparaiso

FOEN (2007) Environment Switzerland 2007. Federal Office for the Environment and the Federal Statistical Office, Bern/Neuchatel

Häberli W, Beniston M (1998) Climate change and its impacts on glaciers and Permafrost in the Alps. Ambio 27(4):258–265

Häberli W, Huggel C, Paul F (2005) Gletscherschwund im Hochgebirge. Alpenvereinsjahrbuch Berg 2005 des DAV

Horton P, Schaefli B, Mezghani A, Hingray B, Musy A (2005) Prediction of climate change impacts on Alpine discharge regimes under A2 and B2 SRES emissions scenarios for two future time periods. Bundesampt für Energie (BFE), Bern

Kundzewicz ZW, Mata LJ, Arnell NW, Döll P, Kabat P, Jiménez B, Miller KA, Oki T, Sen Z, Shiklomanov IA (2007) Freshwater resources and their management. In: Parry ML, Canziani OF, Palutikof JP, van der Linden PJ, Hanson CE (eds) Climate change 2007: impacts, adaptation and vulnerability. Contribution of working group II to the fourth assessment report of the Intergovernmental Panel on Climate Change. Cambridge University Press, Cambridge, pp 173–210

Mauch C, Reynard E, Thorens A (2000) Historical profile of water regime in Switzerland (1870–2000). Working paper de l'IDHEAP 10/2000:IDHEAP, Lausanne

OcCC (2008) Das Klima ändert – was nun? Der neue UN-Klimabericht (IPCC 2007) und die wichtigsten Ergebnisse aus Sicht der Schweiz. Organe consultatif sur les changements climatiques, Bern

Pellicciotti F, Burlando P, Van Vliet K (2007) Recent trends in precipitation and streamflow in the Aconcagua River basin, central Chile Glacier mass balance changes and meltwater discharge. Selected papers from sessions at the IAHS assembly in Foz do Iguaçu, Brazil, 2005, IAHS Publication 318

Pellicciotti F, Helbing J, Rivera A, Favier V, Corripio J, Araos J, Sicart JE, Carenzo M (2008) A study of the energy balance and melt regime on Juncal Norte Glacier, semi-arid Andes of central Chile, using melt models of different complexity. Hydrol Process 22:3980–3997

Reyes Carbajal L (2007) Analysis of hydroclimatic trends in the Aconcagua river basin, Central Chile. Institute of Environmental Engineering, ETH Zurich

Ribbe L, Gaese H (2002) Water management issues of the Aconcagua Watershed, Chile. Technol Resour Manage Dev 2:79–100

Rogers P, Hall AW (2003) Effective water governance. GWP TAC background papers no. 7. Global Water Partnership Technical Committee, http://www.gwpforum.org/servlet/PSP?iNodeID=215&itemId=197

Solanes M, Gonzalez-Villareal F (1999) The Dublin principles for water as reflected in a comparitive assessment of institutional and legal arrangement for integrated water resources management. Global Water Partnership: Technical Advisory Committee (TAC). http://info.worldbank.org/etools/docs/library/80638/IWRM4%5FTEC03%2DDublinPrinciples%2DSolanes%26Gonzales.pdf. Accessed 9 Dec 2008

Solomon S, Qin D, Manning M, Chen Z, Marquis M, Averyt KB, Tignor M, Miller HL (2007) Summary for policymakers. In: Climate change 2007: the physical science basis. Contribution of working group I to the fourth assessment report of the Intergovernmental Panel on Climate Change. Cambridge University Press, Cambridge/New York

Souvignet M, Gaese H, Ribbe L, Kretschmer N, Oyarzún R (2008) Climate change impacts on water availability in the Arid Elqui Valley, North Central Chile: a preliminary assessment. Paper presented at the IWRA World Water Congress, Montpellier, France

Valais (2009) Informations Statistiques: Office de statistique du canton du Valais. http://www.vs.ch/Navig/navig.asp?MenuID=16556&RefMenuID=0&RefServiceID=0. Accessed 9 June 2009

Varone F, Reynard E, KisslingNäf I, Mauch C (2002) Institutional resource regimes: the case of water management in Switzerland. Integr Assess 3(1):78–94

Vicuña S, McPhee J, Garreaud RD (2011) Agriculture vulnerability to climate change in a Snowmelt Driven Basin in Semiarid Chile. J Water Resour Plan Manage, URL: http://dx.doi.org/10.1061/(ASCE)WR.1943-5452.0000202

Viviroli D, Archer DR, Buytaert W, Fowler HJ, Greenwood GB, Hamlet AF, Huang Y, Koboltschnig G, Litaor MI, Lopez-Moreno JI, Lorentz S, Schadler B, Schreier H, Schwaiger K, Vuille M, Woods R (2011) Climate change and mountain water resources: overview and recommendations for research, management and policy. Hydrol Earth Syst Sci 15(2):471–504. doi:10.5194/hess-15-471-2011

Von Arx J (2009) Unterhalb der Kraftwerke sind viele Flüsse tot. Pronatura Magazine, Feb 2009, pp 8–9

Wiegandt E (1977) Inheritance and demography in the Swiss Alps. Ethnohist 24:133–148

Wiegandt E (2008) Framing the study of mountain water resources: an introduction. In: Wiegandt E (ed) Advances in global change research 31: mountains: sources of water, sources of knowledge. Springer, Dordrecht, pp 3–13

Chapter 7
Water Governance in the Context of IWRM: Switzerland

Abstract This chapter outlines the work completed for the ACQWA project on the governance assessments for the Chilean and Swiss case area and is used to provide vital background to the water governance situation and challenges in the Swiss case area. In the Swiss case, despite the fulfilment of accountability, transparency and participation indicators, the assessment suggested that there is a significant gap between the conceptual strands of IWRM in federal laws and policies and their translation at the regional and local levels. The complex institutional framework, legislative provisions and levels of sovereignty which govern water resources in the Canton Valais implied a lack of coordination and long term planning amongst the different politico administrative levels and sector groups. These challenges are linked to concerns that the ramifications of climate change and expanding water uses are not adequately reflected in the current governance framework.

Keywords Rhône, Canton Valais, Switzerland • Water governance assessment • Legislative and policy challenges • IWRM • Sectoral and subsidiarity challenges

7.1 Introduction to the Assessment

The STRIVER/BRAHMATWINN governance assessment represents a systematic methodology to measure governance in the specific context of IWRM, which has been applied in a number of basins within Portugal, Spain, Vietnam and India as part of the STRIVER (www.striver.no) and BRAHMATWINN (www.brahmatwinn.uni-jena.de/) projects (Allan et al. 2007). Its application in the Rhône Basin in the Valais, Switzerland and the Aconcagua Basin in Chile provides a baseline assessment of the governance context as it specially relates and addresses water governance in river basins.

M. Hill, *Climate Change and Water Governance: Adaptive Capacity in Chile and Switzerland*, Advances in Global Change Research 54,
DOI 10.1007/978-94-007-5796-7_7, © Springer Science+Business Media Dordrecht 2013

The indicator approach to water governance is based on three core elements of good governance:

- Accountability: holding governments responsible for their actions; contestability of political power.
- Transparency: right to information; availability and clarity of information.
- Participation: involvement of citizens in decision making.

A comprehensive set of indicators was developed in the style of a questionnaire comprising 18 key questions, and 60 sub-questions testing both commitment (degree to which the governance context adheres to accepted standards of good governance) and process (extent to which this governance context is implemented in reality) (Allan and Rieu-Clarke 2007). The methodology provides a tool to assess existing laws, policies and formal institutions as well as how the relevant governance arrangements have been implemented in practice. The output provides an important benchmark for identifying potential gaps and barriers to implementing IWRM (Rieu-Clarke et al. 2008, p 2) and baseline from which to better understand core governance mechanisms that may affect adaptive capacity in water governance arrangements. Research was conducted initially through a desktop study to gather all relevant information on the laws as well as policies and institutions related to governance and IWRM. Interviews then ascertained the degree to which the law has been implemented. The following chapters provide an overview of the findings from the governance assessments, as a useful baseline from which to move deeper into the adaptive capacity assessment work that will be presented in Part III.

7.2 Swiss Water Governance Assessment

There are a number of laws with water elements at both the federal and cantonal level in Switzerland as shown in Table 7.1. The main federal provisions for property rights concerning water are set out in the Swiss Civil Code (CC) 664, 704 and 705. If a deed holder wishes to use these waters, they are required to follow federal law governing use and protection. Public property is deemed as rivers, streams, lakes, glaciers, as well as springs arising from uncultivable land. However, in Switzerland, federal provisions refer or imply additional cantonal legislation, which differs across the cantons, some of which have not passed additional regulation at all. In Valais, a more decentralised canton, while the lateral rivers of the Rhône are property of the communes, the Rhône itself is the property of the canton.

7.2.1 Accountability

Switzerland represents a direct democratic system and high level of decentralisation for decision making and planning. The primary system to challenge laws is through its citizens' constitutional rights to petition (FC Art. 33), initiative and referendum

Table 7.1 Legal provisions concerning water and environment in Switzerland

English	Source/Original Text
Section 1: Federal level	
Federal Constitution (see below)	Bundesverfassung der Schweizerischen Eidgenossenschaft (BV), 2000, SR 101, http://www.admin.ch/ch/d/sr/sr.html
Swiss Civil Code (see below)	Schweizerisches Zivilgesetzbuch (ZGB), 1912, SR 210, http://www.admin.ch/ch/d/sr/c210.html
Federal Judiciary Act	Bundesgerichtsgesetz (BGG), 2007, SR 173.110, http://www.admin.ch/ch/d/sr/c173_110.html
Federal Administrative Procedure Act	Verwaltungsgerichtsgesetz (VGG), 2007, SR 173.32, http://www.admin.ch/ch/d/sr/c173_32.html
Federal Law on the Principle of Administrative Transparency (LTrans)	Bundesgesetz über das Öffentlichkeitsprinzip der Verwaltung (BGÖ), 2006, SR 152.3, http://www.admin.ch/ch/d/sr/c152_3.html
Environmental Protection Act (EPA)	Bundesgesetz über den Umweltschutz (USG), 1985, SR 814.01, http://www.admin.ch/ch/d/sr/c814_01.html
Protection of Nature and Landscape/Cultural Heritage Act (PNLA)	Bundesgesetz über den Natur- und Heimatschutz (NHG), 1967, SR 451, http://www.admin.ch/ch/d/sr/c451.html
Water Protection Act (WPA)	Bundesgesetz über den Schutz der Gewässer (GSchG), 1992, SR 814.20, http://www.admin.ch/ch/d/sr/c814_20.html
Environmental Impact Assessment (EIA)	Verordnung über die Umweltverträglichkeitsprüfung (UVPV), 1989, SR 814.011, http://www.admin.ch/ch/d/sr/c814_011.html
Federal Fishing Act (FA)	Bundesgesetz über die Fischerei (BGF), 1994, SR 923.0, http://www.admin.ch/ch/d/sr/c923_0.html
Federal Land Planning Act (FPA)	Bundesgesetz über die Raumplanung (RPG), 1980, SR 700, http://www.admin.ch/ch/d/sr/c700.html
Federal Forest Act (FFA)	Bundesgesetz über den Wald (Waldgesetz, WaG), 1993, SR 921.0, http://www.admin.ch/ch/d/sr/c921_0.html
Ordinance on the Cleanup of Contaminated Sites	Verordnung über die Sanierung von belasteten Standorten (Altlasten-Verordunung, AltlV), 1998, SR 814.680, http://www.admin.ch/ch/d/sr/c814_680.html
Use of Water Power Act (UWPA)	Bundesgesetz über die Nutzbarmachung der Wasserkräfte (Wasserrechtsgesetz, WRG), 1918, SR 721.80, http://www.admin.ch/ch/d/sr/c721_80.html

(continued)

Table 7.1 (continued)

English	Source/Original Text
Law on Hydraulic Engineering (i.e. management of waterways)	Bundesgesetz über den Wasserbau, 1991 (WBG), 1993, SR 721.100, http://www.admin.ch/ch/d/sr/c721_100.html
Ordinance on Environmentally Harmful Substances	Verordnung zur Reduktion von Risiken beim Umgang mit bestimmten besonders gefährlichen Stoffen, Zubereitungen und Gegenständen (Chemikalien-Risikoreduktions-Verordnung, ChemRRV), 2005, SR 814.81, http://www.admin.ch/ch/d/sr/c814_81.html
Polluter Pays Principle	Bundesverfassung der Schweizerischen Eidgenossenschaft (BV), 2000, SR 101, http://www.admin.ch/ch/d/sr/sr.html
Precautionary Principle	Schweizerisches Zivilgesetzbuch (ZGB), 1912, SR 210, http://www.admin.ch/ch/d/sr/c210.html
Principle of Sustainable Development	Bundesgerichtsgesetz (BGG), 2007, SR 173.110, http://www.admin.ch/ch/d/sr/c173_110.html
Section 2: Federal Constitution	
Bundesverfassung der Schweizerischen Eidgenossenschaft (2000), SR 101 is available on the website of ‚Die Bundesbehörden der Schweizerischen Eidgenossenschaft': http://www.admin.ch/ch/d/sr/101/index.html	
An English version of selected parts of the Federal Constitution is available at: http://www.servat.unibe.ch/icl/sz00000_.html	
Art. 2 Purpose, SR 101.2	[1] The Swiss Federation protects the liberty and rights of the people and safeguards the independence and security of the country. [2] It promotes common welfare, sustainable development, inner cohesion, and cultural diversity of the country. [3] It ensures the highest possible degree of equal opportunities for all citizens. [4] It strives to safeguard the long-term preservation of natural resources and to promote a just and peaceful international order.
Art.33 Right of Petition, SR 101.33	[1] Every person has the right to address petitions to authorities; no disadvantages may arise from using this right. [2] The authorities have to take cognizance of petitions.
Article 73 Sustainable Development, SR 101.73	The Federation and the Cantons are engaged to establish a durable balanced relationship between nature, particularly its renewal capacity, and its use by human beings.

(continued)

Table 7.1 (continued)

English	Source/Original Text
Art. 74 Environmental Protection, SR 101.74	[1] The Federation adopts rules on the protection of human beings and their natural environment against harmful or irritating effects. [2] The Federation provides for the fact that such effects are avoided. The costs of such avoidance and removal carry the causers. [3] The execution of the regulations falls to the Cantons, as far as the law does not reserve it for the Federation. This clause is further supplemented by other environmental decrees such as the Ordinance on: Environmentally Harmful Substances (9th June, 1986), which provisions that the law should protect men, animals and plants from harmful or long lasting impacts from harmful substances on their communities, living space or environment.
Art. 76 Water, SR 101.76	[5] On rights concerning international water resources and therewith connected duties [6] In the fulfilment of its tasks
Art. 138 Right to Popular Initiative for a Total Revision of the Federal Constitution, 2003, SR 101.138	[1] 100,000 eligible voters can suggest a total revision of the Federal Constitution from 18 months of the public launch of their Initiative.
Art. 139 Right to Popular Initiative for a Partial Revision of the Federal Constitution, 2009, SR 101.139	[1] 100 000 eligible voters can request a partial revision of the Federal Constitution 18 months from the public launch of their initiative.
Art. 140 Right to Referendum, 2003, SR 101.140	[1] The population and states are call to vote on: (a) a change to the Federal Constitution (b) the entry to organisations of collective security or to super-national communities. (c) urgently declared federal laws, which have no constitutional foundations and have exceeded their: 1 year period of validity; these federal laws must be submitted to a vote within 1 year from when they were taken on through the federal convention. [2] The population are called to vote on: (a) popular initiatives for a total revision of the Federal Constitution (b) popular initiatives for a partial revision of the Federal Constitution in the form of a general challenge, which was rejected by the federal convention. (c) questions on whether a total revision of the Federal Constitution is to be put into effect, where there is disagreement between both councils.

(continued)

Table 7.1 (continued)

English	Source/Original Text
Section 3: Swiss Civil Code (1912)	
SR 210, Schweizerisches Zivilgesetzbuch (ZGB) http://www.admin.ch/ch/d/sr/c210.html	
Art. 664 Public	Abandoned sites and the property of the public domain are subject to state policing on the territory on which they are located. In the absence of evidence to the contrary, public water bodies, as well as regions unsuitable for cultivation, boulders, masses of fallen rocks, névés (firns), glaciers and their springs shall not be considered private property. Cantonal legislation regulates those things which are free, such as the exploitation of common use public properties, such as roads, open spaces, water courses and river beds.
Art. 704 Private	Springs are components of the property and can only be owned in conjunction with the ground from which they arise. The law of spring waters from external property is to be established as subservience through registration in the land register. Groundwater is on equal terms with spring waters.
Art 709	Cantonal legislation can reconcile use laws between neighbours or other persons, notably for the extraction of water, the watering of livestock, water sources, springs, and streams which are private property.
Art. 711	The title bearer of sources, springs or streams which are not useful for him, or which have an unreported use with their worth, is required to divest against full indemnity (with compensation?) for the drinking water services, hydrants or other public good services in general.
Art. 712	Title bearers of drinking water can demand the relinquishing of the surrounding ground, in the instance of expropriation, so far as the protection of their water sources against contamination is necessary.
Section 4: Legal provisions at the Canton level in the Valais	
A full listing of Cantonal Acts, Ordinances and Decision is available at: http://www.vs.ch/Navig/navig.asp?MenuID=4609&RefMenuID=0&RefServiceID=0	

(continued)

Table 7.1 (continued)

English	Source/Original Text
Law on Hydraulic Engineering	Gesetz über den Wasserbau, 2007, 721.1, http://www.vs.ch/Navig/navig.asp?MenuID=4628&Language=de&RefMenuID=0&RefServiceID=0&link=
Law on the Utilisation of Hydropower	Gesetz über die Nutzbarmachung der Wasserkräfte, 1990, 721.8, http://www.vs.ch/Navig/navig.asp?MenuID=4628&Language=de&RefMenuID=0&RefServiceID=0&link=
Law on the Protection of Nature and Landscape/Cultural Heritage	Gesetz über den Natur- und Heimatschutz, 1998, 451.1, http://www.vs.ch/Navig/navig.asp?MenuID=4610&RefMenuID=0&RefServiceID=0
Canton Fishing Act	Kantonales Fischereigesetz, 1996, 923.1, http://www.vs.ch/Navig/navig.asp?MenuID=4610&RefMenuID=0&RefServiceID=0
Law on Official Surveys and Geo-information	Gesetz über die amtliche Vermessung und Geoinformation, 2006, 211.6, http://www.vs.ch/Navig/navig.asp?MenuID=4610&RefMenuID=0&RefServiceID=0
Law on Agriculture and the Development of Rural Land	Gesetz über die Landwirtschaft und die Entwicklung des ländlichen Raumes (Landwirtschaftsgesetz), 2007, 910.1, http://www.vs.ch/Navig/navig.asp?MenuID=4610&RefMenuID=0&RefServiceID=0
Regulation concerning leisure cruising on motorboats on Valaisanne waterways	Reglement betreffend die motorisierte Vergnügungs- Schiffahrt auf den Walliser Wasserläufen, 1990, 747.201, http://www.vs.ch/Navig/navig.asp?MenuID=4610&RefMenuID=0&RefServiceID=0
Ruling concerning Drinking Water Installations/Facilities	Beschluss betreffend die Trinkwasseranlagen, 1969, 817.101, http://www.vs.ch/Navig/navig.asp?MenuID=4610&RefMenuID=0&RefServiceID=0
Ruling concerning the Use of Groundwater, Lakes or Waterways for Thermal Energy	Beschluss betreffend die Nutzung des Grundwassers, der Seen oder Wasserläufe zur Gewinnung thermischer Energie, 1982, 730.102, http://www.vs.ch/Navig/navig.asp?MenuID=4610&RefMenuID=0&RefServiceID=0
Ruling on the Draining of Dams and Reservoirs and the Purification of Waterways	Beschluss über die Spülungen, die Entleerungen von Stauanlagen und Speicherstollen und die Reinigung der Wasserläufe, 2002, 721.805, http://www.vs.ch/Navig/navig.asp?MenuID=4610&RefMenuID=0&RefServiceID=0
Further provisions on water also exist in the laws and decisions concerning land protection: e.g. Entscheid betreffend den Schutz des Auengebietes Gletschboden sowie des Gletschervorfeldes des Rhônegletschers in Oberwald (from 10 March 1999) 'Decision concerning the Protection of glacial floodplains such as the glacial forefield of the Rhône Glacier in Oberwald'.	

(FC 138, 139, 140). A number of provisions are present in the Swiss federal constitution (FC), which guarantee access to legal proceedings and the courts (Arts. 29–33 and 64FC). Switzerland's judiciary is independent of the executive and the legislature, with the Federal Court (*Bundesgericht*) being generally viewed as being an effective and independent institution. Rights in legal and judicial proceedings for civil society as well as organisations are embedded in the FC as well as in a number of Federal Acts, namely, the Federal Judiciary Act (FJA), Federal Administrative Act, Environmental Protection Act (EPA) and the Federal Law on the Principle of Administrative Transparency (LTrans). The right of appeal for interested civil society/environmental organisations (*Verbandsbeschwerderecht*) to challenge decisions in court, which may affect their members' interest, was initially introduced in 1983 in the referendum on the EPA (Art. 55).

The complex governing process, while it ensures a consensus is built, does mean that laws can be difficult to implement, and the process of change or implementation is very slow and potentially difficult (Uhlmann Brögli and Wehrli 2008). In enforcing the law, the federal government tends to utilise a hands off approach, but can wield some form of soft enforcement power in terms of financial incentives and subsidies for the implementation of certain principles in projects at the canton and commune level. Under the terms of the *Neuefinanzausgleich* (NFA 2008), communes can receive greater subsidies for projects from the cantons and federal funds if they meet certain criteria (participative planning, integrated risk management, ecological aspects and technical aspects).

7.2.2 Transparency

Transparency indicators generally score well in the assessment, though issues were raised in terms of quality, quantity and coherence of certain hydrologic data set across regions, particularly in the Alpine areas. Despite Switzerland's reputation as the nation of banking secrecy, legal provisions for access to environmental information preceded those in the rest of Europe. In general, access to environmental information is perceived to work effectively in Switzerland, although concerns were raised with aspects of implementation in some of the more remote Alpine areas. The 30-day notification period was seen to be limiting when dealing with some communes that may be far away in the mountains, and/or adverse to environmental organisations, and therefore do not want to give them the report, no matter what rights are provided in the law to the Environmental Impact Assessment (EIA). Additionally, EIAs are outsourced to private bureaus, which can undermine the objectivity of the report. Instances where such bureaus simply do not know thoroughly the detail of the Water Protection Act (WPA), and mistakes are made in how the law should be interpreted and implemented, have also transpired. However, in the time period 1997–2007, there were no court cases or judicial proceedings relating to provisions for environmental information (Kölz and Brunner 2007).

7.2.3 Participation

Swiss citizens maintain extended political rights through the specific constitutional rights for referendum (Art. 140 FC), petition (Art. 33 FC) and initiation of a referendum (Arts 138 and 139 FC). These rights of participation are a fundamental part of the Swiss Constitution not only in law, but also in practice. Therefore, a large number of legislative acts in most policy fields are subject to referendum, requiring ratification by a majority of the electorate and the cantons. This also applies to water policy issues, allowing NGOs, trade unions and professional associations to exert a considerable influence on political decision-making processes (Mauch and Reynard 2002). The right to referendum and the resulting people's initiatives concerning water policy[1] (SFV 2006) show that public participation has been key in moving forward the ecological agenda in water governance in Switzerland. However, it has been well documented that since the 1970s, voter turnout has started to decline (IDEA 2009).

In the practice of water management, participation takes place at the different institutional rather than individual levels. Within the Valais, the implementation phase of the major flood protection project, the Third Rhône Correction (TRC), (Valais 2009) is highly participative, with the different segments of each project having its own local planning commission (*Commission régionale de pilotage*) that includes the different interested parties. However, the level of participation is highly dependent on local factors with inclusion and collaboration in some areas functioning very well, but not in other communes. Further difficulties have been detailed in the problems that arise from the participative process, namely in the slow progress of the project as well as in attempting to align conflicting interests, specifically agricultural stakeholders, who have set up a lobbying group to force the project to follow a more technical approach (Arborino 2009), and the environmental considerations bound by law (WPA) into the project. Other than the TRC, there are not that many other opportunities for participation or where participation is demanded.

7.2.4 IWRM

The sub-categories of the IWRM indicator suggest that while law and policy are certainly more integrated today and legal provisions for different element of IWRM are generally strong, complexity is high in that they are found in a number of separate federal and cantonal acts and ordinances.

[1] 1991 Save our Waterways (*Volksinitiative: Zur Rettung unsere Gewässer*) and 2007 Living Waters – Renaturalisation Initiative (*Lebendiges Wasser – Renaturierungs-Initiative*). More information available in Sect. 9.6 (Appendix) at: http://www.admin.ch/ch/d/pore/vi/vis164.html and at http://www.parlament.ch/d/suche/seiten/geschaefte.aspx?gesch_id=20070060 respectively.

7.2.4.1 Adopting a Basin Approach

Notably there is no framework agreement, convention nor cooperative institution for the Rhône basin as a whole. The River Rhône is the border between the cantons of Valais and Vaud, and is therefore shared 50:50 by each canton. There are federal and cantonal legislative clarifications for managing cross-cantonal waters (Ordinance on the Hydraulic Engineering, 2007), and generally, the Federal State is responsible for those water courses that flow across multiple cantons. However, the responsibility for implementation lies at the cantonal or commune level. The International Commission for the Protection of Lake Geneva (CIPEL; www.cipel.org) does provide a coordinating role for environmental protection across different cantons (Geneva, Vaud and Valais) as well as countries (Switzerland and France) affecting the quality of the Lake of Geneva.

In practice, issues of coordination amongst different communes over one basin are particularly problematic in Valais (Clivaz and Reynard 2008), since the communes have a large degree of autonomy, while the canton has low financial capacity. This decreases the canton's ability to implement federal legislation or a common hydropower policy (Staatsrat 2008) across different communes (Clivaz and Reynard 2008). Since all lateral rivers are under the sovereignty of the communes, some stakeholders suggested that it does not in fact matter what the canton says, as the communes have the end decision about how the water is used, and which projects are implemented. However, others commented that it is in their interests to comply to avoid litigation or losing out financially. Art. 7(3) WPA states that "the cantons provide a communal and where necessary a regional drainage plan (*regionale Entwässerungsplanung* (REP))", a provision only binding for built-up areas, but other aspects have to be included in spatial planning tools (*Sachplan* and *Richtplan*) (Heller 2009). However, very few cantons have completed an REP.

7.2.4.2 Water Allocation and Prioritisation Measures

There are currently no overarching principles on how to manage user conflicts in periods of water stress that address international, national and local actors all together. Provisions for allocation and prioritisation measures can be categorised into two groups: concessions and residual flows. Concessions are administrative agreements allowing exploitation of natural resources. For the exploitation of water power and irrigation, they are subject to the general provisions of the Use of Water Power Act (UWPA). The act provides regulations and guidelines for instances where water courses run through more than one canton (Arts. 6, 7, 61 and 68). In Valais, most concessions were granted by the communal administration for an 80-year time period and. Residual Flows (*Mindestrestwassermenge*) are provisions in both WPA (Arts. 31, 33, 34 and 36) and the Federal Fishing Act (FA), which require that sufficient quantities of water should be either left or returned to watercourses, whatever the water use.

While Swiss legal provisions recognise both economic and ecological water uses, implementation of these provisions has been difficult (Petitpierre 1999), as a 2006 *Eidgenössische Anstalt für Wasserversorgung, Abwasserreinigung und*

Gewässerschutz (EAWAG) study on the implementation of Art. 80 WPA shows (Uhlmann Brögli and Wehrli 2008). Interviews also suggested that it has led to increased challenges for how to best manage the different user groups, with the key issue being the development of the growing micro-hydro sector. There are very few provisions in the law that concern the management of scarcity situations and no overarching principles on how to manage user conflicts in periods of water stress that address international, national and local actors all together.

7.2.4.3 Protection of Aquatic Ecosystems

Aquatic ecosystems are protected in both qualitative (Art. 1 WPA) and quantitative (Arts. 29–36 WPA) terms through provisions in the WPA, as well as Protection of Nature and Landscape/Cultural Heritage Act (PNLA) and Federal Forest Act (FFA) (preservation of natural diversity of riparian species). Within the legislation of the Canton Valais, the Law on Hydraulic Engineering (LHE) (Arts. 5g and 39, 15 March 2007) provides protection of aquatic ecosystems, as does a 1999 law protecting the floodplain of the Rhône. In practice, a number of water courses and aquatic ecosystems have been severely impaired and federal targets are not being met (e.g. nitrate concentrations) (FOEN 2009a). Hydro-peaking regularly impacts rivers, while some periodically dry up from over extraction. Environmental lobby groups have expressed concern with the fact that the legally binding provisions for residual flows are too weak for effective nature protection (Bonzi 2009a), raising questions as to whether an effective and efficient instrument exists for coordinating water's protection and use (Bonzi 2009b). Enforcement of protection provisions are seen to be hampered by resource limitations in staff numbers at the canton level.

7.2.4.4 Flood Risk Management and Response Systems

Since the 1970s there has been a shift from technical building and a hard canalisation approach to a more integrated and eco-system based flood management philosophy (Zaugg 2002), which has meant that implementing flood protection projects (such as the TRC) requires a more complex negotiating process. While federal and cantonal law (WBG, Valais) state that the natural condition of the river must be improved, other stakeholder groups, such as agriculture, have rallied against the impacts this would have for their own resources. However, financing mechanisms are perceived to effectively assist the federal government in implementing current philosophy of the law. Recurring issues of sovereignty and capacity were raised in interviews across the cantonal departments. Hazard maps are a key requirement of the flood protection concept due for completion by the end of 2011. Progress is recorded in the ShowMe maps (FOEN 2009b). It was noted that better coordination across different departments was required to reduce duplication of effort (Fig. 7.1).

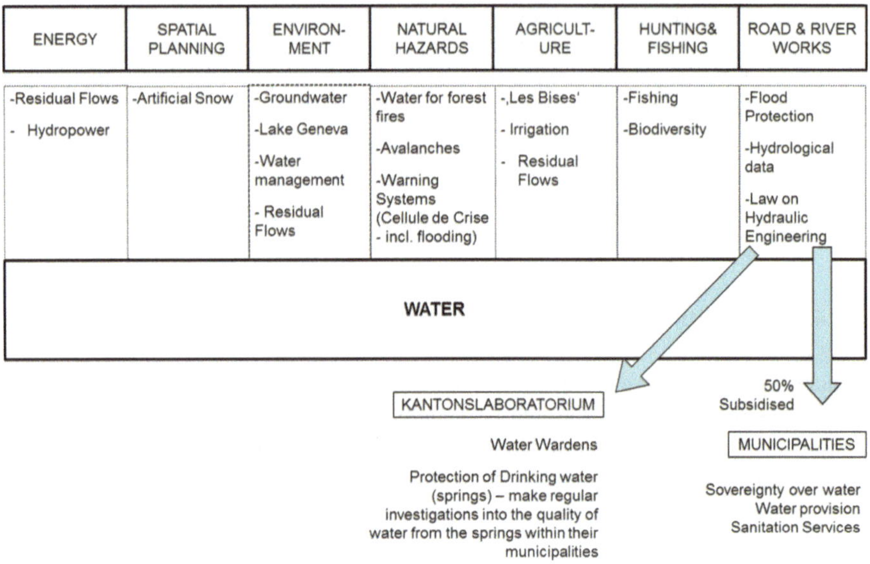

Fig. 7.1 Organigram of water resource management across Canton Valais

7.2.4.5 Institutional Arrangements and Challenges Related to IWRM

Swiss water management is driven from the local level up, thereby reducing the impact of IWRM policies proposed by FOEN at the cantonal and communal level. Organisations such as Wasser Agenda 21 (WA21) are more focused on the theory of IWRM, so have very little impact in practice. Despite policy briefs from FOEN (SAEFL 2002, 2003), so far there is no cooperative institution at the basin level in the Canton Valais or the Rhône basin in general, instead a complex and segregated approach (Fig. 7.1) makes coordination across the different actors difficult. There are many institutions that focus on the different elements of water, spread across the federal, canton and communal level, leading to weak internal policy coherence within the federal administration (Varone et al. 2002) and in the country as a whole.

Many interviewees raised the issue of professionalism and lack of capacity at the commune and canton level. Those responsible for a water management component often do not have the time, training or both. The autonomy from federal control (provided by Art. 50 FC) is another facet of the principle of subsidiarity and one that has significant consequences in water management (Aschwanden et al. 2008). It makes it particularly difficult to establish appropriate geographical units for water management, since their areas are too small to represent either natural or technical territorial units of water courses. Over the years each canton has developed their own brand of water management, along with their own institutions, leading to a lack of overall vision for Switzerland (Chaix 2008). Some question whether this decentralised approach to water management is compatible with the goals of IWRM (Chaix 2008). While limited and independent examples of integrated watershed management have been documented across different cantons (WA21 2007), the ability for IWRM to move more comprehensively from theory to practise is yet to be seen.

7.3 Conclusion

Although Switzerland scores well on accountability, transparency and participation indicators, the assessment has shown that there is a significant gap between the conceptual strands of IWRM in federal laws and policies and their translation at the regional and local levels. In many instances, reality in implementation of more integrated principles of water related law is still divorced from commitments in federal and cantonal law. Most significantly, it shows that integrated watershed management has yet to be fully defined in Switzerland.

Key findings in the assessment can be grouped under the headings listed below:

- Sectoral approach & demarcation by political boundaries for water management
- Suitability of the 'lowest possible level' concept of subsidiarity & *Kantonligeist* Syndrome with ramifications for resource constraints at the municipal level
- Complexity of water sovereignty at different levels of government
- Conflicts on the Horizon: artificial snow, climate change, long term hydropower concessions, growth of micro-hydropower
- Balancing of protection and use provisions in the different laws concerning water

Swiss water law, although progressive, maintains a focus on sectoral and end of pipe regulation. To date, Swiss water management has been described as an over layering of more or less sectoral coordinated plans and management processes. The various tasks on the protection or use of water are often separate and administered in geographically very small areas (namely the communes). Most water associations are still organised by sector, and therefore management remains driven by sectoral interests. It has been commented that the decentralised approach to water management (with the duty of implementation designated to the communal level) is incompatible with the goals of integrated water resources management, and watershed management (Chaix 2008). Communes tend not to be able to establish appropriate geographical units for water management – most commune areas are too small to represent either natural or technical territorial units of water courses.

Furthermore, over the years each canton has developed their own brand of water management, along with their own institutions, leading to a lack of overall vision for Switzerland (Chaix 2008). An optimistic development has been the emergence of WA21, which recognises these issues and challenges and is attempting to implement its agenda to achieve a more sustainable and integrated approach to water management (WA21 2008). However, in the highly decentralised political climate in Switzerland, national and federal programmes are often viewed with suspicion and are unwelcome at the local and regional level. While limited and independent examples of integrated watershed management have been documented across different cantons (WA21 2007), the ability for IWRM to move more comprehensively from theory to practise is yet to be seen.

The complex institutional framework, legislative provisions and levels of sovereignty which govern water resources in the Canton Valais imply a lack of coordination and long term planning amongst the different politico-administrative levels and sectoral groups. These issues impede the implementation of a more integrated water management framework. Future management should be far better coordinated at the watershed level, yet given the Swiss politico-administrative order, sole management at the watershed level is unlikely, since the logical political boundaries are the communes (Clivaz and Reynard 2008). In the past some communes in the Valais have shown interest in the creation of a Master Water Plan within the framework of the Environment and Health Action Plan (Clivaz and Reynard 2008). However, more recently the suggestion of the *Centre de Competence d'Eau Valais* is a positive development. At the time of interview, only an initial planning phase was underway, therefore the form and shape of the institution is yet to be seen. The extent to which it will embrace IWRM principles and make strides towards a basin approach management style is also yet to be seen.

The tendency to plan and manage water at the lowest political level implies a lack of oversight and raises questions as to what really is the 'lowest suitable level' in the principle of subsidiarity. The limitation of this concept therefore requires far closer investigation in the discourse on water governance in Switzerland. Resource and professionalism issues at local institutional levels were a source of concern, not just for current management issues, but in light of future challenges as well. Even in the Swiss Alpine region, where water is plentiful, multiple uses of water has caused a degree of stress in supply, due to the non-management of demand.

A further challenge is to create and integrate new institutions that can manage not only sectoral uses but also cross sector problems, within a climate of change. The lack of an oversight institution in the Rhône basin is a situation that may lead to issues amongst stakeholders (at local, regional and international levels) in the future. Other external factors such as the intensification of the energy market between Switzerland and Europe (Von Arx 2009) also raise questions as to the ability of such a devolved and un-coordinated governance setting to manage change and uncertainty.

Finally, there are concerns that the ramifications of climate change and expanding water uses are not adequately reflected in the current governance framework. Some have suggested that a lack of urgency to address this issue in an integrated manner and vision is due to fact that there has historically been a low level of pressure on water management in the region, and Switzerland in general (Heller 2009). However, even within the water tower of Europe, the prognosis of rising conflicting demands on a water system facing uncertain changes from climate impacts, suggest that it may be time for a change of speed.

References

Allan A, Rieu-Clarke A (2007) Mapping of the governance framework. Report for EC project BRAHMATWINN. no. GOCE –036952. University of Dundee, Scotland

Allan A, Rieu-Clarke A, Borja R, Dekens J, Deng A, Magsig BO (2007) Identification of the legal and policy framework in Upper Brahmaputra an Danube River Basins. Report for EC project BRAHMATWINN no. GOCE –036952. University of Dundee, Scotland

Arborino T (2009) Ihre Fragen an Rhone.vs. Rhonevs: Informationsmagazin über die dritte Rhonekorrektion 15 (April 2009):4

Aschwanden H, Pflaunder M, Vollenweider S (2008) Wasserwirtschaft Schweiz 2007: Eine Auslegeordnung mit Thesen zur Weiterentwicklung Bundesamt für Umwelt. Ittigen, Bern

Bonzi C (2009a) Gewässerschutz darf nicht weiter verwässert werden. pronatura magazin February 2009:4–5

Bonzi C (2009b) Kein Leben ohne Restwasser. Pronatura Magazin. February 2009:6–7

Chaix O (2008) Einzugsgebietmanagement: Koordination der Bereiche und wasserwirtschaftliche Planung. Umweltr Prax 5:527–544

Clivaz C, Reynard E (2008) Crans-Montana: water resources management in an alpine tourist resort. In: Wiegandt E (ed) Advances in global change research 31: mountains: sources of water, sources of knowledge. Springer, Dordrecht, pp 103–119

FOEN (2009a) Environment Switzerland 2009. Federal Office of the Environment (FOEN), Bern

FOEN (2009b) Stand Gefahrenkartierung Januar 2009: Gefahrenkarte vorhanden in % der Flächen. Federal Office for Environment. http://www.bafu.admin.ch/naturgefahren/01922/01926/01930/index.html?lang=de. Accessed 22 July 2009

Heller M (2009) Insights of policy instruments in integrated river basin management from the Dutch practice for the Swiss context: a comparative study of the policy instruments in the sub-catchments regge (NL) and birs (CH). Wageningen University, Wageningen

IDEA (2009) Voter turnout: country view, Switzerland. International Institute for Democracy and Electoral Assistance, Stockholm, Sweden

Kölz M, Brunner RM (2007) Umweltrecht in der Praxis: Register der Jahrgänge 1997–2007. Vereinigung für Umweltrecht, Zürich

Mauch C, Reynard E (2002) The evolution of the national water regime in Switzerland. Institut de Hautes Études en Administration Publique, Lausanne

NFA (2008) Geltender Finanzausgleich – NFA. http://www.efv.admin.ch/d/themen/finanzausgleich/index.php. Accessed 20 June 2009

Petitpierre A (1999) Environmental law in Switzerland. Kluwer Law International/Stämpfli, The Hague/Bern

Rieu-Clarke A, Allan A, Magsig BO (2008) Assessing governance in the context of IWRM. Striver policy brief: strategy and methodology for improved IWRM – an integrated interdisciplinary assessment in four twinning river basins. Issue No. PB6.2. http://www.dundee.ac.uk/water/projects/striver/

SAEFL (2002) Sustainable management of water resources: the need for a holistic ecosystem approach running out of freshwater or maintaining freshwater through a ecosystem based approach – an easy choice. World Summit on Sustainable Development 'Sustainable management of water resources': The Ramsar Convention on Wetlands, Ramsar

SAEFL (2003) Guiding principles for Swiss watercourses: promoting sustainable watercourse management. SAEFL/FOWG, Bern

SFV (2006) Eidgenössische Renaturierungsinitiative eingereicht; Mehr als 160'000 Unterschriften für «Lebendiges Wasser». Schweizerische Fischerei-Verband (SFV), Bern

Staatsrat (2008) Bericht des Staatsrats zur kantonalen Energiepolitik. Staatsrat, Sion

Uhlmann Brögli V, Wehrli B (2008) Sichere Restwassermengen gegen uneingegeschtränkte Wasserkraftnutzung – ein Vollzugsdilemma? Umweltr Prax 5:469–486

Valais (2009) Plan d'aménagement de la 3ème correction du Rhône – PA-R3. http://www.vs.ch/Navig/navig.asp?MenuID=16520&RefMenuID=0&RefServiceID=0. Accessed 28 July 2009

Varone F, Reynard E, Kissling-Näf I, Mauch C (2002) Institutional resource regimes: the case of water management in Switzerland. Integr Assess 3(1):78–94

Von Arx J (2009) Unterhalb der Kraftwerke sind viele Flüsse tot. Pronatura Magazine February 2009:8–9

WA21 (2007) Beschreibung und Analyse von Fallbeispielen zum integralen Einzugsgebietsmanagement (IEM). Wasser Agenda 21: Teilprojekt 'Auswertung und Evaéuation von IEM-Ansätzen'. Im Auftrag des Bundesamtes für Umwelt BAFU, Abteilung Wasser

WA21 (2008) Vision Wasser-Agenda 21. Wasser Agenda 21

Zaugg M (2002) More space for running waters: negotiating institutional change in the Swiss flood protection system. GeoJournal 58:275–284

Chapter 8
Water Governance in the Context of IWRM: Chile

Abstract This chapter outlines the work completed for the ACQWA project on the governance assessments for the Chilean case area and is used to provide vital background to the water governance situation and associated challenges. In the Chilean case, significant challenges persist across the governance indicators, in particular in relation to transparency and accountability. While water governance at the political level is driven through a centralised approach, water management happens in the private sphere and is driven by private interests. Despite the strong codified nature of water governance through the Water Code, the weakness of enforcement and capacity in the DGA means that provisions relating to protection of aquatic ecosystems can effectively be ignored at the basin level. The market focus on water management has meant that public institutions responsible for water rights management or water and environmental issues have very limited capacity to address water issues.

Keywords Aconcagua • Region V • Chile • Water governance assessment • Legislative and policy challenges • IWRM • Transparency and enforcement challenges

8.1 Development of Water Rights in Chile

Water rights in Chile have undergone a number of evolutionary steps over the past century. The following section provides a brief overview of the development of water rights in Chile according to the different periods of change and reform from the Agrarian Reform of 1960 to the latest changes to the radical 1981 Water Code that took place in 2005 (Bauer 2004; Carruthers 2001). Prior to the Agrarian Reform in 1968, water had been a constitutional right of the state, with water users able to obtain a right of use for this water. Water rights were linked to land rights, which meant that a separate registry of water rights did not exist, but instead were assumed to be part of the deed registries recording land ownership. In 1968, the agrarian reform effectively divided land rights from water rights, yet during this period of

M. Hill, *Climate Change and Water Governance: Adaptive Capacity in Chile and Switzerland*, Advances in Global Change Research 54,
DOI 10.1007/978-94-007-5796-7_8,

transference, there was no requirement to register the new water rights that were now separated from the land they had originally belonged to. Within the Aconcagua, the majority of land (or parcel) owners originate from the period of agrarian reform, yet while the land rights were transferred, the water rights followed without registration.

Beyond the basic nature of Chilean water rights, there are important distinctions concerning different kinds of water rights, namely whether they are permanent, eventual, provisional, continuous or non-continuous, consumptive or non-consumptive (Water Code, Art. 12), which allows the water rights holder different levels of security and timing. *Permanent* water rights are an expression of volume per time that are granted within the guaranteed flows of the water body (Water Code, Art. 16, 17). *Continuous* rights can theoretically be used 24 h a day of the 365 days a year (Water Code, Art. 19) while *discontinuous* rights are only allowed to be used during predetermined periods. However, the official assumption for calculating these rights is that they are not used permanently, but according to different factors of use, which are used to calculate the level of 'real' exploitation with the surface and ground waters of a basin. For surface water, *eventual rights* are those rights which are granted beyond the limit of secure water in the river. These rights are less secure than permanent because if flow goes below the average level of the river (e.g. in times of drought) then these rights cannot be used. The difference in qualification between permanent and eventual rights is the time that a rights holder is allowed to use it. Within the Water Code, there is an express article (Art. 18) that prohibits water from reservoirs (*las aguas lacustres o embalsadas*) being subject to eventual rights.

In restricted zones (such as a depleted aquifer or groundwater zone) all new water rights are allocated as 'provisional' *rights*, with a set of conditions (Water Code Art. 62; Resolution 425, Art. 30-41) that restricts how they can be used for the proceeding 5 years. After 5 years, if the rights holder can show that their rights did not impact the aquifer, then these rights can be converted into permanent rights (DGA source). A further classification is between consumptive and non–consumptive water rights (Water Code, Art. 13, 14, 15). *Consumptive* rights allows for the total consumption of the allocated water by the water rights holder (Water Code, Art. 13), while *non-consumptive* rights requires rights holders to use their water rights in a non-consumptive fashion, returning the water rights to the water body for usage by other water rights holders according to predetermined standards of quantity and quality (Art. 14).

8.2 Chilean Assessment

In 1981, during the early years of the Pinochet regime, a new water code was passed (1981 Water Code), that was heavily informed by neo-liberal economic doctrine. After an acrimonious 15 year struggle to amend the 1981 Water Code, a set of minor reforms were passed in 2005 that aims to address issues concerning externalities caused by water market transactions, the hording of water rights from non-consumptive water users, a lack of transparency and information concerning the market and registered water rights as well as mounting challenges from environmental externalities and the lack of river basin management (Bauer 2004). Due to the limited scope of 2005 amendments, water governance remains dominated by the provisions set in the 1981 Water Code, but is also impacted by the Energy and Mining Codes (Tables 8.1 and 8.2).

8.2.1 Accountability

In Chile, the Constitution (Art 12 (2)) provides for the right of equality before the law, as well as the more recent right to a clean environment and protection of the environment (Framework Environmental Law 19.300). Despite provisions for clear timeframes in court procedures (Water Code, Art 129 bis 11), a number of stakeholders alluded to the time and financial barriers that the court process entailed for effective and expedient conflict resolution. Instances of corruption were also reported, such as the circumvention of the EIA process during the government project to build motorway, resulting in damaging impacts on the Aconcagua River. The weak enforcement capability was also brought up in interviews, in terms of the DGA's inability to stop illegal extractions of groundwater unless it is denounced by a water user. Furthermore, interviewees and literature alike point to a lack of agency for stakeholders involved in day to day water issues to influence the system or challenge decisions by governmental bodies. Some studies have pointed to the numerous situations where an individual or company's personal bargaining power outweighs any legal norms in conflict resolution or interest coordination (Bauer 1997; Carruthers 2001).

8.2.2 Transparency

Article 31 bis (Environmental Law) provides for the right to access to environmental information[1] held by authorities, in accordance with the Constitution (Art. 19 (12, 14)) as well as Law 20.285 concerning access to public information. With respect to water resources information, water rights owners are required to register water rights with the '*Conservador de Bien Raices*' in the '*Registro de Aguas*' (Water Code, Art 112). However, it is the overall responsibility of the DGA to maintain a consolidated information system on the water rights through the '*Cadaster Publico de Aguas*' (Water Code, Art 120–122). However, the water market was referred to by a number of stakeholders as 'dark' market, with a complete lack of

[1] The process for access to environmental information is described in the Law of Access to Public Information (20.285, Art. 10–30).

Table 8.1 Legislation concerning water and environment in Chile (including rights to information on the environment and participation in decision making concerning the environment) English

English	Source/original text
Constitution of the Chilean Republic	Constitución Política de la República de Chile (1980), Santiago, 17 de septiembre de 2005. http://www.leychile.cl/Navegar?idNorma=242302&buscar=Constituci%C3%B3n+Pol%C3%ADtica+de+la+Rep%C3%BAblica+de+Chile
Water code	El Código de Aguas, Ley N°1.122, Santiago, 13 de agosto de 1981. http://www.leychile.cl/Navegar?idNorma=5605&buscar=codigo+de+aguas
	Ley 18.450 Aprueba Normas para el Fomento de la Inversión Privada en Obras de Riego y Drenaje, 30 de octubre de 1985
	Decreto Supremo N°285 – Reglamenta Procedimiento para Aplicación del D.F.L. N° 1.123, de 1981, sobre Ejecución de Obras de Riego por el Estado.
Environmental Framework Law	Ley N° 19.300 sobre Bases Generales del Medio Ambiente, 1994. http://www.leychile.cl/Navegar?idNorma=30667
Law on Indigenous Rights	Ley 19.253, Establece Normas sobre Protección, Fomento y Desarrollo de los Indígenas, y creas la Corporación Nacional de Desarrollo Indígena, 28 de septiembre de 1993. http://www.leychile.cl/Navegar?idNorma=30620&buscar=Desarrollo+de+los+Indigenas
Law on Access to Public Information	Ley N° 20.285 sobre Acceso a la Información Pública, 11 de agosto de 2008. http://www.leychile.cl/Navegar?idNorma=276363&buscar=INFORMACI%C3%93N+P%C3%9ABLICA
Administrative Procedure Law	Ley N° 19.880 sobre Bases de los Procedimientos Administrativos que rigen los actos de los órganos de la Administración del Estado, 22 de mayo de 2003. http://www.leychile.cl/Navegar?idNorma=210676&buscar=Bases+de+los+Procedimientos+Administrativos+que+rigen+los+actos+de+los+%C3%B3rganos+de+la+Administraci%C3%B3n+del+Estado
General Law on City Planning and Construction	Ley General de Urbanismo y Construcciones, 16 de abril de 1992. (FIJA NUEVO TEXTO DE LA ORDENANZA GENERAL DE LA LEY GENERAL DE URBANISMO Y CONSTRUCCIONES) http://www.leychile.cl/Navegar?idNorma=8201&idVersion=2010-01-16#área (currently being revised)
Chilean Civil Code	Norma DFL 2, DFL 2-95, Código Civil de Chile, 21 de septiembre de 1995. http://www.leychile.cl/Navegar?idNorma=3551&idParte=8896843&idVersion=

Tribunal Code	Código Orgánico de Tribunales (Ley N° 19.665), 25 de febrero de 2000. http://www.leychile.cl/Navegar?idNorma=160254&buscar=C%C3%B3digo+Org%C3%A1nico+de+Tribunales
General Norms of Citizen Participation	Norma General de Participación Ciudadana, Santiago, 17 de abril de 2009. http://www.leychile.cl/Navegar?idNorma=1001531&buscar=cambio+clim%C3%A1tico
Law on Legal Authority and Legal Aid	Ley N° 17.995 sobre Concede Personalidad Jurídica a los Servicios de Asistencia Jurídica que se Indican en las Regiones que se Señalan, 8 de maya de 1981. http://www.leychile.cl/Navegar?idNorma=29425&buscar=CONCEDE+PERSONALIDAD+JURIDICA+A+LOS+SERVICIOS+DE+ASISTENCIA+JURIDICA+QUE+SE+INDICAN+EN+LAS+REGIONES+QUE+SE+SE%C3%91ALAN
Law on Public Defence	Ley N° 19.718 sobre Crea la Defensorio Penal Publica, 27 de febrero de 2001. http://www.leychile.cl/Navegar?idNorma=182755&buscar=CREA+LA+DEFENSORIA+PENAL+PUBLICA
Law on Freedom of Speech and Information and Practice of the Journalism	Ley N° 19.733 sobre Libertades de Opinión e Información y Ejercicio del Periodismo, 19 de mayo de 2001. http://www.leychile.cl/Navegar?idNorma=186049&buscar=LIBERTADES+DE+OPINION+E+INFORMACION+Y+EJERCICIO+DEL+PERIODISMO

Table 8.2 Legal provisions of particular relevance to discussion in this book: excerpts taken from the full text of the Chilean Governance Assessment (Torres 2010)

Code/law	Articles and descriptions/original text
Chilean Constitution Article 24 (Right of Property)	Article 24° recognises and guarantees the right of property. The right of ownership in its diverse aspects over all classes of corporeal and incorporeal property. Only the law may establish the manner to acquire property and to use, enjoy and dispose of it, and the limitations and obligations derived from its social function. Said function includes all the requirements of the Nation's general interests, the national security, public use and health, and the conservation of the environmental patrimony
	In no case may anyone be deprived of his property, of the assets affected or any of the essential faculties or powers of ownership, except by virtue of a general or a special law which authorises expropriation for the public benefit or the national interest, duly qualified by the legislator. The expropriated party may protest the legality of the expropriation action before the ordinary courts of justice and shall, at all times, have the right to indemnification for patrimonial harm actually caused, to be fixed by mutual agreement or by a sentence pronounced by said courts in accordance with the law
	The rights of private citizens over waters, recognised or constituted in conformity with the law, shall grant proprietorship to the owners thereof
Water Code Article 5	Article 5 states that water is a national good of public use, for which particular rights of use can be awarded in compliance with the regulations set out in the current Code (refer to Chilean Constitution, Art 18 (24) & Chilean Civil Code, Art 589)
Article 122	The DGA is required to establish a Public Cadaster of Water
Article 129 bis 1	The Water Code has limited scope for protection of water related ecosystems. In issuing new water rights, the DGA has to protect nature and the environment by establishing a 'minimum ecological flow' which will be applicable only to future allocation of water rights. Such minimum water flow should not surpass the 20% annual mean flow of the superficial water course. In exceptional cases, such limit if increased until the 40% (Art 129 bis 1). Another rule considered in the Water Code is its article 58 forbids explorations -and thus, exploitation- for water in headwater zones for wetlands in the north of Chile (Regions of Tarapacá and Antofagasta), unless authorized by DGA. Moreover, its article 63 declares the same zones as 'prohibition areas', where no further exploration and exploitation will be permitted, unless authorised by DGA
Article 129 bis 3	The DGA must establish a network of stations in each basin to monitor water quality, quantity and levels both in superficial and subterranean waters. The information must be available to the general public
Article 299	The DGA must fulfil certain functions, including: development planning of natural water sources, including recommendations on how they are used; observation and monitoring of the resource, including maintaining and operating the national hydrology service and supplying and publishing related information; commission technical studies for resource and infrastructure planning; police water resources to avoid un-authorised destruction or modification of hydrological work; in cases where a Junta de Vigilancia is not legally constituted, to impede the extraction of water without rights or over-extraction

Article 32	The Water Code forbids construction on the river bed, unless an administrative authorization is issued (Art. 32). Furthermore, the Construction Law ("Ordenanza General de Urbanismo y Construcción") bans or limits edification on flooding prone terrain, and in protected areas of natural value resources (e.g. wetlands). Such restriction areas should be contemplated in the City's Zoning Instrument ("Instrumentos de Planificación Territorial")
Prioritisation of Uses	In Chile there is no legal preference of one use over another, as the rule "first come, first served" is applied (Vergara, 1998). Notwithstanding, according to the Law 19.253 on Protection, Promotion and Development of Indigenous People, article 64 provides on protection of the water of Aymaras and Atacameña Communities. The law granted property rights over the water of those communities, such as rivers, canals, irrigation ditch, slopes notwithstanding the right granted for third parties according with the Code of Water. No new water rights will be granted on to lakes, ponds, streams, rivers and aquifers that supply water owned by various indigenous communities established by law without ensuring in advance the normal water supply to the affected communities
Regulations on the Control of Water Contamination	The most overarching rule is the statutory instrument to control water pollution ("Reglamento para el Control de la Contaminacion Acuatica", Decreto 1, Ministry of National Defense, 06 January 1992). It establishes the general regime for prevention, control, and combat of pollution in sea water, ports, rivers, and lakes. Most of its rules are related with water related activities, such as navigation, fishing, etc.
Law 19.300 Environmental Framework Law	The environment legal framework does provide for protection of water. However, general legislation is scarce and scattered. The main pillar of protection is given by the Environmental Impact Assessment system considered in the Environmental Law (Ley N° 19.300), which assess the environmental impact of particular projects
Article 31 bis (Access to Environmental Information)	Provides for the right of every person to access environmental information obtained by the Administration, as required by the Constitution and Law N° 20.285 on Access to Public Information
Article 7 (Strategic Environmental Assessment)	According to article 7 bis, the President of the Republic shall decide on the policies and plans subject to the strategic environmental that have an impact on the environment or the sustainability, on proposal of the Council of Ministers. The development of policies and plans should include the design and approval stages. In doing that, the design stage of a policy or plan another relevant bodies should be incorporated to ensure a coordinated action of public entities affected by the policy or plan
Article 4 & 26 (Public Participation)	According to Article 4, the State is required to facilitate citizen participation and allow for access to environmental information as well as promote educational campaigns on environmental protection. Article 26 provides for the establishment of mechanism to allow for public participation in the Environmental Impact Assessment process, which itself is provided for in Article 27 (EIA)

clarity and oversight on the DGA's part on the trading and transactions of water rights. Stakeholders also alluded to the lack of transparency in decisions and data from *Conservador* or even judges, in terms of the information that judicial decision making was based on.

Criticism was also directed at the lack of regular monitoring and assessment of water quality and quantity (which are carried out by MMA and DGA respectively), as well as at challenges arising from a lack of impartiality and objectivity in research and monitoring studies to provide accurate scientific information upon which to base decisions. It has been estimated that over two thirds of water rights are based on the return flow of other rights in the upper watersheds (Pena 1997, 2001). In areas of the country such as the Copiapo, the legal over exploitation of groundwater has reached 18,000,000 m^3 per year (MMA, Expert). Again, this issue can partly be assigned to the DGA, which set a very wet period (just 1 year as opposed to a longer time period) when water availability was high as the baseline from which to calculate abstraction levels (MMA, Expert), meaning that abstraction levels are simply too high for more recurrent dry periods.

8.2.3 Participation

Article 4 of the Environmental Law requires the State to facilitate public participation, access to environmental information and promote educational campaigns to protect the environment. Article 186 of the Water Code provides for the establishment of water user communities (Junta de Vigilancia) where more than two rights owners share the same watercourse. However, while the deregulated approach places water management in the hands of the water rights owners, participation in the broader issues that affect water resources in the basin is very limited. The EIA has been the principle mechanism since 1997 through which public participation is envisaged, with the MMA and its regional agencies responsible for its implementation (Environmental Law, Art, 4). However, many authors have criticised the EIA in the Chilean deregulated neoliberal model because a meaningful enforcement power of provisions for the process is practically non-existent (Carruthers 2001). In reality it has been seen to have been used as a pre-emptive tool to demobilize conflict and local opposition to mega projects, such as controversial hydro-electric projects (Bio Bio, Baker), and has been described as 'inherently cautious and exclusionary where environmental concerns might challenge economic priorities' (Carruthers 2001, p 350).

After decades of citizen exclusion from public debate and destruction of Chilean social fabric, citizen culture is returning. There is now a growing movement of activism and unrest on issues related to environmental and social injustices, as well as a growing consensus that there is a need for a new constitution, that enables a fairer balance of power between citizens and corporations (see Guardian 2011; Patagonia 2011; Nacion 2010). More recently, a new public participation law has been brought into effect (Law N° 20.500), that provides for broader public participation than just the EIA through the right of the people to participate in policies, plans, programmes and actions. Art 70 requires that each body of the Civil Service should establish a formal and specific method of participation for people and organisations, which it must update and publish through electronic or other means.

8.2.4 *IWRM*

8.2.4.1 Adopting a Basin Approach

The Treaty between Argentina and Chile on the Environment (DS N°67) states that both parties are concerned about the deteriorating state of the global environment, recognising the need for joint action on environmental protection. Specifically, the Protocol on Shared Water Resources provides for the integrated management of watersheds on shared waters that drain across or overlap the national borders of the two countries (Art 1, 4). Beyond this, the legislative framework in Chile does not take account of IWRM[2] or even water resources management, with the Water Code providing for the allocation of water rights. During the Bachelet period a set of IWRM projects were piloted in three different basins (Dourojeanni 2010), which were criticised as being weak at the time, but have not been prioritised in the Piñera government. However, the Water Code does provide for the establishment of Junta de Vigilancia (Art 186), which although it does not prescribe a basin approach, does allow for the formation of user communities sharing canals, reservoirs or aquifers (though as separate sources).

8.2.4.2 Water Allocation and Prioritisation Measures

Water is treated as any other property right, for which there can be a one off payment, and then allocation of this resource will be corrected through the market for it. There is, therefore, no legal preference for one use over another, as a 'first come, first served' rule is in place. The only exception is during official periods of drought (Water Code, Art. 314), when there is an official intervention in a river basin (but not as soon as a scarcity zone has been declared). However, the water market is deemed to be inactive outside of the southern regions of Chile and therefore inoperative (Dourojeanni and Jouravlev 1999). At present, water rights holders pay a one off fee for the initial purchase of a water right, and afterwards pay for the costs of distribution, operation and maintenance of the infrastructure, according to their amount of rights (whether to a utility as a domestic consumer, or as a farmer to the Junta or the Canal Association). However, beyond these costs, there is no on-going

[2] Except for the last amendment of the Environmental Law (Law N° 19.300) in January 2010 that introduced the Strategic Environmental Assessment. It establishes that it is subject to, among other requirements, integrated watershed management. Therefore, the Chilean counterpart is obliged always to make a Strategic Environmental Assessment according to article 7 bis.

volumetric cost for water.[3] After the initial transaction costs, there is no payment for the water itself (in comparison to the water market in Australia where there is a volumetric fee), nor is there a basin authority setting priorities for different uses (as in areas of the USA).

The application of usage factors, or 'factor de uso' in Chile has been complex.[4] The application of this principle exists as an internal DGA estimate, and therefore comes with no legal obligation for the water rights holder to respect it. In some areas of Chile, this legislative gap has led to disastrous consequences for the over-exploitation of basins such as the Copiapo and more and more water rights being relegated to 'derechos de papel' (paper rights) in that granted use rights far exceed availability. According to interviewees (irrigators and administrators), there is little incentive for farmers to reduce the amount of water they use, or sell it on, as most feel that as scarcity situations mount, it is better to keep use rights for years in which there is proportional reduction in the basin (currently most years).

8.2.4.3 Protection of Aquatic Ecosystems

Broadly, the environmental and ecological impacts of the market are not considered, despite the fact that ecological flows are now in the Water Code (Art 129 bis 1). However, these provisions are only in effect for new rights, of which there are none in most of the regions in central and northern Chile, including the Aconcagua. While the environmental legal framework (Environmental Law, 19.300) does provide protection for water resources and aquatic ecosystems (on a case by case basis), the main pillar of protection is provided through the EIA (Environmental Law, Art 8-25[5]), which has already been highlighted as a highly flawed instrument in Chile. Furthermore, quality rules are subject to a cost benefit analysis (Environmental Law,

[3] A positive outcome of the market has been seen to be the highly developed and well-connected water services provision. Yet in recent years, the tariffs have been rising quite considerably, with a corresponding fall in consumption per person. While the costs of water have been rising considerably for domestic consumers, the cost of water for companies (utilities) is deemed to be relatively inexpensive and water losses are rising (Lentini 2011). A similar perversity in the market system is that efforts to improve irrigation efficiency tend not to lead to water savings, since the incentive is to reduce water use for increased profits rather than decrease water to decrease costs.

[4] Most water rights are registered as permanent (DGA, Expert), yet in reality these rights are not permanent as rights holders do not tend to permanently use their water 24 h a day for the duration of a year. This common contradiction in water rights terminology led to a new classification, introduced in 2005, for 'effective use' (*factor de uso*). For example, a rights holder whose right is 10 l/s, 24 h a day, is unlikely to use that amount throughout the year, so the DGA introduced an 'effective use' calculation to estimate the amount of water used, as opposed to the exact amount of the water right. However, farmers are still allowed to sell the amount that they don't use from their total water right. So, if out of that 10 l/s, a farmer only uses 6 l/s, then the farmer would be able to sell the remaining 4 l/s. Or if a farmer only uses the water right 4 days a week, the remaining 3 days per week could also be sold.

[5] http://www.prodiversitas.bioetica.org/doc56.htm

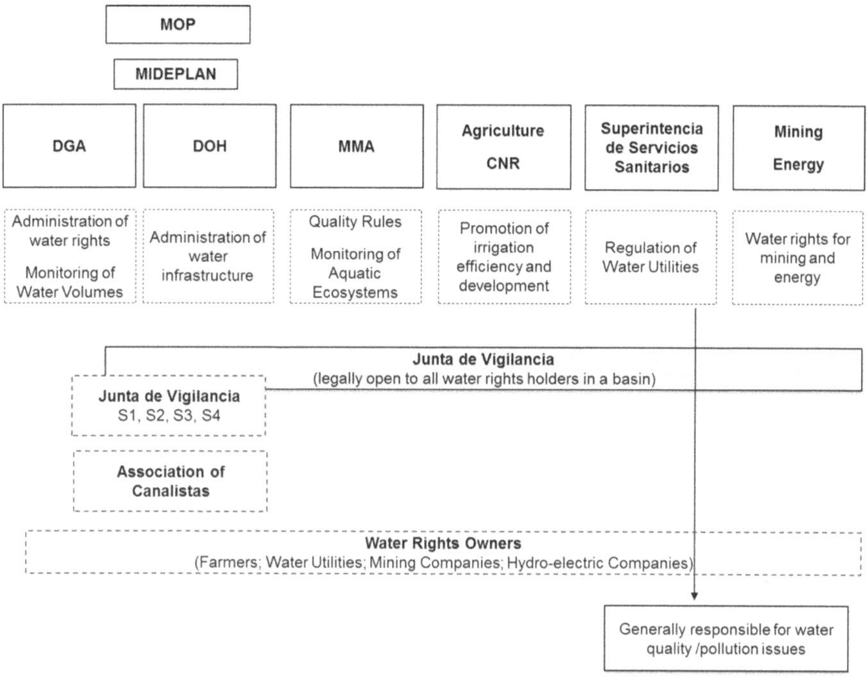

Fig. 8.1 Chilean water institutional framework (with reference to local situation in Aconcagua Basin)

Art 32–39). While it may not have direct implication on the short term protection of aquatic ecosystems, the recent elevation of CONAMA (Environment Commission) to ministry status (MMA) and the establishment of three environmental courts (*Tribunales Ambientales*)[6] are positive signs that the issue of environmental protection might begin to hold more sway.

8.2.4.4 Institutional Arrangements and Challenges Related to IWRM

Officially, the DGA is the primary administrative body with responsibility for water resources, through its administration of water rights (Water Code, Art 130). However, Fig. 8.1 depicts the complex institutional arrangement across which different aspects of water are managed. Water is therefore fragmented across the different Ministries of Mining, Energy, and Public Works. The Ministry of Public Works further separates aspects relating to water rights to the DGA, water infrastructure to the DOH, while any projects plans need to be passed by MIDEPLAN. Monitoring of water

[6] However, these environmental courts have not yet come into effect as the implementation law is yet to be passed (see http://www.mma.gob.cl/1257/w3-propertyvalue-16001.html).

quality is the responsibility of the MMA (Environmental Law, Art 70 (u)), while monitoring of water quantity is fragmented across the DGA and the Junta de Vigilancia (Water Code, Art 122 & 146). Power imbalances across the different ministries and institutions further complicate the fragmented water institutional landscape (i.e. MMA is weak in comparison to MOP, which is less powerful than Ministries for Mining and Energy).

Moreover, Map 6.3 in Chapter 6 shows the set of boundaries that separate the basin in four different sections. The division of the resource in the basin along the administrative boundaries of the juntas is echoed in the legislative framework through the separation of subterranean and surface waters (Water Code, Art 2). The Water Code (Art 186) does provide for multi-sector participation where two or more parties own water rights in the same canal, reservoir or aquifer in the Junta de Vigilancia. However, non-agricultural stakeholders do not take part in the different Junta de Vigilancia in the Aconcagua, nor are there any such organisations for groundwater. In general, hydroelectric companies have been reluctant to become Junta de Vigilancia members, leading to issues in how users co-habit basins across the country.

8.3 Conclusions

In Chile, the informal approach to water management is driven through its conception as an economic good. While water governance at the political level is driven through a centralised approach, water management happens in the private sphere and is driven by private interests. Despite the strong codified nature of water governance through the Water Code, the weakness of enforcement and capacity in the DGA means that provisions introduced to build some resilience in the system (i.e. residual flows and sustainable use of aquifers) can effectively be ignored at the basin level. The governance approach has also produced a number of blind spots, including one on the ecological impacts of the market driven approach. It is a common saying, that what you measure you manage. While water rights are supposed to be recorded and administrated by the DGA, quality issues and ecosystem impacts are not being consistently measured or managed.

Another blind spot in the system are issues that will arise from increased uncertainty and climate change impacts. Investments are being made now, which do not incorporate any sense of uncertainty or climate impacts in the basin. It is clear that the market commodity definition of water rights in Chile has impaired a holistic view of water resource management that looks beyond the limited definition of water as an economic input for agriculture, mining or energy production. It is only during presidentially declared drought periods that water is prioritised for human consumption, and the even then, the declarations of drought periods also allow for the exploitation of underground water to which one does not have constituted rights.

In times of such extreme drought in other areas, water extractions are limited to protect fragile ecosystems, in Chile, it seems the opposite. The economic and market focus on water management has meant that public institutions responsible for water rights management (DGA) or water and environmental issues have very limited capacity to address water issues, yet the DGA is expected to step in at the most extreme time of drought to manage water conflict as soon as the drought is formally declared. Unsurprisingly, this has tended not to end well.

8.4 Summary of Chilean and Swiss Governance in the IWRM Context

The descriptions and analysis in Chap. 7 provide a baseline understanding of the governance systems relationship with sustainable water management, and can be taken as a point of departure from which to develop a better understanding of adaptive capacity. The aim of the STRIVER/BRAHMATWINN indicator based approach is to better understand how the governance systems can assist in the implementation of IWRM. The full governance assessments provide a rich and detailed picture of the governance framework in relation to IWRM and highlight the core challenges in each case area to the development and implementation of an IWRM based approach. While these full reports were developed for use in the ACQWA project, abbreviated versions included in this book provide a useful baseline from which to better understand the governance context and associated challenges in adaptive capacity that must be developed and mobilised. In addition, as highlighted in Chap. 4, some adaptation assessments have utilised these indicators as determinants of adaptive capacity.

Despite the very contrasting legislative frameworks across the two cases, significant challenges persist in each according to the indicators of IWRM. In the Swiss case, despite better fulfilment of accountability, transparency and participation indicators, there remain institutional fragmentation and the challenges of implementing federal policy at the local level of implementation. Furthermore, findings from the Swiss case suggest that there may be a limit to the level of devolvement, and that it can only be effective when combined with requisite levels of experience and resources as well as a propensity for stakeholders to work across the other levels of decision making. In Chile the key institutional challenges relate to the lack of data and information on the market, challenges in holding water users to account due to the lack of enforcement capacity and informality of the governance approach, and finally significant barriers to integrating environmental and social considerations in the water governance framework.

The analysis provides a useful baseline from which to explore factors affecting adaptive capacity. Results from the governance assessment in the Swiss case area showed that despite the water governance system performing well under the indicators of accountability, transparency, participation, there are concerns about ability to cope and adjust to a changing climate and rising competition for water use, mainly through the challenges that the strength of local autonomy generates for ross scale integration and collaboration, effective decision making in the face of new challenges and the acceptance of innovation from higher administrative levels. Issues also arose in interviews which suggested that a correlation between ‘participation’ and ‘decentralisation’ and greater adaptive capacity should not be taken as a normative assumption. Chile performs less well according to the indicators, which would suggest that governance challenges in relation to IWRM are likely to be further exacerbated by issues relating to climate change impacts.

References

Bauer CJ (1997) Bringing global markets down to earth: the political economy of water rights in Chile, 1976–95. World Dev 25(5):639–656

Bauer CJ (2004) Siren Song: Chilean water law as a model for international reform. Resources for the future. Johns Hopkins University Press, Washington, DC

Carruthers D (2001) Environmental politics in Chile: legacies of dictatorship and democracy. Third World Q 22(3):343–358

Dourojeanni A (2010) El Contexto de GRRH por Cuencas: Primera Parte. Fundacion Chile

Dourojeanni A, Jouravlev AS (1999) El Código de Aguas de Chile: entre la ideología y la realidad, vol LC/R.1897. Comisión Económica para América Latina y el Caribe, Santiago de Chile

Guardian (2011) Chile riot police clash with student protesters. The Guardian, Wednesday, 10 Aug 2011

Lentini E (2011) Servicios de agua potable y saneamiento: lecciones de experiencias relevantes. United Nations Economic Commission for Latin America and the Carribean, Santiago de Chile

Nacion (2010) Suicidios, el calvario pehuenche. Nacion.cl, Sunday 21 Nov, pp [online]. http://www.lanacion.cl/suicidios-el-calvario-pehuenche/noticias/2010-2011-2020/190404.html

Patagonia (2011) Patagonia Chilena: Sin Represas: El Problema – Monopolio y Concentración. http://www.patagoniasinrepresas.cl/final/contenido.php?seccion=problema. Accessed 15 Sept 2011

Pena H (1997) Modificaciones al Codigo de Aguas y su Aporte a la Gestion del Agua. Direccion General de Aguas, Ministerio de Obras Publicas, Santiao de Chile

Pena H (2001) 20 Anos del Codigo de Aguas: Vision Desde la Administracion. Paper presented at the 4th Jornados de Derecho de Aguas, Facultad de Derecho, Pontificia Universidad Catolica de Chile. Santiago de Chile, 19–20 Nov

Chapter 9
Converging Threats: Driving Pressures for Adaptive Capacity

Abstract Previous studies of adaptive capacity have shown the value in utilising past climate events to explore the experiences in mobilising adaptive and coping responses. The case events used in each case study area served as reference points of climate variability and as useful indications for the impact of extremes in a future, warmer climate. The exploration of past experiences in relation to climate related extreme events acted as a means to understand and assess adaptive capacity. This chapter details the extreme events used as focussing events for the exploration of adaptive capacity, as well as the attenuating management challenges which frame the context within which adaptive capacity must be mobilised. This chapter provides an in depth account of the focussing events, in the context of climate change impact projections, as based on interview data, archival data and primary research utilising meteorological and climate model data.

Keywords Rhône, Canton Valais, Switzerland • Aconcagua, Region V, Chile • Exploration of past climatic events • Coping and adaptation actions • Drought • Flooding

9.1 Switzerland

The climate of the Alpine region is characterised by Beniston (2004) as having a high degree of complexity due to the interactions between the mountains and the general circulation of the atmosphere. Resulting features of this complexity are the aforementioned rain shadow effect in the Valais as well as blocking highs and föhn winds. Precipitation patterns vary according to altitude, sun exposure (which is greater in the Southern Alps), and dryness of climate (Weingartner 2007). To date, the warming experienced in the Alps since the early 1980s has been roughly three times as strong as the global climate signal (Beniston et al. 2003). Broadly, climate change impacts in the European Alps will lead to higher winter temperatures and a

M. Hill, *Climate Change and Water Governance: Adaptive Capacity in Chile and Switzerland*, Advances in Global Change Research 54,
DOI 10.1007/978-94-007-5796-7_9,

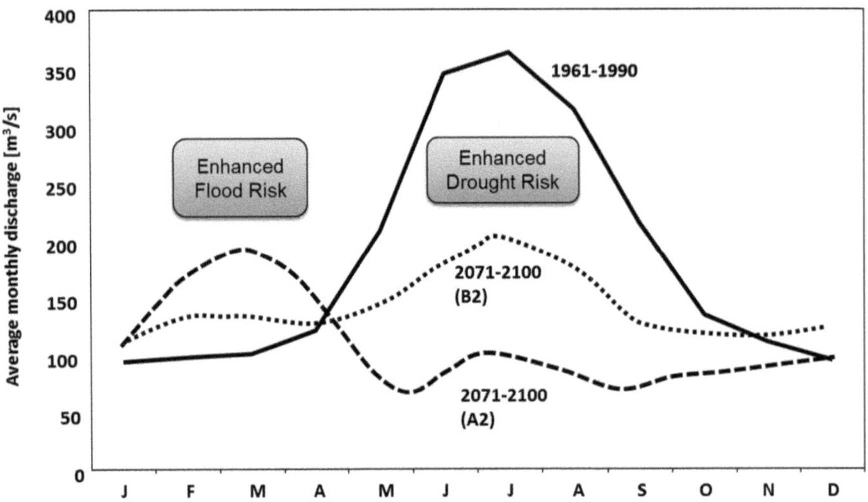

Fig. 9.1 Taken from Beniston et al. (2011) showing projected developments of flooding and drought instances for the Rhône Basin

more marked increase in summer temperature, higher precipitation that is more intense during winter, but likely to be much reduced in summer months (Beniston 2006). Increased temperatures will continue to have a significant impact on the melting of glaciers, whose thickness has already decreased by roughly five times more in the period 1980–2000 than during average loss (Büchler et al. 2004).

It has been suggested that the largest source of vulnerability from climate change is likely to come from changes in the intensity or frequency of extreme events, such as heat waves (winter and summer), heavy precipitation events and drought (Beniston et al. 2007). Increased glacial melt also is leading to an increase in flood risks and other natural hazard events (OcCC 2008). Figure 9.1 represents the difference in seasonal distribution of run off for the River Rhône between baseline values (1961–1990) and projected values for A2 SRES scenario (800 ppm) by 2100. The dark black line represents baseline run off, showing typical seasonal flows (high run off in summer, low in winter) for an alpine regime. The dashed line shows the 2100 projection, with increased run off during early spring and decreased run off in mid to late summer, while the dotted line indicates the intermediate alterations projected. The graph suggests that summer months could experience enhanced drought situations through reduced glacial mass and precipitation, while in winter months increased intense precipitation periods could not only impact on flood risk, but a wide range of geomorphological processes such as landslides and rock falls.

Increased flooding and extreme precipitation events are compounded by an increase in risk exposure due to infrastructure/housing development in vulnerable areas which are currently seen as 'safe' due to technical interventions. Temperature increases at alpine elevations raises demand for water uses such as artificial snow making and summer cooling and drinking water leading to complex management shifts, compounded by changes in seasonality. There have already been examples

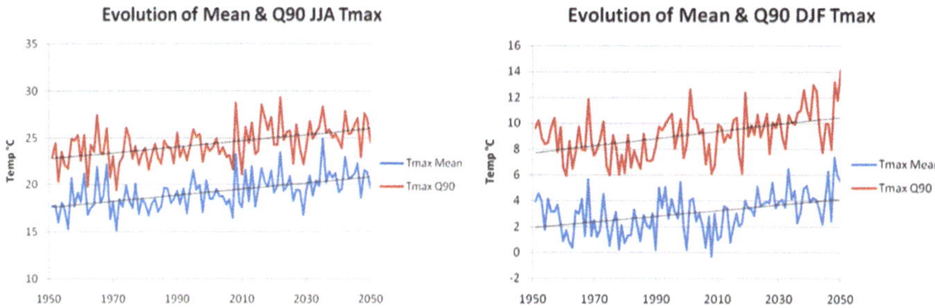

Fig. 9.2 Evolution of temperatures at Visp according to A1B scenario for summer temperatures (*left*) and winter temperatures (*right*)

where a lack of planning for drinking water supply has led to sector issues between hydropower use, tourism use and drinking water supply (Reynard 2000a), as well as tensions between the different sectors that arose in the 2003 summer heat wave, despite a paradoxical increase in surface flows in the River Rhône due to enhanced glacier melt.

The impacts of climate change may also compound the reduction in ecological status of many surface waters in Switzerland. Of 65,300 km of surface waters in Switzerland, 10,600 km have been considerably altered through technical projects, thereby impairing their ecological functions (FOEN 2009). Hydro-peaking (artificial high and low flow phases) also impacts rivers, in that they regularly dry up from over extraction of water, with damaging impacts on aquatic ecosystems. It is also worth mentioning the potential impacts of climate change on mountain forests, which have a variety of highly significant environmental benefits, including the protection against natural hazards (Gautam et al. 2004). Increasing temperatures and lower precipitation levels during the hottest periods are likely to lead to greater risk of forest fires, further increasing the vulnerability of mountain communities to the mounting hazards (rock falls, flash floods, landslides) from which they normally provide protection (Gautam et al. 2004).

Climate model data from the ACQWA project was used to calculate return periods as per A1B and B2 emissions scenarios. A synthesis of results from the stations at Visp and Zermatt are detailed below (Fig. 9.2). The figures show an evolution of higher mean temperatures for both summer and winter periods, with corresponding impacts on glacier melt that will influence the run off regime of the Rhône based as depicted in Fig. 9.1 above.

9.1.1 *Focusing Events*

To date, the Valais has been more seriously impacted by extreme precipitation events and flooding than by situations of drought and scarcity. Drought impacts were viewed by stakeholders as being relatively minor while extreme weather associated with high precipitation is seen as a much more significant issue. The issues associated with hydro-climatic extreme events have therefore been too much water, rather than too little water. This has meant that stakeholders have generally been preparing for increased periods of flooding, and little effort has gone into assessing relevant measures for water stress. Additionally, it was also noted, that it is also easier to work out how to cope with more flooding than with more drought.

It is worth mentioning the unique weather patterns that the Valais experiences; its relative dryness in comparison to the rest of the country (due to the rain shadow effect, e.g. Grächen has the lowest levels of precipitation in Switzerland), yet the extreme precipitation it often undergoes when warm air from the Mediterranean rises and releases huge amounts of precipitation in the area surrounding Simplon. This phenomenon of extreme dryness to extreme volumes of precipitation sets the backdrop for the focussing events that shall be discussed next. Valaisanne farmers have a long history of adapting to climate variability and dry conditions through the system of water canals (*Die Suonen/Les Bisses*). This system of canals that lead glacier melt water from the high alps to the alpine meadows continues to buffer farmers from the most extreme impacts of very dry periods such as the 2003 heat wave, since as glacier melt increases more water can be exploited through the irrigation infrastructure.

Generally, stakeholders concluded that in 2003, despite the very low precipitation levels (30%[1]), water in the streams and rivers was in fact plentiful from the record glacier melt (Huss 2011)[2], meaning that the Valais experienced the opposite problem to the rest of Europe because of this increased melt water. However, drier summers have been leading to lower recharge levels in the springs, which did require certain non-essential domestic uses to be stopped in places such as Visp (e.g. watering the garden, swimming pools, washing cars). Farmers were also required to stop using drinking water supplied from the utility to irrigate their meadows and fields in the area of Visp, and instead pumped water from the Rhône, as a one off adaptation to the extremely low precipitation levels. The commune is now implementing measures that would reduce the amount of water used for irrigation purposes from the local utility.

The extreme heat wave in 2003 did lead to some tensions between farmers who needed water to irrigate their fields, but mostly on the 'right' side of the valley (i.e. the northern alpine side), where the glaciers are not as high, extensive, or as numerous, and the distance to the valley shorter. Issues of scarcity also tend to hit at commune rather than cantonal level. The amount of water farmers receive is traditionally co-regulated, with each farmer knowing exactly how much water should be received over a certain number of hours. Problems have started arising after very dry winters, or winters with low amounts of snow,

[1] MétéoSuisse Data show that precipitation deficits range from 20% at Montana to 38% at Gd St Bernard. Therefore a median figure of 30% is seen to be representative of Valais as a whole.

[2] In August 2003, a recent study has calculated ice melt to have been over three times the mean (Huss 2011).

where traditional irrigation is not possible from the end of July onwards. For example, there are a number of communes (Bralanch, Gutthed, Faessil) above Leuk that in 2003 could not maintain irrigation from the end of July after a snow-poor winter, halting the second growth of grass for hay making due to lack of water in the streams.

For the hydropower operators, the 2003 summer heat wave meant a period of increased melt, and therefore increased production. For example, hydropower operators reported production that was about 20–25% higher than normal. However, some operators are concerned that maximum levels of run-off have now been reached and are unlikely to increase further. Between the 1960s and later 1990s to early 2000, stakeholders referred to the steady augmentation in melt, which has been seen to stabilise since 2000, mirroring studies that suggest a peak might have already been reached and thus the transition from a glacial to a nival run off regime may have already commenced (Huss 2011; Huss et al. 2008). In general, the operators tend to receive more water earlier in the summer melting, even if precipitation (as rain or snow) is reduced.

Low flows in winter exacerbate the already high pressure on multi-use rivers and streams in the Valais, where many rivers have already experienced some form of drying up because of over-extraction, in particular during peak vacation periods in this region, highly dependent on tourism for a significant part of its income. The drying up of rivers becomes apparent after August, where uses compete over less melt water in the late summer period (e.g. La Reche River) causing tensions between the fishery and environmental lobby and on the other side the agricultural and hydropower lobby. Peak periods of over demand (Christmas, New Year and Easter) occur at the lowest periods of flow (also when hydropower plants are capturing much of the water that would flow into the streams), but climate impacts can also aggravate these lower flows in winter, further damaging micro-organisms and fish.

From the mid-1980s, there were a series of heavy precipitation events that occurred in relatively short intervals. In 1987, Muster and Goms were heavily impacted, and then in 1993, heavy rain for a number of days resulted in destructive flooding in the Saastal down to Brig, where the damage from the debris flows through the heart of the city generated damage costs of close to CHF one billion (FOS 2011). The winter of 1999/2000 became known as the avalanche winter, and in the autumn of 2000 more major flooding events impacted Stalden, Baltschieder, Gondo and Brig as well as the lower Valais at Riddes. In Gondo, the event resulted in 13 dead, with practically the whole village being washed away (Amweg 2011). While in Baltschieder, about 80% of the sewage infrastructure was affected, and it took 5–6 years to repair. Stakeholders noted that while impacts on water provision from flooding events, impacts on sewer system and drainage infrastructure has tended to be much worse. While in all of these cases, it was the valley communes that were most significantly damaged, the initial increases in river water volumes started much higher up, at around 3,500 m.

Stakeholders allude to the shock at the increasing volumes of water that came down during those periods, the increasing frequency of events across the two decades as well as the increasing amounts of damage and destruction that they caused.[3] Some stakeholders suggested that the increased damage has been partly assigned to the relative failure of infrastructure that had not been as well maintained. The relative lack of experience of such flooding events in the previous decades (living memory of the hydraulic engineering or flood protection managers) had meant that they had allowed the built up areas to encroach on the river's space, giving the Rhône only as much room as it was thought necessary (about half as much as was sufficient for adequate protection). As soon as quantities did increase, and dramatically so in the series of flooding events, it became clear that high flows needed about twice as much room as currently was allocated. The increasing intensity of precipitation also has negative impacts on water quality and damages the catch points for water in the streams.

In 2000, damage was limited by the remediation enacted after the 1993 event, despite the fact there was about 25% more water, but those projects which were built around 20 years ago were already not enough to have the capacity to hold back future quantities. Some 350 m^3/s fell in the Vispa in 1993, which was exactly the limit that they could cope with. The first and second Rhône Correction projects were 100 and 50 years ago respectively, and the events showed how the dams were no longer stable enough nor had sufficient capacity for heavier precipitation events, therefore were insufficient for protecting the settlements and industrial areas bordering the Rhône. The new measures that have now been implemented are prepared for 550 m^3/s.

In addition to the impacts felt within villages and towns themselves, hydropower operators also experienced impacts on their reservoirs and operations. In 1993, the Mattmark dams amongst others overflowed, while the power station at Stalden was also exposed as being vulnerable to the flooding. The overflow from the dams exposed the fact that the storage volume of the reservoirs was no longer large enough to contain the volumes of water, in fact worsening the impact of the 'natural' flooding. In the 1993 case, the road was severely damaged between Mattmark and the valley as far as Saas Almagell and the village itself was also severely impacted. In the 2000 event, which took place over 3 days, there was a significant impact on operations. During this period, issues in how to regulate the hydropower reservoirs occurred between the canton and the hydropower operators, in that while they needed to discharge water from the reservoir, the overall management of the river was unclear. During a period of 3 h, around 30 m^3/s needed to be let out of the Mattmark reservoir, which is likely to have further impacted the intense flooding of the Rhône in Saxon.

It is not just the volume of water that affects the hydropower operations, but the extra volumes of material, such as sand, gravel, dirt and bed load which affect water quality, blocking or damaging the turbines. While from a profit perspective, the worst case scenario is that the installations need to be de-activated or are damaged, from an operational and river management perspective, it is that the reservoirs or facilities overflow and aggravate the flooding that impacts the villages further downstream.

[3] For example in the 1993 event, about 80 m^3/s fell, and then in the 2000 event, 125 m^3/s fell.

With the Mattmark reservoir, extra storage capacity can be created by pumping water out of the reservoir to prepare for higher precipitation volumes as and when is needed. In 2000, a debris flow (*murgang*) caused extensive damage to the Vispa in Stalden. During this event the extra storage volume was used to ensure that excessive amounts of water did not flow into Visp, hence avoiding more serious flooding impacts, although some damage still occurred.

The Third Rhône Correction plan was outlined in the aftermath of the earliest of these events, and agreed upon after the later events, as the impacts of the floods highlighted how the earlier remediations on the Rhône (the first and second Rhône corrections) could no long ensure sufficient security for the Rhône flood plain. In the central Valais, there are a number of priority measures (e.g. at Alcom) because it is an industrial zone where they produce aluminium, and the damage potential is millions of franks worth. The 2000 flood was a prime example of the increasingly aggressive autumn floods that occur between September and October, when a cold snow spell is followed by higher than average temperatures.

While the elevation of the snow line has significant implications on the alpine tourism industry, there are also severe ramifications for water quantity and timing and the impacts from heavy precipitation events. For example, in the Valais above 2,000 m the terrain tends to be mainly glacial or rock and cliff. When the 0°C isotherm is only at 4,000 m as opposed to about 2,000 m, then most of the precipitation falls as rain, rather than snow, with repercussions for the amount that is stored for later melt (Beniston 2006), and that which flows down directly through the streams and off the cliffs (a 2,000 m difference can increase or reduce the volume by half).

Increasing flows of water from rapidly receding glaciers is not only an influencing factor on the increasing flooding events, but also provides certain benefits to the hydropower operators, who have more seasonal production than when the 0°C isotherm is lower. Changing patterns in glacial melt have had repercussions on spring levels, which are fed by seasonal glacier and snow melt. At the commune level, the largest source of water for domestic supply is from glacier melt (in that it feeds the springs and groundwater from which domestic use is supplied). In the Zermatt region, the rapidly reducing Findler Glacier is negatively affecting spring recharge, which the communes rely upon for domestic water supply.

In other areas of the Valais, there are situations where levels of melt water are insufficient to meet demand from mid-August (but mainly on the northern alpine side). Low snow levels in winter and periods of low precipitation in summer, also negatively impact spring levels, which can lead to an increasing exploitation of groundwater sources. Spring and groundwater levels are dependent on a number of variables, including precipitation levels in winter, whether precipitation falls as rain or snow, and when the melt period begins. In general, if the months after March are very dry, then the dual impact of less melt and less precipitation reduces the replenishment of the springs (April 2010 and 2011 were both extremely dry). These compounding impacts have reportedly led to situational increases in competition at the commune level, for which the canton has no oversight. However, while utilities have diversified supply from precipitation (rain/snow) and melt water to recharge springs, hydropower operators rely solely on glacier melt.

To date, winter heat waves have not been observed to have a major impact on river levels, partly because the glaciers and water capturing points in the Valais are at high enough altitudes, that temperatures remain negative. Additionally, rivers in the Alps are at their lowest levels during the winter months, therefore any increase of melt would have been adequately absorbed by relatively dry waterways to avoid sudden flooding. Stakeholders also referenced noticeable changes in invasive and damaging species such as the Colorado Beetle, which is now recorded at higher altitudes, affecting different crops such as potatoes. Additionally, melting permafrost has also led to increasing problems of landslides and rock falls where previously there had been none. However, permafrost is not a climate impact that will be explored further in this chapter.

9.1.2 Converging Threats: Non-climatic Drivers

In addition to the specific impacts from climate related stresses as detailed above, there are a number management related challenges that were identified through the coding exercise as converging threats within the basin. Most of these issues have already been addressed in Chaps. 1 and 4 and were presented through the broad governance context in relation to indicators of good governance and IWRM (Chap. 5). The issues highlighted in this section (and in the following Chilean section) relate mainly to specific geographical, demographic and infrastructural issues that interact with the climate driven issues and so in some respects are difficult to separate from the climatic drivers of adaptation responses (see Sect. 6.2.1 for a definition of response).

Perhaps the biggest management challenge for the mountain municipalities is the issue of periodic rivalries. Peak period demand, when water flows are at their lowest, are precisely when water demand is at its highest. For example, in Zermatt and Les Bagnes, 90% of demand is during winter and notably at specific points during winter. This requires a commune with a population of 3,000–6,000 to be able to cover water supply for an intermittent population of 30,000 over the course of a few weeks during the winter season (December–April), when the springs are at their lower. This has caused some local water managers to suggest that the communes will in the future need to rely more heavily on exploiting groundwater sources (Zermatt is currently fully dependant on spring water), if demand keeps rising, and their ability to recharge the springs during summer diminishes. A related impact of tourism peaks is the steady rise of electricity demand, which has risen by 3% per year for the previous 5 years, the majority in winter (EWZ 2010).

Despite the strong principles of decentralisation that defines the Swiss governance framework, concentrations of power have gradually been shifting from lower to higher levels of government as well as across private and public sector responsibility. Certain services that used to be the responsibility of private actors at the commune level have now been transferred to the public realm either at the commune or cantonal level, most notably after the 2000 flooding events. For example, reconstruction and repair of damages from extreme weather events is no longer managed privately as remediation work was so extensive it required the structure and support of the commune or cantonal authorities. In the aftermath of the 2000 event in Baltschieder, the commune and canton collaborated to take over the clean-up operation, rebuilding the streams and repairing the canalisation network. After the 1993 events, there was a realisation that the damage costs were so high, that the communes did not have the financial capacity to cover the bill. The canton therefore took responsibility for remediation costs, later sourcing percentages of repayment from both federal and commune authorities.

Despite these shifting demographic and political structures, traditional associations for the water irrigation channels have tended to remain, yet in a weakened form. As agriculture and the number of full time farmers decreases,[4] the number of members of these common property resource regimes (CPRs) is in decline. While water is linked to property rights in these regimes, many people who now own the relevant property are no longer farmers, but perhaps holiday or second home owners, and therefore no longer use the associated water rights or consider themselves responsible for paying for the upkeep of irrigation infrastructure they no longer use. Where these CPRs are dwindling, many of the activities are being transferred to the commune. Likewise, similar institutions and associations for more relevant needs are being discussed (domestic water, artificial snow) and additional financial support sought through organisations such as Berghilfe.

9.2 Chile

There has been very little documentation of the potential impacts of climate change in semi-arid watersheds in subtropical South America (Vicuña et al. 2011a), despite most climate models projecting a strong future climate change signal on the Western side of the Andes (Mata and Campos 2001; Souvignet and Heinrich 2011). Climate projections based on GCM's for central Chile[5] consistently demonstrate

[4] Alpine farmers have traditionally played an important role in the upkeep of the 'alpine cultural landscape', including the maintenance of traditional defences and infrastructure that play a role in water management and associated protection mechanisms against avalanches, flooding and landslides. As more and more alpine farmers move in either a full time, or part time, capacity to other economic sectors and financial resources are more constrained, infrastructure upkeep becomes more of a challenge at the same time as hazard recurrence is increasing (Kantonszentrum für Landwirtschaft; lack of upkeep on the irrigation system of Vispa, Saasvispa, Mattervispa meant that water was not as well transported through the canal system, and instead flowed wildly off the slops intensifying damage from the flooding event). In the Oberwallis for example 85% of farmers now are part-time, and often work either in tourism or in the Lonza factory (Kantonszentrum für Landwirtschaft).

[5] However, it should be recognised that 'because of the special physiographic characteristics of watersheds on the western slope of the Andes cordillera (steep, short river lengths, with a 3+ km elevation gain from the Pacific Ocean in less than 200 km), the spatial scale of current GCM modelling grids is inadequate to assess local effects on the hydrologic regime and downscaling approaches (statistical or dynamical) introduce an additional layer of uncertainty' (Vicuna et al. (2011a, b, JWRM).

both a warming and a drying trend throughout the rest of the twenty-first century (Christensen et al. 2007). Winter months (June–August) feature both minimum temperatures and maximum precipitation, namely about 80% of annual total precipitation between May and August (Vicuña et al. 2011a). Summer months (Dec–Feb) feature minimum precipitation and tend to be snow or glacier melt dominated, with the main proportion of stream flow taking place in late spring and summer (Sep–Jan) (Vicuña et al. 2011a). This leads to almost total reliance on glacier and snow pack melt for water during the growing season, in areas where there is no storage capacity. Climate change associated reductions in run-off, hydrograph timing and enhanced evapo-transpiration will have significant impacts on agriculture in the semi-arid areas of northern and central Chile (Vicuña et al. 2011b).

Furthermore, precipitation and temperature are both strongly influenced by different large scale natural phenomena such as ENSO as well as, the Pacific Decadal Oscillation (PDO) (Garreaud et al. 2009; Souvignet and Heinrich 2011; Verbist et al. 2010), leading to high inter-annual variability (Vicuña et al. 2011a). ENSO is a coupled ocean–atmosphere phenomenon, tied to the tropical Pacific Ocean, that is characterised by fluctuations (periodicity between 2 and 7 years) between a warm phase (El Niño), generally associated with higher than average precipitation in central Chile, and a cold phase (La Niña), associated with lower than average precipitation (Garreaud et al. 2009). While ENSO is observed as the primary driver of inter-annual variability, PDO has been suggested to force decadal and inter-decadal variability, with temperature and precipitation anomalies related to ENSO, but with smaller amplitude (Garreaud et al. 2009). In the preceding decades, ENSO events have become increasingly frequent, but high levels of uncertainty mean that projecting its development according to climate change scenarios is still poorly understood (Kim and An 2011).

While glacier shrinkage in the Dry Andes (generally between 20 and 35 S) has been relatively well captured (Le Quesne et al. 2009), the impacts on stream flow have been less well documented, in part due to the challenges of data collection (Gascoin et al. 2010). Despite high uncertainty and general lack of data on climate change impacts in the central Chilean region, studies and observation show that in the Aconcagua Basin, there has been a significant decrease in the annual and seasonal trend of streamflow from the Aconcagua basin glaciers, related to decreasing contributions from glaciers and snow cover (Casassa et al. 2009; Pellicciotti et al. 2007). Pellicciotti et al. (2007) suggest that melting rates tend to be higher in the Central Andes, since the glacier ablation area occurs at lower elevations and so higher temperatures in summer have increased melting. Furthermore, simulations in northern central Chile suggest that the Dry Andean mountain range is likely to encounter warmer winters, decreasing precipitation, changes in snowpack, changes in snow and glacier melt and generally increasing dry periods, though as mentioned earlier, this is still poorly modelled by GCMs (Souvignet et al. 2008; Vicuña et al. 2011b). As Vicuna et al. (2011b, p 482) clarify 'increase in temperature leads to a reduction in snowpack accumulation during the rainy season and an earlier, faster snowmelt process during spring and summer'.

The changes in amounts and timing of hydrological resources converge with enhancing levels of water demand from growing urban populations, irrigation areas and mining activities (Reyes Carbajal 2007). Beyond the challenges implied by decreasing amounts of water resources for economic inputs, increasingly dry conditions in the spring and summer months would also have severe consequences for the farmers in the agricultural belt that is situated in central Chilean areas, through impacts in the biological productivity of ecosystems (Vicuña et al. 2011b). Changes in stream flow timing and amounts already have begun to impact the different economic sectors in the Aconcagua, and shall be further discussed in relation to the focussing events below.

9.2.1 *Focusing Events*

Over the past 15 years, while there have been a few flooding events associated with increased snow melt and heavy precipitation events, drought has been a far greater preoccupation of water stakeholders with far reaching impacts for the SES. Andean watersheds generally experience low precipitation in summer and rely heavily on storage of winter precipitation within the snow pack and glaciers of the high Andes. Climate change impacts on water quantity have already led to increased water stress, which compounded by increasing water abstractions, has led to a reduction in surface water recharge that tends to impact water rights in medium and lower segments of the basin more severely (Desmadryl 2010). Melting glaciers and reductions in water availability have also been observed to have exacerbated impacts on water quality, ecosystems and overexploitation of certain aquifers in the northern and central regions. With around three quarters of Chilean economic produce and activity seen as water intensive, the repercussions from climate change impacts on water reserves with far reaching potential consequences (Desmadryl 2010).

In the preceding 15 years, there have been a few incidents of the river overflowing, such as the 2009 event in the Panquehue region, primarily from increased flow due to ice melt in spring time. DOH has undertaken a number of remediation works, and in the past 6–7 years, there have been no major issues from flooding, but also no major substantial overflows either. Vulnerable areas are situated at La Calera, Panquehue and Los Andes. Increasing variability is seen to reduce the former predictability and the innate knowledge of precipitation volumes, snow fall and flow behaviour. In May 2010, a combination of precipitation in the high mountains and ice melt during summer led to increased run off and thus an overflow of the river.

The natural reservoir of the region, 'La Cordillera de los Andes' has already been exposed to a rise in the zero-degree isotherm, reducing the capacity of snow storage and thus further aggravating over exploitation of surface waters during the dry summer months. Combined with the potential diminution in run off contribution from melting glaciers, the scarcity situation of the basin has been deteriorating in the past years. While there have been a number of droughts in recent years in the

Aconcagua region, 1996/1997, 2002, 2008 and most recently 2010/2011, the most severe were those in 1996/1997 and 2010/2011. While stakeholders have known drought periods to occur about twice per decade, recently, an increase has been observed. In 2002, 2008 and then in 2010, an official drought zone was declared by the President with the support of the DGA. In 2010, the drought zone was declared on 25 November, while in 2008, it was January 1. The earlier the declaration, the more the DGA can try to mitigate impacts of the drought, which affect agricultural stakeholders and utilities most severely. In 2010, a significant reduction in snow fall, despite rain volumes remaining constant, was seen as the primary cause of the drought.

Not all sections of the basin are evenly affected by the drought impacts, due to the high hydrological and geographic diversity in the basin (see Chap. 4). While the third section is most sensitive to drought and the second section the least sensitive. Petorca and La Ligua are some of the worst affected areas, as they do not contain any surface water, but instead rely on groundwater rights extracted through deep wells. The impact of severe droughts, such as 1996, 2002, 2008 and the most recent 2010 droughts have impacted drinking water distribution. In response, the DGA was called on, through the official drought declaration, to identify the most important needs and distribute water in equal quantities for drinking water and agriculture. In the 2008 drought, there were water transfers by truck from Algibe to Ligua, and from Cabildo to Ligua, which is the capital of Petorca province. At the time of interview, the expectation was that this would need to be repeated in the 2010/2011 drought. There were also transfers to Limache, and a transfer from Quillota to Marga-Marga. The summer droughts are also aggravated by the large population surges in the coastal cities of Valparaiso, Viña del Mar and Reñaca during the warmer months.

Drought impacts are exacerbating issues of general over-exploitation of water resources in northern and central areas of Chile that had already led to decreasing levels of water availability. For *regantes* (farmers with water rights to irrigate) each drought period has meant a significant reduction in allocated rights. A number of instances of illegal abstractions from the canals were reported during interview. These illegal abstractions can take multiple forms, including the placing of pumps in the canals to extract their full rights allocation, or even position glass sheets in the canals so that the water can be 'invisibly' siphoned off. Often during such drought periods, the water might not reach the last farmer in a canal (as in the case of 2010 drought), even after interventions have been made. In addition to climate-related drought impacts, there are a number of aggravating factors that in themselves intensify drought related impacts, and are themselves aggravated during drought periods. Firstly, irrigators, utilities and water managers alike refer to the amount of water that is lost to the ocean due to the lack of storage infrastructure in the basin.

The Aconcagua is one of the only basins in Chile not regulated by a major dam (the only major dam in the region is in Los Aromas, in Section 4). This is partly attributed to the fact that the region has been historically known for sufficient hydrological resources and highly suitable climatic qualities for agricultural production, a situation that over-exploitation and diminishing contribution from snow pack and glacier melt has now reversed into a hydrological deficit. As climate change impacts further reduce the capacity of the Andean natural storage (i.e. glaciers and snow pack), irrigators and water managers have stepped up their demands for the increasingly drought-prone Aconcagua River to be more regulated, with the construction of at least two new dams, and a battery of wells.

9.2.2 *Converging Threats: Non-climatic Drivers*

While the focus has been on water quantity, a number of issues were raised by agricultural and utility stakeholders about the combined impacts of mining, urbanism and drought for water quality. ESVAL's water rights in Estero Riecillos, in the high Andes in the upper part of the watershed, have become increasingly impacted by the expansion of Mina Andina. Mina Andina is not only using water from Riecillos, but there are additional reports on the impacts of mining activities on the glacier itself, thereby further exacerbating the increasingly stressed situation. Transparency over mining activities or planned activities is difficult, but within the basin, it is common knowledge that CODELCO are constructing some of the largest covered mines in Chile (reportedly to be larger than Coquichamaca), but this evidence must be taken as hearsay since public knowledge is limited due to the secrecy of the company itself, and the lack of transparency concerning approved project plans, as well as the fact that water rights for glaciers do not exist.

Another aggravating factor on water quality is the increasing growth of urban areas, in particular the associated littering of vulnerable waterways (canals) with urban waste. Irrigators from the third section noted that they had to remove between 1 and 9 m^3 of domestic waste from the canals that passed near Quillota. The impacts from this pollution were intensified during each period of drought. Another issue that affects water resources across the central and northern parts of Chile is the over exploitation and illegal extraction of water resources, both surface and ground (including aquifers). Hydropower companies also have extraction points in the river (*compuertas*), higher up in the watershed, where water is pumped from one point in the river, used for electricity generation, and then pumped in at another part, creating relatively drier sections at the extraction points. Illegal extraction is not just an issue for water resources, but also in terms of ground bed of the rivers. Irrigators reported that the Aconcagua has recently been extensively, and illegally, mined, with stones and gravel removed from the river bed during recent construction of the state highway.

For example, between Punta del Rey and San Felipe in section one, reportedly 4 million m^3 of sand in the last 3 years, gravel and rocks had been removed from the river by different companies under the auspices of the state government. The law states that if you remove more than 100,000 m^3 of gravel from the river (Water Code, Art 32; Environmental Law, Art 27) then an EIA should be completed. In order to circumvent the law once a company removed 99,000 m^3, the extraction company was allegedly changed every 99,000 m^3 until the 4,000,000 m^3 reached for the project.

Irrigators believe that this removal has punctured the river's seal leading to a drop in water level, thereby further exacerbating low river flows during the present drought period of 2010/2011.

In the Aconcagua Basin, the different sections are characterised by different hydrological contexts. While Sector 1 is at the top of the watershed, Section 2 enjoys relative water abundance due to the upwelling in the area (groundwater meets surface water), allowing for comparatively simple illegal groundwater use. Sector 3 on the other hand is in a more resource scarce area, while Sector 4 sits at the end of the catchment at the mouth of the river and not only has low coverage of agriculture, but less water availability than the rest of the basin. These contrasting hydrological characteristics are attenuated by the fact that upstream water right owners are generally able to abstract their full allocation, while downstream rights holders then enjoy less valuable water rights due to their inability to guarantee flows being released to their sections.

Finally, the Aconcagua basin is characterised in part as one of the only in Chile not to have any major regulation works. Irrigation security due to lack of dams, wells and other major regulation works is therefore seen as one of the biggest issues in the basin, resulting in a 'loss' of water (mainly in winter – in the non-irrigation season) to the ocean. Additionally, a lack of investment and maintenance in the water supply channels is blamed for the general lack of impermeable irrigation channels across the different sections of the basin, particularly in the areas where the length of the channels exceed 40–50 kms. For example, the 140 km long Waddington channel in the basin experiences about 50% loss, which means that farmers at the end of the channel can on occasion receive no water at all (but instead just rely on groundwater abstraction). The Waddington was created in the early nineteenth century, with a soil bed that has settled over the years because of natural condition. Today, leaks have to be constantly adjusted and losses dealt with, yet it is considered too expensive to invest in the restoration and enforcement of the channels.

References

Amweg (2011) Unwetter im Wallis, Oktober 2000. http://amweg.ch/d/unw2000.html. Accessed 15 July 2011

Beniston M (2004) Climatic change and its impacts. An overview focusing on Switzerland. Kluwer Academic, Dordrecht/Boston

Beniston M (2006) The August 2005 intense rainfall event in Switzerland: not necessarily an analog for strong convective events in a greenhouse climate. Geophys Res Lett 33. doi:10.1029/2005GL025573

Beniston M, Keller F, Koffi B, Goyette S (2003) Estimates of snow accumulation and volume in the Swiss Alps under changing climatic conditions. Theor Appl Clim 76:125–140

Beniston M, Stephenson DB, Christensen OB, Ferro CAT, Frei C, Goyette S, Halsnaes K, Holt T, Jylha K, Koffi B, Palutikof J, Scholl R, Semmler T, Woth K (2007) Future extreme events in European climate: an exploration of regional climate model projections. Clim Change 81:71–95. doi:10.1007/s10584-006-9226-z

Beniston M, Stoffel M, Hill M (2011) Impacts of climatic change on water and natural hazards in the Alps: can current water governance cope with future challenges? Examples from the European "ACQWA" project. Environ Sci Policy 14:734–743

Büchler B, Bradley R, Messerli B, Reasoner M (2004) Understanding climate change in mountains. Mt Res Dev 24(2):176–177

Casassa G, Lopez P, Pouyaud B, Escobar F (2009) Detection of changes in glacial run-off in 25 alpine basins: examples from North America, the Alps, central Asia and the Andes. Hydrol Processes 23:31–41

Christensen JH, Hewitson B, Busuioc A, Chen A, Gao X, Held I, Jones R, Kolli RK, Kwon RT, Laprise R, Magaña V, Mearns CG, Menendez CG, Raisanen J, Rinde A, Sarr A, Whetton P (2007) Regional climate projections. In: Climate change 2007: the physical science basis. Contribution of working group I to the fourth assessment report of the Intergovernmental Panel on Climate Change. Cambridge University Press, New York

Desmadryl M (2010) La visión del gobierno chileno sobre el futuro del agua en Chile. Paper presented at the Seminario Internacional de Fundación Copec-Universidad Católica/Agua: Desafíos de su escasez, Santiago de Chile, 16 Nov 2010

EWZ (2010) Geschäftsbericht 2010. Elektrizitätswerk Zermatt AG. http://ew.zermatt.ch/pdf/GB_2010.pdf

FOEN (2009) Environment Switzerland. Federal Office of the Environment (FOEN), Bern

FOS (2011) Dangers naturels et accidents majeurs. Federal Office for Statistics. http://www.vd.ch/fileadmin/user_upload/themes/securite/protection_population/fichiers_pdf/opam_dangers_naturels.pdf

Garreaud RD, Vuille M, Compagnucci R, Marengo J (2009) Present-day South American climate. Palaeogeogr Palaeoclim Palaeoecol 281:180–195

Gascoin S, Kinnard C, Ponce R, Lhermitte S, MacDonell S, Rabatel A (2010) Glacier contribution to streamflow in two headwaters of the Huasco River, Dry Andes of Chile. Cryos Discuss 4:2372–2413

Gautam AP, Shivakoti GP, Webb EL (2004) Forest cover change, physiography, local economy, and institutions in a mountain watershed in Nepal. Environ Manage 33(1):48–61

Huss M (2011) Present and future contribution of glacier storage change to runoff from macroscale drainage basins in Europe. Water Resour Res 47:1–14

Huss M, Farinotti D, Bauder A, Funk M (2008) Modelling runoff from highly glacierized alpine drainage basins in a changing climate. Hydrol Process 22:3888–3902

Kim BM, An SI (2011) Understanding ENSO regime behavior upon an increase in the warm-pool temperature using a simple ENSO model. J Clim 24:1438–1450. doi:10.1175/2010JCLI3635.1

Le Quesne C, Acuna C, Boninsegna JA, Rivera A, Barichivich J (2009) Long-term glacier variations in the central andes of argentina and chile, inferred from historical records 5 and tree-ring reconstructed precipitation. Palaeogeogr Palaeoclim Palaeoecol 281:334–344

Mata LJ, Campos M (2001) Latin america. In: McCarthy JJ, Canziani OF, Leary NA, Dokken DJ, White KS (eds) Climate change 2001: impacts, adaptation, and vulnerability. Contribution of working group II to the third assessment report of the intergovernmental panel on climate change. Cambridge University Press, Cambridge, pp 693–734

OcCC (2008) Das Klima ändert – was nun? Der neue UN-Klimabericht (IPCC 2007) und die wichtigsten Ergebnisse aus Sicht der Schweiz. Organe consultatif sur les changements climatiques, Bern

Pellicciotti F, Burlando P, Van Vliet K (2007) Recent trends in precipitation and streamflow in the Aconcagua River basin, central Chile glacier mass balance changes and meltwater discharge. Selected papers from sessions at the IAHS Assembly in Foz do Iguaçu, Brazil, 2005, IAHS Publication 318

Reyes Carbajal L (2007) Analysis of Hydroclimatic Trends in the Aconcagua river basin, Central Chile. Institute of Environmental Engineering, ETH Zurich, Switzerland

Reynard E (2000a) Gestion patrimoniale et inégrée des ressources en eau dans les stations touristique de montagne. Les cas de Crans-Montana-Aminon et Nendaz (Valais). Travaux de Recherches, no 17. Institut de Géographie, Université de Lausanne 17

Souvignet M, Heinrich J (2011) Statistical downscaling in the arid Central Andes: uncertainty analysis of multi-model simulated temperature and precipitation. Theor Appl Clim 106:229–244

Souvignet M, Gaese H, Ribbe L, Kretschmer N, Oyarzún R (2008) Climate change impacts on water availability in the Arid Elqui Valley, North Central Chile: a preliminary assessment. Paper presented at the IWRA World Water Congress, Montpellier, France

Verbist K, Robertson AW, Cornelis WM, Gabriels D (2010) Seasonal predictability of daily rainfall characteristics in central northern Chile for dry-land management. J Appl Meteorol Clim 49(9):1938–1955

Vicuña S, Garreaud R, McPhee J (2011a) Climate change impacts on the hydrology of a snowmelt driven basin in semiarid Chile. Clim Change 105:469–488

Vicuña S, McPhee J, Garreaud RD (2011b) Agriculture vulnerability to climate change in a snow-melt driven basin in semiarid Chile. J Water Resour Plann Manage. http://dx.doi.org/10.1061/(ASCE)WR.1943-5452.0000202

Weingartner R (ed) (2007) Hydrologie – Im Wasserschloss Europas. The World of the Alps, Heritage of the World (Welt der Alpen, Erbe der Welt). Haupt Verlag, Bern

Part III
Applying the Assessment

Chapter 10
Governance in the Face of Uncertainty and Change

Abstract This chapter presents the first stage of findings in the multi-pronged approach, used to triangulate towards a more nuanced and empirically based set of adaptive capacity indicators. First, the results from the assessment and categorization of the adaptive responses are presented, elucidating the adaptive mechanisms across the different governance scales. Next, those mechanisms are characterised according to the categories of transformation, persistent adaptation and passive responses. Finally, the different categories of adaptive outcomes are discussed in more depth in relation to the specific governance mechanisms associated with them. Results indicate a higher concentration of transformative and persistent adaptive responses in the Swiss case area than in the Chilean case area.

Keywords Rhône, Canton Valais, Switzerland • Aconcagua, Region V, Chile • Assessing adaptive mechanisms • Adaptive outcomes • Transformative adaptation • Persistent adaptation • Passive change • Governance scales

10.1 Adaptive Mechanisms Across Scales

The following section and table records the different institutional and governance mechanisms that were mobilised, drawn on, or relevant to preparing for or navigate the case events in each case area. The different mechanisms recorded in Tables 10.1 and 10.2, are categorised by the different scales in which they are invoked. For the purposes of this study, adaptive mechanisms are termed as a response, institutional or governance mechanism (law, regulation, policy, institution) that are undertaken at the national, regional or local (both community and individual) level in order to prepare for or respond to different scales of environmental change (i.e. inter-annual variability, drought, floods, climate change impacts).

M. Hill, *Climate Change and Water Governance: Adaptive Capacity in Chile and Switzerland*, Advances in Global Change Research 54,
DOI 10.1007/978-94-007-5796-7_10, © Springer Science+Business Media Dordrecht 2013

Table 10.1 Overview of adaptive actions and major institutions, governance and management mechanisms affecting adaptive capacity identified in the Chilean case

Case area	Scale	Institution/governance mechanism/adaptive action	Explanation/example
Aconcagua basin, Chile	National	Presidential Declaration of Drought Zones (Water Code, Art. 314) Provides for the declaration of a drought zone for a maximum of 6 months, allows the General Directorate of Water (DGA) to intervene on request in the management of the river and redistribute the water (as in 1996), allows the extraction of superficial or subterranean waters where users do not have rights, prioritises human consumption of water (but liable to pay compensation to users adversely affected by decisions during drought situations).	Article 314 (Water Code) allows prioritisation of uses in a river basin, for the construction of hydraulic engineering works to abstract water without water rights, for the potential intervention of the DGA to manage water distribution, as well as for the prioritisation of water uses, namely for drinking water and agriculture. The declaration aims to mitigate harmful effects of drought partly by reducing the amount of irrigation, as opposed to suspending it altogether, allowing farmers to reduce the losses of their crops at least partially. However, in allowing the abstraction of water for which users do not have rights, the DGA is liable to pay compensation to rights holders whose water rights are affected by the increased extraction of groundwater.
		Resolution 39 (1984) Resolution establishes the 3 criteria for the DGA regional office to call a period of drought. The resolutions refers to Regions I, II, III, IV, V, metropolitan and VI. The three parameters are set as follows, and require that: (1) Stored/accumulated rainfall, as from April is less than 70% of the average statistical value; within the same point (2) Accumulated rainfall as from September should have a return period of 10 years (3) Monthly average flow of rivers is less than 70% from the average statistical flow of the that same month.	These three parameters are the criteria set to declare a drought event in Region V. The resolution dates back to 1984, and has not been updated since. The regional DGA office is unclear as to whether the hydrological scarcity decree can be issued based on 1 or the 3 criteria. Additionally, the hydrological parameters have remained consistent with 1984 hydrological records.

Commision Nacional del Riego (National Commission for Irrigation) – Law on Promoting Irrigation Efficiency (Ley 14.450) Law 14.450 had provided financing and subsidies to farmers to improve irrigation efficiency. The CNR was responsible for implementation of the law, and managing the legal instruments that funded projects aimed at increasing agricultural output as well as efficiency.	Law 14.450 had focussed mainly on increasing agricultural output, but ended in 2010. It was renewed with an updated focus, in that farmers must now focus on water efficiency in their new projects. Within the submission criteria for projects, different points are allocated for different criteria, water efficiency being one of them. Promoting efficiency is one of the main lines of intervention with farmers for the CNR.
Direcion General de Aguas/General Directorate of Water The DGA is responsible for the allocation of water, the monitoring of water quantity in surface and groundwater, and the maintaining of sustainable levels of abstraction according to a set of internal directives.	Major focus at the moment is to improve the transparency, availability and quality of the water rights information system (Cadaster de Agua Publica).
Direcion de Obras Hidrologicas/Ministerio de Obras Publicas The General Directorate of Public Works and Directorate of Hydraulic Works coordinate plans and proposals on water infrastructural works that aim to increase efficiency and improve irrigation security in response to increasing water losses and drought events.	Los Aromos Dam is situated in Section 3 of the basin at Limache, with a volume of 35 million m^3. As the government body responsible for the construction of major water infrastucture, the DOH is the main istitutional partner for the Aconcagua Project.
Aconcagua Project Major infrastructural project to improve irrigation security in Region V.	The Aconcagua Project was first officially presented in 2001, with the purpose of constructing a dam at Puntilla del Viento, along with two other dams, and a battery of wells that would recharge from the dam in order to improve the irrigation security and allow for water transfers to La Ligua. For a more detailed discussion of the project refer to Box 10.4 in Sect. 10.2.3.

(continued)

Table 10.1 (continued)

Case area	Scale	Institution/governance mechanism/adaptive action	Explanation/example
	Basin	Mesa Technica de Aconcagua Regional round table for interested stakeholders for the Aconcagua Project.	Informal group of stakeholders in the basin, mainly agricultural actors, that is lobbying for the construction of the Aconcagua Project. The group consists of hydrologists, farmers, representatives from the DOH, as well as CODELCO and COLBUN.
		Turno between Sections The turno is the process employed during drought periods to manage a proportional reduction of water rights and distribution of water to different users on a specific daily schedule.	The system is a management mechanisms to reduce the amount of water each rights holder receives, while attempting to ensure that all the rights holders do receive a proportion of their water rights. The turno is divided into shifts, which last a certain number of days. Turno occur between for example, if there are five zones to a canal, each zone will receive water for 7 days. So the first zone is allocated water for 1 week, and then the next 4 weeks they receive no water. During the time that it doesn't receive water, the zone is dry, and the other zones are allocated water in their shifts.
	Local	Junta de Vigilancia (Vigilance Committee) and comprising Association de Canalistas (Canal Association) These user based associations manage water allocation within and between sections of the river according to property rights and availability. They are responsible for coordinating proportional reduction amongst the members of the junta or canal association based on daily monitoring of the level of river flow.	Private negotiation is common between user associations across different sections of the river, and amongst the different sectoral actors (agriculture & water supply). The managers of the Juntas negotiate with the DGA to intervener not intervene in the management of the basin.

The turno is the process employed during drought periods to manage a proportional reduction of water rights and distribution of water to different users on a specific daily schedule.	receive water for 7 days. So the first zone is allocated water for 1 week, and then for the next 4 weeks they receive no water. During the time that it doesn't receive water, the zone is dry, and the other zones are allocated water in their shifts.
Farm/Company level adaptations	
Crop Choice – At the farm level, some of the Juntas encourage diversification of crops during more extreme periods. Once planted, the avocado trees are permanent features of the agricultural landscape and have a larger requirement for irrigation. About 70% trees are avocado, the rest are orange, lemons and mandarins. A smaller percentage of farmers grow vegetables, which are much more flexible.	One method is to warn farmers in advance of the planting season, that the next summer is likely to be dry allowing them to not plant certain vegetable and plants and so reducing their overall demand for water.
Extraction of underground water, for which farmers may or may not have rights.	Farmers also adapt to the reduction in supply (from the turno or not) by relying on wells used to extract underground water (whether or not they have rights to it). Therefore, when surface water dries, they can just dig down to extract the groundwater.
Water Transfers from Aconcagua River to drier areas of the region.	Water transfers have been conducted between the Aconcagua River to La Ligua, Limache and Marga-Marga using trucks to transport the water across during drought periods.

Table 10.2 Overview of adaptive actions and major institutions, governance and management mechanisms affecting adaptive capacity identified in the Swiss case

Rhône basin, Canton Valais Switzerland	Federal	1992 Federal Water Protection Act (WPA, Art 29) Residual flows now quantitatively protect flowing waters; efforts to reduce impacts from hydro-peaking; legislative guidance on sustainable management of water resources	The updated water law passed in 1992 provides for the enhancement of ecological services and the sustainability of waterways. It led to new policy guidance on flood management, that has been implemented in the plan for the TRC. However, there are no overarching principles on managing water conflict or stress.
		Bundesverordnung für Trinkwasserversorgung in Notzeiten/ Federal Ordinance on Drinking Water Provision in Emergency Periods (1995) The Federal Government proposes an organisational structure to deal with drinking water in any kind of crisis, i.e. drought, earthquakes, war, floods. It is an ordinance of the WPA.	This has yet to be implemented in the Canton Valais. The canton is supposed to develop a 'wasserkarte/atlas' (a water map), which is a separate requirement to the hazard maps, but the Valais has not done this yet. The ordinance provides guidelines as to how to maintain water provision during disaster periods, but does not set guidelines as to priorities during drought periods.
		Wasser Agenda 21 (WA21) Informal cross sector collaborative body on water issues in Switzerland with multiple committees	This informal institution represents a cross-sector collaboration that aims to provide guidance and support for adapting to the increasing stresses facing water resources management in Switzerland.

	NFA – Subsidies	
	Subsidies are linked to the implementation of federal legislative principles (environmental and participative provisions) through the Neugestaltung des Finanzausgleich (Reorganisation of Financial Equalisation/Compensation between the federal and cantonal governments)	Communes can receive greater subsidies for projects from federal and cantonal funds if they meet certain criteria: Participative planning (2%), Integrated Risk Management (4%), Ecological Aspects (2%) and Technical Aspects (2%). A building or river works project for example could request higher funding if the communes show that the project was planned in a participative manner and that ecological improvements have been made over 5 years.
Regional (Cantonal)	Cantonal Law on Utilisation of Hydropower (Art 42) Legal provision for irrigation rights to take priority over residual flows and hydropower concessions from 1 April–30 September	Art. 42 On the Use of Irrigation Canals: The diversion of irrigation water through water canals and conduits that belong to the 'Geteilschaften' or 'Grundeigentümern' rights remain reserved. Where the particular use or concession does not stipulate otherwise, these diversions (i.e. non irrigation) are restricted between 1 April and 30 September.
	Third Rhône Correction (TRC) 169CHF million project to reinforce flood security in the Rhône valley, with a number of sub-objectives to enhance social, ecological and economic security and well-being.	New integrative and uncertainty based approach to flood management, increased space for the Rhône, emergency evacuation channels, linked to spacial planning.

(continued)

Table 10.2 (continued)

	MINERVE The project stands for 'Modélisation des Intempéries de Nature Extrême, des Retenues Valaisannes et de leurs Effets' which is the initiative by the canton to improve the modelling of extreme events and natural hazards, their retention and effects. The project represents a public – private partnership between the Canton Valais, l'Ecole Polytechnique Fédérale de Lausanne (EPFL), Météosuisse, as well as a selection of hydropower companies.	EPFL and Météosuisse were engaged in order to develop the information management system, while the hydropower operators have been in negotiations with the canton to secure extra storage in hydropower reservoirs as a means of limiting the impact of flooding events and other natural hazards. At the prior knowledge of a heavy precipitation event, hydropower companies are warned so that the reservoirs (Speicherbäche) can be automatically lowered to buffer the volume of precipitation that comes down.
	Cantonal subsidies and support Investment in infrastructure and irrigation technology by the cantonal authority for agriculture (Amt für Strukturverbesserungen, Dienstelle für Landwirtschaft).	The cantonal office for agriculture is encouraging regional farmers to move irrigation technology towards drop irrigation, to enable a 30–40% reduction in water use. Shifts in technology in the vineyards (where drop irrigation is increasingly used) has already reduced water use.
	Modeling and observation networks: Canton authorities are responsible for monitoring and evaluation networks in the Valais, but are vertically connected to MétéoSuisse as well as to the communes and hydropower companies to coordinate information exchange prior to and during extreme events.	Multi-stakeholder working groups supported through cantonal authorities and national level agencies and associations (WA21, Cantonal Agricultural Centre, Association of Utilities) to initiate scoping exercises on climate change impacts in the Valais. Integration of climate information into planning by hydropower companies but general apathy to future impacts in other sectors

Hazard Mapping Canton authority for river course management (DFSB) provide support, financial incentives and expertise to the communes so that the canton's hazard mapping exercise can be completed.	DFSB are responsible for coordinating the communes and the engineering offices in the production of these hazard maps, and underwriting their validity and publishing them (Art. 3 (1)) since the results of hazard mapping must be transposed into spatial plans (according to FPA). The local authorities are responsible for completing the mapping exercise and implementing its findings.
Cellule scientifique de crise (CERISE) The scientific cell for crises is composed of representatives from the cantonal office for forests and landscape, the DFSB/SRCE and the service for hydropower (service des forces hydrauliques – SFH) in order to aid the authorities in their decision making ability during flood and avalanche hazards. Communes are legally required to ensure the safety requirements, and the canton natural hazards division subsidises their work.	CERISE is split into summer and winter competencies and comprises canton geologists, river work engineers and the natural hazards division and is linked to MeteoSchweiz. Of the 23 regions in the Valais, so far, there are 3 regions where the process already functions. In heavy precipitation events, they need to measure the rainfall quantity in certain streams, but up until now, they have not done this. (http://www.vs.ch/Navig/navig.asp?MenuID=2410&Language=fr)

(continued)

Table 10.2 (continued)

	Local (Commune)	Municipal Provisions for use prioritisation Ad hoc municipal level directives for reducing water use in periods of drought. Municipal provisions prioritise historic flow rights to agricultural users as per the Law on the Use of Hydropower, Art. 42 (see above). Local agreements for use transference and priority setting (informal Bewilligung). Commune regulations (reglemente) determine when the utilities are able to decide to supply farmers with domestic water. While the priority is the population, when there is 'superfluous' water, farmers are able to be supplied during these dry periods.	The communes are responsible for prioritising the use of drinking water. In Visp and Les Bagnes directives were issued to stop non-essential domestic uses during 2003 heat wave. Fully sometimes impose restrictions (e.g. 1 day per week) during periods of water stress, when precipitation has been low, and run off is reduced. In extremely dry summers, water suppliers have supplied drinking water to the farmers through the fire hydrants as a temporary measure. The utilities may grant, when there is enough supply for the population, an informal authorisation or licence (Bewilligung) that allows farmers to use domestic supply for irrigation between 18.00 and 08.00 (e.g. in Zermatt, this has happened about every 4–5 years). In the high ski resorts, the ski piste operators have agreements with different hydro-power operators to supply water for the snow cannons. In Zermatt, the commune water supplier has supplied water (spring water) for snow making from the end of October to December, but this amount only stands at about 5% currently. In Verbier, the commune utility (Les Bagnes – SIB) supplies water for snow-making from the Lac de Louvie only during the period of October/November.

	Infrastructural Diversification & Integration Development of artificial lakes at higher altitudes to transport water down to villages during periods of high demand and manage scarcity areas. In addition, a number of 'Bergbahnen' have now also constructed their own large storage reservoirs (Speicherbecker), to be able to supply their own needs for artificial snow production, no longer needing drinking water for this purpose. Diversified sources: Domestic use comes mainly from aquifers, groundwater and springs, while hydropower use relies on glacier and snow melt, as do farmers, either from the irrigation canals or pumping from the rivers to supply the fields and meadows.	Currently, the utility in Zermatt can supply water for snow production between mid October to end of December as that is the time when there is sufficient water availability. After mid-December water needs to be supplied to the growing tourist population. In the preceding 10 years, more investment has been channeled into the construction of extra reservoirs and storage for snow production to avoid reliance on utility water. Examples of this can also be found in Montana, with the construction of Lac de Tseuzier. More and more water is needed for ensuring coverage of increasing numbers of slopes by mid/end of November. Competing uses rely on different sources of the hydrological system, so at present avoid direct competition. However, some farmers do irrigate their fields with water supplied from utilities, but in the 2003 summer drought, swapped from using drinking water to pumping water from the Rhône instead. In certain communes, there are measures to reduce the amount of irrigation water that comes from drinking water.

(continued)

Table 10.2 (continued)

	Commune level utilities Consolidating more communes into shared water utility service provision (increasing integration and connectivity as in Les Bagnes and Martigny) to better manage periods of high demand or scarcity – spreading water reserves across hydrologically diverse communes.	30–40 years ago the water resources of many of the alpine villages were separate, meaning that if there were villages with fewer springs or more vulnerable sources of water, they would have to deal with this alone. By connecting resources, and linking vulnerable villages (with perhaps only one spring) with other, more water abundant, villages, supply was secured across the geographic area.
	Water Canals (Suonen/Les Bisses) These irrigation canals form part of common property resource regimes that were initially developed and constructed in response to drier climate. The irrigation channels and associated water rights are governed through user groups and canals.	In certain communes, the original 'Geteilschaft/Gnossenschaft' are used to manage rivalries and govern the water system amongst farmers, but these tend to be more important in the mountain communes.
	Krisenstab (Emergency Working Group) to respond and plan for emergency situations.	Each commune has their own emergency working group that consists of police, firefighters, military and commune council who are responsible for planning and coordinating emergency situations in collaboration with the canton.

This definition therefore takes into account both proactive and preparatory adaptation as well as reactive and autonomous adaptation (Dovers and Hezri 2010; Engle 2010; Tompkins and Adger 2005). Actions and mechanisms included in this table represent legislation, or particular articles, regulation, policy frameworks or institutional actions (i.e. decisions or rules of user group associations) that provide guidance or mechanisms for drought or flood management, the prioritisation of users during particular peak periods (scarcity or high demand) and infrastructural adaptation to shifting hydrological patterns. While the Swiss case area covers adaptive mechanisms relating to both flooding and scarcity situations, the Chilean examples pertain only to drought and scarcity. The definition is deliberately broad and evades an exclusive linkage to climate change impacts since other studies have highlighted the difficulty in separating 'pressures exerted as a result of climate change from other economic, environmental or developmental pressures' (Tompkins and Adger 2004, p 564).

Across the two cases areas, adaptive actions ranged from historical coping techniques to legal prescriptions for prioritising uses in periods of scarcity to more radical policy reform. Unsurprisingly, the mechanisms for dealing with drought and flooding were very different, but lessons can be drawn from the institutional processes that allow for these mechanisms to be implemented. Other studies (e.g. NeWater) comparing adaptation across case studies experiencing flood or drought impacts have noted that flooding tends to illicit more advanced strategies (Huntjens et al. 2011).

The NeWater project suggests that this may be explained partly by different risk perceptions (Green et al. 2007) and the difference in available solutions to the two extremes, which itself is related to the unique natures of the different extremes. Huntjens et al. (2011) posits that flooding is primarily a safety concern, while drought management concerns water scarcity and allocation management problems. The suggestion seems to be that drought and scarcity issues can be seen as more polemic and divisive than flood management issues, with less potential technical and management fixes available. While the adaptive actions across the Chilean and Swiss studies are quite different, interestingly, the nature of the Swiss flood management solutions can be seen to be as polemic as those of the drought issues within the Chilean studies, on which the following sections will go into more detail.

10.2 Characterising Adaptive Responses

Adaptive responses in each of the case areas were categorised according to the concepts of transformation, persistent adaptation and passive change, as discussed in Part I. By categorising the responses in terms of these categories, it allowed a linkage to be established between governance mechanisms that allowed for more sustainable and resilient approaches to water management solutions and those that fostered responses that might not build adaptive capacity or even degrade resilience in the face of increasing stresses and uncertainty. In addition, in order to characterise the governance elements that were associated with different categories of adaptive responses, a mixed methods analysis was conducted in MAX QDA to identify the intersections between the different response categories and the governance related indicators (under three broad categories of *Regime*, *Knowledge* and *Networks*) to establish which governance mechanisms were most associated with different categories of response.

The following tables represented in Figs. 10.1 and 10.2 present the analysis and subsequent results of this coding exercise, which show a higher concentration of transformative and persistent adaptive responses in the Swiss case area than in the Chilean case area. The results of the coding exercise of the adaptive responses will be presented and discussed below, in conjunction with the governance indicators related to them.

10.2.1 Transformative Adaptation

Transformational responses were classed as those that exhibited examples of *innovation, and possibly transformation of SES into trajectories that sustain and enhance ecosystem services, societal development and human well-being* (Folke et al. 2010). Transformability has also been described as the 'capacity to create a fundamentally new system when ecological, economic, or social structure makes the existing system untenable' (Walker et al. 2004). Adaptive responses were coded as 'transformation' if they exhibited traits of managing for uncertainty (i.e. practices and policies that prepare for uncertainty in context of climate change or inter-annual variability, including unanticipated changes), or if they showed that policy makers for/and water managers were searching for alternative governance or management practices that integrate ecological and social consideration, or had signs of innovation and development of new strategies that enhance ecological and social aspects as well as economic. A full list of criteria is given in the tables in Figs. 10.1 and 10.2, which show that of the adaptive responses, very few exhibited characteristics of transformation. Within the Chilean case, none of the responses exhibited transformative characteristics, while in Switzerland only the TRC and MINERVE had transformative attributes, and only the TRC could be categorised as 'transformation' (albeit with limits).

Initially, a scoping study was conducted to assess the most vulnerable areas, in terms of potential flooding events and damages (e.g. industrial sites and residential zones). In addition, the concept of 'residual risk' has been applied to the scoping studies for the project in order to meet the challenge of designing a project that incorporates the uncertainty of climate impacts (modifications in flows) so that the management plan can adapt to changing hydrological parameters. Engineers recognised that the project would need to find a means of incorporating hydro-climatic uncertainties to ensure that the project could statistically calculate projected levels of flow for individual sections under climate change conditions. Engineers are currently assessing the possibility of the calculated levels of flow being exceeded, and

Analytical categories		Adaptive Action / Policy Response						
		Presidential Declaration of Drought Zones	Irrigation Efficiency	Water Data System Imprövement	Aconcagua Project	Water Transfers	Turno	Farm level adaptations
Descriptive	Responsible Institution	DGA	CNR	DGA	DOH	ESVAL	Junta de Vigilance	
Descriptive	Description	Declares of a drought zone for a maximum of 6 months; DGA may intervene on request in the management of the river and redistribute the water; extraction of superficial or subterranean waters without rights allowed; human consumption of water prioritised.	Provides financing and subsidies to farmers to improve irrgiation efficiency; administered and implemented by the CNR.	New government prioritised the improvement of the public information system for water rights, including transparency; better information is seen to enable better managemnet of resources in increasing scarcity.	Initated in 2001, with the purpose of constructing a dam at Puntilla del Viento, along with 2 other dams, and a battery of wells that would recharge from the dam in order to improve the irrigation sectutity and allow for water transfers to La Ligua.	Water transfers have been conducted between the Aconcagua River to La Ligua, Limache and Marga-Marga using trucks to transport the water across during drought periods.	Process employed during drought periods to manage a proportional reduction of water rights and distribution of water to different users on a specific daily shedule	Crop Choice Extraction of underground water
Descriptive	Legal Baseline	Water Code, 1981 (Art. 314) Resolution 39 (1984)	Law or Promoting Irrigation Efficiency (Law 14.450)	Water Code (Cadaster Public de Agua)	Water Code (Infrastructure development)	Water Code	Traditional Method	Traditional Methods
Descriptive	Status of Implementation	In effect	In effect - but loosing institutional support	Policy priotiy, not yet completed	Not yet implemented	Ad hoc adaptations	Ad hoc adaptations	Ad hoc adaptations
Transformative Adaptation	Paradigm Shift: Altered way of thinking incorporated into management measures or regulatory regime	No	No	No	No	No	No	No
Transformative Adaptation	Climate change & uncertainty being addressed (inter-annual variability; unanticipated changes)	No - Resolution 39 establishes the 3 criteria for the DGA regional office to call a period of drought based on data from 1984.	No	No	No	No	No	Yes
Transformative Adaptation	Incorporation of criteria for sustaining and enhancing ecosystem services, societal development and human well being	Limited - domestic supply prioritised	No	No	No	Only in that it prioritses water for domestic use	No	No
Persistent Adaptation	Recognise uncertainty and climate related drivers in adjusting responses to changing circumstances	No	No	Yes		No	Yes	Yes
Persistent Adaptation	Social Infrastructure: Development of Actor Network; Knowledge base	No	No	Yes	No - but it has led to the development of the Mesa Technica de Aconcagua	No	Partly - initiates period of increased and intense communication and cooperation between sections and canals for risks sharing	No
Persistent Adaptation	Efficiency & Conservation	No - focus on exploiting groundwater resources in drought	Yes	No	No	No	Partly - signals to farmers period where water resources need to be conserved	Yes
Passive	Technical Infrastructure	Yes	Yes	No	Yes	Yes	Yes	Yes
Passive	Focus on up-scaling past techniques that may still degrade the SES	Yes	Yes	No	Yes	Yes	Yes	Yes
	Category	PASSIVE	PASSIVE with elements of P.A.	PERSISTANT ADAPTATION	PASSIVE	PASSIVE	PERSISTANT with elements of P	PERSISTANT with elements of P

Fig. 10.1 Overview coding of Chilean adaptive responses

Analytical categories		Adaptive Action / Policy Response						
		TRC	MINERVE	KRIZ/CERISE Krisenstab	Suonen / Les Bisses	Municipal Priority Setting	Informal Use Transfers	Integration/Consolidation
Descriptive	Responsible Institution	TRC Coordination Commission	Canton du Valais	Commune Council	Water Associations	Commune Council	Commune Council / Utilities	Commune Council / Utilities
	Description	169CHF million project to reinforce flood security in the Rhone valley, with a number of sub-objectives to enhance social, ecological and economic security and well-being.	Canton iniative (linked to TRC) to improve the modelling of extreme events and natural hazards, their retention and effects; includes EPFL, MeteoSchweiz, and a group of hydropower companies	Early warning system to respond to flood & avalanche hazards; comprises canton geologists, river work engineers, natural hazards division, with MeteoSchweiz; links with commune krisenstab	Common property resource regimes that were initially developed and constructed in response to drier climate; still used to manage rivalries and govern the water system amongst farmers	Ad hoc municipal level directives for reducing water use in periods of drought; municipal provisions prioritise historic flow rights to agricultural users.	Local agreements for use transference and priority setting; commune regulations determine when the utilities are able to decide to supply farmers with domestic water	Consolidating more communes into shared water utility service provision to better manage periods of high demand or scarcity by spreading water reserves across hydrologically diverse communes.
	Legal Baseline	721.100 Federal Law on management of watercourss (1991); 721.1 Cantonal Law on management of watercourses (2007)		Art. 31 kRPG	Commune rights registers	Art. 42 Law on the Use of Hydropower; Commune private contracts and agreements	Commune contracts and agreements	Inter-commune contracts and agreements
	Status of Implementation	First Implementation Phase	Implementation Phase 2008-2011	3 out of 23 regions in VS functioning	Traditional Method - In effect since 13th Century	In effect	In effect	
Transformative Adaptation	Paradigm Shift: Altered way of thinking incorporated into management measures or regulatory regime	YES Structural change in watercourse management legislation	No	No	No	No	No	No
	Climate change & uncertainty being addressed (inter-annual variability; unanticipated changes)	Yes - buffers for future flows, and iterative planning process in light of uncertainty in climate impacts.	Yes	No	No	No	Yes	No
	Incorporation of criteria for sustaining and enhancing ecosystem services, societal development and human well being	Yes Ecological resilience of the floodplain incorporated into implementation plan	No	No	Yes	No	No	No
Persistent Adaptation	Recognise uncertainty and climate related drivers in adjusting responses to changing circumstances	Yes	Yes	Yes	Yes	No	Yes	Yes
	Social Infrastructure: Development of Actor Network; Knowledge base	Yes - develops knowledge based, while participative process is expected anyway	Yes - New stakeholders and networks being developed	Yes	Yes		No	Yes
	Efficiency & Conservation	No	No	No	Yes	Yes	No	Yes
Passive	Technical Infrastructure	Yes	No	No	Yes	No	Yes	Yes
	Focus on up-scaling past techniques that may still degrade the SES	Mixed - compromise has been made in implementation phase that still employs less innovative techniques	No	No	No	No	Yes	No
	Category of change	TRANSFORMATIVE Policy priorities are potentially diluted in implementation	PERSISTANT with elements of T	PERSISTANT Focus is on developing knowledge base and networks	PERSISTANT	PERSISTANT WITH Elements of P	PASSIVE With elements of P A	PERSSISTANT

Fig. 10.2 Overview coding of Swiss adaptive responses

Box 10.1 A Brief History of the Rhône Corrections in Canton Valais (Source: www.vs.ch/Rhône)

The Rhône Valley has historically been a site of numerous natural hazards most notably flooding, avalanches and landslides. Prior to the first correction, there had been numerous attempts to control the river in order to protect both inhabitants and infrastructure. However, an overall co-ordinated plan had escaped these efforts, limiting their efficacy across the broader region. After a devastating flood in 1860 that impacted on the entire valley, the first programme of corrective interventions took place on the Rhône River between 1863 and 1893. The results were, however, disappointing. Despite the dredging of the river, levels continued to rise. The second correction took place between 1930 and 1960, reinforcing and raising the level of the dikes that had been constructed in the first correction. The works were accompanied by an increased exploitation of the river bed, with companies beginning to remove the gravel for commercial purposes but also assisting in flood protection.

The third correction planning phase was announced in 2008, in a process that is set to take between 25 and 30 years. After a series of high flooding events in the 1980s and 1990s (1987, 1993, 2000), it became clear that the engineering corrections of the first and second corrections were no longer enough to protect the growing towns in the Valais from the increasing amounts of precipitation and melt that were having devastating impacts, and that a new tactic was needed. While the 2000 flooding event was a turning point, cantonal authorities had identified the need for a new set of corrections well before that catastrophe. The 1992 change to the federal law signaled a move to a more integrated approach to water management, including flood management. The 169 million CHF provided by the federal government for the TRC represents 65% of the project's costs, more than they are legally bound to by the WBG (45%), because the costs of potential damage and required measures are so comprehensive. The federal guidelines for the project's financing require it to integrate climate change uncertainties and to take account of social-ecological benefits, rather than just technical security and economic priorities.

where supplementary corridors of evacuation (*couloirs d'evacuation*) may be situated in order to manage this eventuality. The TRC project team therefore proposed an interpretation of the legal basis to the federal office, in which they proposed that for the project to be fair and balanced (i.e. fulfilling its security function and enhancing the ecology of the Rhône floodplain) the Rhône should be enlarged to roughly twice its current size.

However, while the TRC has transformational characteristics in its policy formation, the challenges of passing its implementation plan at the local level has led to a dilution of those attributes, which aim to enhance the ecological and social benefits of the projects. While the highly participative process in itself would be considered to be requisite for a transformational governance approach, in this case it also shows the deep challenges that it may also bring, to the detriment of innovation. Once the priority measures were identified, an interpretation of the legal baselines was conducted to draw up an implementation plan which identified the level to which it was possible to achieve a two times enlargement of the Rhône, or where the enlargement needed to be scaled back in order to take account of occupied land by urban zone or industry. The compromise was to complete a security enlargement of 1.6 times the current river size, with a more consequential 2 times enlargement in other areas. It is hoped that this will meet both federal demands, but reduce disruptions to more heavily urbanised and industrialised areas.

Windows of opportunity generated by successive extreme events that surpassed past management practices and technologies have limited currency when policy and engineering innovation meets the reality of implementation within a physically (urban and industrial reality) and socially (land rights and perceptions) constrained reality. The participative process of project implementation allows for the integration of these different voices, to ensure a socially equitable solution, but yet may result in a dilution of the principles that allow for greater resilience to climate uncertainty.

In Chile, despite the availability of scientific information on climate change impacts and sector impact studies, adaptive actions did not account for uncertainty in the context of climate change (particularly given that the development of inter-annual variability in relation to ENSO – El Niño/La Niña events is currently one of the areas of climate science with the largest amount of uncertainty). Environmental impacts and the ecological integrity of the social-ecological system are not considered outside of economic parameters, and innovation is generally low, with a reliance on classic technical fixes of large scale dam storage and increased groundwater exploitation. While water conservation and efficiency improvements are active or within the scope of governmental bodies (DGA focus on efficiency), the CNR is at present scaling back its irrigation efficiency programmes (notably to focus more on improving information and transparency of the water market). Yet, within the current framework conditions, these programmes do not lead to reductions in water use, but the expansion of supply or irrigation with the water that is conserved.

10.2.1.1 Associated Governance Mechanisms

Regime

A mix of federal legislation (WBG, Federal Policy Directives) and cantonal legislation (Valais WBG) set the framework for the most transformational elements of the TRC (integration of uncertainty and climate information, integrated risk management based on social-ecological resilience). Environmental provisions within these laws and environmental goals integrated into subsidy programmes such as the NFA, attempt to direct water resources management projects and agreements on water resources in a in a direction that would allow transformation of the SES onto a trajectory that could sustain and enhance ecosystem services, societal development and human well-being (WBG, Art 4 (2), Valais WBG, Art 14)[1] (Valais 2009).

Knowledge

Planning time horizons were shown to be insufficient for current challenges, therefore longer term horizons were set, shifting the planning focus to more iterative and integrative and uncertainty (variable risk) based strategies. This was enabled through a diverse range of impact studies and multi-stakeholder investigations to allow for compromise and balance in the project. The integration of climate change adjusted risk and uncertainty into planning was deemed necessary to ensure that time would not be wasted in the future by having to redo the management plan (reflecting the understanding that present day hydrology might not reflect future patterns). Planning and scoping was therefore forward looking, acknowledging that current levels of flows may be surpassed in the future. Within the project itself, the enlargement (by 2 or 1.6 times) signifies redundancy being built into the system (Valais 2009).

An element of flexibility needed to be incorporated into the TRC plan to deal with this, so that the technical experts, rather than politicians, can define the planning process, but overall objectives are set in a top down manner from federal and canton levels (but their strength and closeness to interpretation is negotiated at the local level). Scientific and technical monitoring and modelling are relied upon to diagnose vulnerabilities, and communication programmes tend to translate the outcome studies into justifications for the project with local level stakeholders. Sustainability criteria are integrated into financial incentive criteria and thereby are positively linked with project objectives.

Networks

The distributed legal structure (i.e. canton and federal law) allows for negotiation between canton and federal levels to find a balance in the implementation of legal provisions which encourage a 'sustainability' led approach that matches both federal guidelines and local realities. Reliance on federal financial support allows the federal (more transformative approach) to have some power, but regional particularities and needs are accounted for through the decentralised implementation structure, which in turn is influenced by local autonomy, land rights holders and water owners. Each scale has its own source of power and agency (federal: legislative provisions, financial capability; canton: subsidiarity of Implementation as a constitutional right, technical

[1] Refer to the Management Plan of the TRC for more discussion on the acceptance that absolute security against flooding was no longer an option, and the development of the legal framework for the management of watercourse in conjunction to this shift in thinking. Available online at: http://www.vs.ch/Navig/navig.asp?MenuID=16521

expertise, some financial power over the communes; local: right to local autonomy, water sovereignty, land rights) leading to an extenuated impasse in passing the implementation plan, but the potential to negotiate a common, integrated solution.

10.2.2 Persistent Adaptation

Responses that allow for the 'persistence of the fundamental properties of the current system through adaptation' (Chapin et al. 2009, p 20) were classified as 'persistent adaptation', to distinguish it from *transformative adaptation*. Adaptive responses were coded as *persistent adaptation* if they exhibited aspects of technical or governance innovation, which while they may not be transformative in terms of fostering SES resilience, it still introduced new, innovative approaches to decision making or water resource management. Examples of governance innovation might be the attempt to generate new or enhanced knowledge or partnerships for addressing resource challenges. Examples of technical innovation could relate to the development of new techniques or improvements to irrigation efficiency (new irrigation technologies or efficiency gains through infrastructure maintenance and repair), or hard path infrastructural solutions for scarcity, drought, rivalries or flooding, that also incorporated aspects of uncertainty relating to climate change.

Adaptation is deemed to be a manifestation of adaptive capacity, notably as a means of reducing vulnerability to present stresses and future impacts (Smit and Wandel 2006). However, this form of adaptive behaviour is more associated with means of 'coping' with climate variability rather than shaping responses to climate change that improves resilience of the SES, and adapting to the changes in physical parameters of the system. Boxes 10.2 and 10.3 below highlight two of the responses that were categorised as *persistent adaptation* responses according to the criteria above: the Turno from Chile and the MINERVE project in the Swiss case. The Aconcagua project also meets certain criteria of *persistent adaptation*, but the focus on steady state hydrology and the lack of integration of climate based uncertainty projections into the scoping plans, means that it was more heavily weighted as a *passive* response, and so shall be discussed later.

MINERVE represents a governance innovation in the knowledge network that frames the cantonal response system to extreme hydrological events. It incorporates a number of transformational characteristics in its fundamental integration of uncertainty based science and cross-sector partnerships for knowledge sharing in the public-private partnership. However, since it represents an innovation in only the information system for improving response to extreme events, it does not have the more transformative characteristics of shaping the broader resilience of the SES.

Box 10.2 Turno in Chilean Irrigation Systems (Source: Interviews; Alvarez (2005))

The turno is the process employed during drought periods to manage a proportional reduction of water rights and distribution of water to different users on a specific daily schedule. The system of turno is a very old traditional Spanish system of proportionally reducing the amount of water each rights holder receives, but attempting to ensure that all the rights holders receive a proportion of their water rights. The turno is divided into shifts, which lasts a certain number of days. For example, if there are five zones to a canal, each zone will receive water for 7 days. So the first zone is allocated water for 1 week, and then the next 4 weeks they receive no water. During the time that it doesn't receive water, the zone is dry, and the other zones are allocated water in their shifts. The turno can take place between a single sub-canal, across a number of canals within one section of the river, or across multiple sections of the river. It tends to highlight the latent power imbalances between upstream and downstream water users, as well as those with stronger and weaker rights within a single canal or section of the river (Alvarez 2005).

Box 10.3 Integrated and Coordinated Action Against Flooding: MINERVE (Source: http://www.aqueduc.info)

The project stands for 'Modélisation des Intempéries de Nature Extrême, des Retenues Valaisannes et de leurs Effets' which is the initiative by the canton to improve the modelling of extreme events and natural hazards, their retention and effects. The project represents a public – private partnership between the Canton Valais, l'Ecole Polytechnique Fédérale de Lausanne (EPFL), Météosuisse, as well as a selection of hydropower companies. In 2002, EPFL and Météosuisse were engaged in order to develop the information management system. Drawing on meteorological forecasts from MétéoSuisse, the system calculates the flow rates up to 72 h in advance, in order to provide water managers with enough time to put into effect anticipatory actions that should minimise flood damage.

The project also takes into account both the impacts on hydropower operations and reservoirs, as well as the potential role that they might play in adaptation to increasing numbers of extreme events. The aim is to develop multi-use infrastructure through innovative partnership techniques. Hydropower operators have been in negotiations with the canton to secure extra storage in hydropower

(continued)

Box 10.3 (continued)

reservoirs as a means of limiting the impacts of flooding events and other natural hazards. When there is prior knowledge of a heavy precipitation event, hydropower companies are warned so that the reservoirs (*Speicherbäche*) can be automatically lowered to buffer the volume of precipitation.
In addition, MINERVE acts to mobilise a number of competencies to respond collaboratively to extreme events. These include meteorologists, hydrologists, information technicians, dam operators, security services (police etc.), as well as vertically coordinating decision making between the canton and communes. Over the 3 days of the 2000 flooding event, there was a degree of uncertainty with the canton over who was responsible for deciding on management steps to protect areas below some of the reservoirs. The lack of oversight and preparatory steps to manage over-flow of the reservoirs was seen as having been potentially counter-productive, perhaps even exacerbating some of the impacts lower down the valley. Now, in a selection of reservoirs across the canton, at the signal that there will be an extreme precipitation event, the reservoirs can be automatically pumped out and lowered, in order to then buffer the volume of precipitation that comes down.

It is also worth mentioning the institutional component of the Suonen/Bisses in this category as well. Cantonal Authorities have recognised that these common property regimes have played an important role in building solidarity and managing conflict resolution (Netting 1981), and therefore have made efforts at both commune and canton levels to support and encourage the maintenance of these organisations since they assist in the upkeep of the infrastructure and minimise costs at the local level. Federal and cantonal administrative levels provide financial support for the CPRs by subsidising infrastructural maintenance projects. The rest of the costs are covered by the commune and whatever then remains must be covered by the landowner, despite some, who are no longer farmers, being hostile to covering the costs for irrigation installations.

10.2.2.1 Associated Governance Mechanisms

Regime

Mechanisms (legal provisions; informal agreements) that allow for emergency drought responses to kick into action provide a clear signal to actors that a different set of parameters have been reached, and so prepare the path to set coping strategies that replace normal 'day to day' management. In Chile, the drought provisions provided in the Water Code, signal that farmers may start negotiating emergency short term exploitation of groundwater to enable irrigation to take place as surface waters diminish. The emergency drought provisions allow for the flexible and provisional use of alternative water sources (wells/groundwater) as a means of short term coping, but also provide protection for groundwater rights holders who may hold the DGA liable for any affectation of third party rights. However, the effectiveness of the declaration is limited to the level of government financial assistance that would allow farmers to actually exploit the water resources to which they've been granted temporary access.

Adaptation at the Junta de Vigilancia and Canalista level is characterised by the Turno, which enables farmers to quickly shift to an alternative water distribution model. The model of temporary coping allows for the proportional reduction of water rights distribution based on different 'shifts' or 'turns', aims to minimise drought impacts across the basin (Box 10.2). In Switzerland, company and association agreements are in place between different actors (e.g. commune utility and farmers; commune utility and cable car companies; hydropower utilities and cable car companies) for short term adaptation of water supply for irrigation and artificial snow production. In addition there are commune level regulations on water provision during emergency times that provide guidelines for supply and sanitation in extreme events, but there are no overarching rules on scarcity or drought.

Knowledge

In Chile, the initiation of declaring the drought zone is guided by an internal technical regulation of the DGA, which sets the hydrological parameters by which drought should be declared. However, the present regional DGA office deems these parameters, and the data that informs them, to be out of date, and no longer relevant to the decision criteria for which it is needed. Despite challenges in the breadth and transparency of state monitoring and assessment, mechanisms are in place at the channel and junta level to evaluate the amount of water every day and proportionally reduce allocations during times of stress. Private actors also are open to learning from other areas and seeking government support for diversification and technical adaptation options as a potential means of coping with climate change impacts. Government actors have the technical capacity to carry out and use research on climate change impact across the water intense sectors. Increasing attention is being paid by government bodies (DGA, DOH and CNR) on improving the state and transparency of hydrological assessment and water rights information to build capacity for managing increasingly scarce water resources, as well as to inform policies such as a National Dam Policy (hard infrastructural adaptation).

In the Swiss case, improvements are being planned and implemented for local level monitoring on run off and water quality from increased precipitation as part of the reaction plans for coping with impacts on quality and quantity (in relation to extreme precipitation events). Flood management planning takes into account the likelihood of increasing water risks from climate change as prescribed by the top down regional planning concept that involves both federal and cantonal levels.

Inventories of water infrastructure take place to inform redevelopment of diversity (Suonen/Canals) as a means of maintaining traditional infrastructure that minimises impacts in heavy precipitation events. Early warning monitoring networks and response systems are already in place for many other hazards across the canton, and are being improved specifically for increases in precipitation events related to changing climatic conditions.

In both cases, the awareness of impending climate change impacts drives actor's perception of the need to find solutions for the challenges it will bring. In the Chilean case, this is however often accompanied by the perception that more water needs to be captured so that less is lost to the sea. In the Swiss case, there is an awareness of quality, quantity and seasonality changes from climate change, as well as dichotomy of extremes (i.e. glacier reduction but more extremes) in the intensification of the hydrological cycle, particularly among the more technocratic hydropower stakeholders. Recent experiences of major floods and precipitation events had led to a high awareness of this intensification and thus implementation of technical protection measures (after 1993 event) that protected them in the 2000 events, but which are already seen as redundant according to current observational data. In both case areas, technical fixes are seen as the main or only means (irrigation efficiency, crop efficiency, irrigation networks, storage capacity/dams) of facing climate change challenges. In Chile, however, attention has turned to the importance of improving market transparency and information so that it may operate better. This is not seen as an adaptation measure, but as a means of improving the baseline administration of water resources management to be better prepared for increasing droughts and pressure on scarce water resources.

Networks

In the Chilean case, cooperation for coping takes place amongst private rights holders through formalised user based institutions. Rights owners are enabled to take responsibility to ensure 'coping' in times of stress, through institutional mechanisms for canal based adaptations (i.e. Turno). While the Presidential declaration of a drought zone provides for increased involvement and connection between user level and administration level, the incentives for cooperation between actors remain fraught. The declaration is seen not to bring the financial capacity for investment in alternative groundwater wells that are needed for increased exploitation during the 6 month period, indicating that without the government's financial assistance, its increased involvement in the management of drought is extraneous.

In the Swiss case, public-private partnerships (government, university, private hydro companies) allow for information and burden sharing to improve protection from flood damage. The partnerships enable collaboration across regional, canton and local (private) and commune (public) actors. Knowledge networks link local and regional managers with research institutions (private and public) and universities, so that scientific information informs watercourse management. Cross sectoral collaborations are in place to improve service and efficiency in the face of novel challenges, for which expertise may not be at hand at the local level. Specifically in MINERVE, there has been a transition from informal collaboration and assistance to a formalisation of the process and agreement. At the local level, communal institutions that are redundant during 'normal periods' (e.g. *Kristenstab*) mobilise quickly to impending extreme events. These flexible institutions contain both private and public actors. While the canton level provides coordination in extreme events, freedom and autonomy persists at the local level.

10.2.3 Passive

In addition to the two categories of adaptive responses, a third category was utilised capture responses that contributed to the degradation of the system to a less favourable state, resulting from either a failure to transform and adapt (Chapin et al. 2009, p 20) or maladaptation. Responses were coded as 'passive' if they adhered to concepts of steady state resource management, impasses in planning and project process with no scope for resolution, or adaptation that further degraded either the social or the ecological system. Responses that were categorised as passive included the Aconcagua Project because although it is a project that has a climate adaptation element to it (managing storage does not necessarily imply maladaptation), it has been proposed purely in the name of irrigation efficiency and its planning is based on steady state principles that do not integrate the potential impacts that climate change may have on the validity of the project.

The Aconcagua project is seen by many agricultural stakeholders as the only means for enhancing the capacity of the system to cope with increasingly dry periods, hence the level of frustration that negotiations have run for 10 years without any resolution. Stakeholders often referred to the loss of water to the sea throughout the winter period and higher periods of precipitation.

Box 10.4 Water Resources Planning: The Aconcagua Project (Source: Presentations by DOH, DGA & Agricultural Stakeholders at Universidad Catolica de Valparaiso, Quillota; (Matta 2011)).

The Aconcagua Project is a major infrastructural project that has been in planning and negotiation for the past 10 years. It is projected to provide irrigation security to existing cropland, while also enabling farmers to increase irrigated area (30 million ha) through enhanced security of their rights. The project is to build a reservoir (Puntilla del Viento) with a capacity of 110 million m^3 together with a battery of wells in the areas of Curimón, Panquehue and Llay Llay. The wells would be relied upon only in order to manage periods of

(continued)

Box 10.4 (continued)

drought. Without the present regulating works, farmers see themselves as losing water to the sea in winter time, which then cannot be used for irrigation purposes in spring time. Currently, the aquifer is being used as an underground reservoir, pumped during periods when river flows are too low to supply rights allocation.

The major impediment to the implementation of the project is a disagreement between agricultural stakeholders with the DOH and the DGA over the availability of water rights for filling up the reservoir. The DGA is under pressure to allow the plan for the dam to be approved, but posits that as there are no more available rights in the Aconcagua Basin, irrigators themselves must use their own rights to stock the dam. The DOH has 400 million m^3 of eventual rights, yet as detailed earlier, these cannot be transformed into permanent rights through infrastructure, but the aim of the dam is to give security to permanent water rights.

10.2.3.1 Associated Governance Mechanisms

Regime

The drought declaration in the Chilean case enables actors to cope in part by allowing the exploitation of 'vulnerable' ground water sources. Additionally, the informality of the Chilean governance approach in 'normal' periods leads to a lack of capacity and knowledge of the river when the 'external' DGA takes over at the most critical moment. This leads to wasted time and conflict possibilities heightened because of the government intervention. However, the intervention of the DGA is still seen as a necessary last resort. In the Swiss case, while legal guidelines exist for the management of increasing flooding issues (governmental policy guidance) there is a void of guidance and rules on scarcity or stress.

Knowledge

The Aconcagua Project is defined by criteria adhering to steady state resource management, since there is no accounting for uncertainty, nor incorporating inter-annual variability (i.e. ENSO), nor the integration of climate change related uncertainties into the project scoping phase. There is a lack of alternative options proposed, and ideological constraints persist, which limit the ability to experiment with alternative solutions. The private adaptations at the canal and river level are reactive measures, and there is a lack of planning that would enable more proactive preparation. The DGA intervention in the river implies a loss of knowledge, since government actors lack the capacity and familiarity of the basin as water management is usually in the hands of private actors.

The lack of agreement and coherence across different evaluations and assessments of the hydrological resources available in the Aconcagua, severely limits the ability of both public agencies and private actors to agree on plans for the development of management and infrastructure in the basin. There is a strong awareness amongst water owners that hydrological patterns are shifting, but as yet this has not translated to enhanced use of technology, monitoring, modelling or integration of uncertainty into the management and planning of water resources in the basin. The ideological rigidity of the water market and Water Code not only informs the adversity to change the framework rules which govern the current system but also constricts and narrows actors' views of how to resolve the complex problems that have been emerging. The Swiss case lacks preparedness and planning for possible scarcity situations in the area of water supply. This is in part due to the perception of climate change as an issue to be taken into account for long term horizon planning (30–40 years) but not yet for operational day to day management. While there is an acceptance and awareness of the inevitability of increasing impacts in flooding and natural disasters, awareness on other impacts of climate change related to water availability remains less engrained. Despite this, there is still awareness amongst technical experts that precipitation patterns are changing and that legal mechanisms for drought are no longer up to date.

Networks

In the Chilean case, the lack of trust between actors is a major impediment towards fostering common integrated solutions to common problems. The impasse over the Aconcagua project has lasted for 10 years for example. Furthermore, the DGA perceives that the agricultural actors have strategically used legal mechanisms such drought provision as a means of forcing the DGA's hand on groundwater exploitation. At the ministerial level, the power imbalances between different ministries and government institutions (mining, energy, agriculture versus environment and water) has so far continued to side line the environment and weaker economic actors in water resource management, limiting the scope for innovation for enhanced SES resilience through cross-sector collaboration and cooperation. At the basin level, public-private sector cooperation has taken place within the realm of the Aconcagua Project, as well as between Junta de Vigilancia and individual companies. Private negotiation of this sort is reported to take the form of financial pay-offs (an alternate version of polluter pays, which does not lead to less pollution, but just an acceptance of it), while multi-sector cooperation in the Mesa Tecnica has not as yet led to a resolution on the project or to a solution being found.

References

Alvarez P (2005) Un itinéraire de l'eau, Approche géographique et agronomique d'une gestion de l'irrigation en zone aride du Chili. Université de Orléans, Orléans

Chapin FS, Folke C, Kofinas GP (2009) A framework for understanding change. In: Chapin FS, Kofinas GP, Folke C (eds) Principles of ecosystem stewardship resilience-based natural resource management in a changing world. Springer, New York

Dovers SR, Hezri AA (2010) Institutions and policy processes: the means to the ends of adaptation. Wiley Interdis Rev Clim Chang 1(2):212–231

Engle NL (2010) Adaptation to extreme droughts in Arizona, Georgia, and South Carolina: evaluating adaptive capacity and innovative planning and management approaches for states and their community water systems. University of Michigan, Ann Arbor

Folke C, Carpenter SR, Walker B, Scheffer M, Chapin T, Rockström J (2010) Resilience Thinking: Integrating Resilience, Adaptability and Transformability. Ecol Soc 14(4):20. [online] URL: http://www.ecologyandsociety.org/vol15/iss24/art20/

Green CH, Tunstall SM, Fordham MH (2007) The risks from flooding: which risks and whose perception? Disasters 15:227–236

Huntjens P, Pahl-Wostl C, Rihoux B, Schlüter M, Flachner Z, Neto S, Koskova R, Dickens C, Nabide Kiti I (2011) Adaptive water management and policy learning in a changing climate: a formal comparative analysis of eight water management regimes in Europe, Africa and Asia. Environ Policy Gov 21:145–163

Matta S (2011) El Proyecto Aconcagua. El Observador, Tuesday, 26 July

Netting RM (1981) Balancing on an Alp: ecological change and continuity in a Swiss mountain community. Cambridge University Press, Cambridge

Smit B, Wandel J (2006) Adaptation, adaptive capacity and vulnerability. Glob Environ Chang 16(3):282–292

Tompkins EL, Adger WN (2004) Does adaptive management of natural resources enhance resilience to climate change? Ecol Soc 9(2):10 [online]. doi:Artn 10. URL: http://www.ecologyandsociety.org/vol19/iss12/art10

Tompkins EL, Adger WN (2005) Defining response capacity to enhance climate change policy. Environ Sci Policy 8:562–571

Valais (2009) Plan d'aménagement de la 3ème correction du Rhône – PA-R3. http://www.vs.ch/Navig/navig.asp?MenuID=16520&RefMenuID=0&RefServiceID=0. Accessed 28 July 2009

Walker B, Holling CS, Carpenter SR, Kinzig A (2004) Resilience, adaptability and transformability in social–ecological systems. Ecol Soc 9 (2):5. [online] URL: http://www.ecologyandsociety.org/vol9/iss2/art5/

Chapter 11
Bridges and Barriers to Adaptive Capacity

Abstract This chapter has presents and discusses the bridges and barriers identified across the different governance scales and cases. The analysis of bridges and barriers to adaptation in relation to extreme events allows for a better understanding of adaptive capacity to be built, primarily through the identification of a set of favourable conditions for fostering adaptive capacity. Becoming aware of what the bridges and barriers are, even those that may seem immalleable can be seen as a first step in defining key factors that can transform barriers into factors that can transform capacity into action. While the analysis focusses on bridges or barriers to specific needs of adaptation and adaptive capacity building, it also drew on findings from the governance assessments and concepts such as IWRM as a starting point from which adaptation to climate related aggravations could be better navigated. Common bridges and barriers related to *autonomy and structure, information and capacity, as well as integration and cooperation.*

Keywords Rhône, Canton Valais, Switzerland • Aconcagua, Region V, Chile • Analysis of bridges and barriers to adaptation • Favourable conditions for adaptive capacity

11.1 Bridges and Barriers in Studies of Adaptive Capacity

In the context of the methodology chosen (i.e. extreme events as proxies for future climate change impacts to which actors and the system responds) context, barriers and bridges were defined not only by coding structural elements of the response according to normative indications, but also by eliciting actors' perceptions on what aspects of the system aided or impeded solutions and adaptive responses to the extreme event, and thus future climate change impacts. Other studies have outlined broad themes for bridges and barriers to adaptation (Crabbé and Robin 2006; Dovers and Hezri 2010; Moser and Eckstrom 2010; Burch 2010). Researchers have

M. Hill, *Climate Change and Water Governance: Adaptive Capacity in Chile and Switzerland*, Advances in Global Change Research 54,
DOI 10.1007/978-94-007-5796-7_11, © Springer Science+Business Media Dordrecht 2013

investigated bridges and barriers at different stages or phases of the adaptation process (Moser and Eckstrom 2010), across both mitigation and adaption response (Burch 2010) or at different scales of adaptation (Crabbé and Robin 2006). Dovers & Hezri (2010) in their summary of barriers included the following categories; physical and ecological, financial, informational and cognitive, social and cultural (e.g. short termism, ignorance).

Often these bridges and barriers are used interchangeably with common indicators of adaptive capacity, by looking at what can stop, delay, hinder, or help actors at different stages of the adaptation process (Moser and Eckstrom 2010). While indicators of adaptive capacity may be framed as bridges in some studies, the inverse may be seen as barriers. A difference in framing specific issues as bridges or barriers rather than indicators is that it allows a more nuanced identification of specific lever points that could be exploited to make targeted interventions to develop more integrative and adaptive responses in water management. Becoming aware of what the bridges and barriers are, even those that may seem imaleable, can also be seen as a first step in defining key factors that can transform barriers into 'enablers of the translation of capacity into *action*' (Burch 2010, p 290).

The investigation of mechanisms for dealing with hydro-climatic extremes allowed an analysis of the factors that enabled or hindered the mobilisation of capacity to respond and adapt to the hydrological events. Other studies that have investigated barriers to mobilising effective responses to climate change have also recognised the role played by shocks to the system, by providing opportunities to re-orient institutional pathways (Burch 2010; Folke et al. 2002, 2010; Herrfahrdt-Pähle 2010; Gelcich et al. 2010). However, according to Pierson (2004), the 'seeds for a particular (i.e. sustainable, or climate change resilient) path must be planted prior to the shock in order for that path to merge during the readjustment period that follows the *shock*' (Burch 2010, p 292). While the analysis focusses on barriers to specific needs of adaptation and adaptive capacity building, it also drew on findings from the governance assessments and concepts such as IWRM as a starting point from which adaptation to climate related aggravations could be better navigated.

11.2 Bridges and Barriers Across Scales

Tables 11.1 and 11.2 present some of the key bridges and barriers to adaptation in each of the case areas with respect to relevant extreme events examined. The results are drawn from the coding and analysing of interviews, which identified points made about the challenges and ability of the systems to respond to both the case extreme events and climate change impacts. The tables represent the bridges and barriers across the different governance scales, rather than according to categories employed by other studies such as regulatory, operational, technical and behavioural (Burch 2009, 2010; Yohe 2001). Since the focus was on identifying the challenges posed at different levels and to different scales of change, rather than across the determinants of adaptive capacity most often cited in the literature (Keskitalo et al. 2010; Smit

Table 11.1 Detailed qualitative analysis of key barriers and bridges to water management solutions for adaptive capacity building across different scales in the Chilean case

	Major barriers and bridges to adaptation: Chile	
Scale	Barriers	Bridges
National level	Informality, weakness and impotency of ministry and regulatory bodies (Ministry of Public Works and General Directorate (DGA) of Water, Ministry of Environment etc.) to regulate water uses and enforce weak protection provisions, leading to environmental degradation and over-exploitation of water resources	Promotion and development of efficient use of water in the agricultural sector, through government support and subsidies
	Informality of governance approach means weakened NGO's tend to take up the role of defencing against environmental destruction, enforcement of protection provisions and developing more long term concepts for sustainable development and watershed management	Through the Water Code, the legislative framework has provided a level of legal certainty on water rights that has successfully funneled investment into both water services and provision (coverage is at between 83 and 100 %) and contributed to the country's economic development, thereby improving the state's ability to finance a number of public services and infrastructural investments related to water
	Lack of institutional capacity to improve levels of transparency and information on water rights, which could serve as the basis of a planning strategy for long term climate impacts	Glacier monitoring programme in place in the DGA represents an initial awareness of the need to improve information and monitoring on climate change impacts
	Challenge in developing a culture of water conservation and awareness around water scarcity	Growing consciousness about climate change as an issue that will need more attention, even if at the moment from a very technical perspective (i.e. no attention paid to institutional adaptation)
	Economic priorities and ideology of water as an economic good for development (commoditised through water rights) trumps environmental concerns and the integration of ecosystem health as a water stakeholder, undermining SES resilience in the face of longer term uncertainty and change.	Awareness and interest in moving towards IWRM, but focus on hard infrastructural mechanisms as a means to do this (i.e. dam building)
		Independent research bodies (e.g. UN ECLAC) conduct studies to isolate factors that have led to successes elsewhere in Chilean water management (e.g. drinking water provision), signalling a potential to link these self –assessing factors to adaptations to climate change

(continued)

Table 11.1 (continued)

Scale	Major barriers and bridges to adaptation: Chile	
	Barriers	Bridges
	Objectivity/Technical versus Political/Judicial/Subjectivity and co-option of the system (Dams, Mining, Factor de Uso)	Government directions and incentives for efficiency improvements in agriculture has been successful – showing that a model of government subsidies and programmes can lead to more optimal water management (just needs a shift from pure economic to SES focussed outcomes)
	Lack of trust between different government institutions – lack of trust in decision making ability of judiciary from a technical/operational perspective	
	Independence/Objectivity/Consistency of research partnerships (independent consultants & universities which are funded by major economic groups), reliance on international experts (WB etc.),	Increasing attention and operational power being given to environmental sector, as CONAMA was recently changed to MMA – 2008/2009 there was a CONAMA strategy to implement a sustainable water resource strategy at the river basin level (however, this has apparently fallen by the wayside); environmental courts now also have a legislative grounding but have yet to be implemented (refer to Chap. 8).
	Legislative Framework (Water Law, Energy Law, Mining Law) is weighted to favour economic exploitation of water resources over striking a balance between different uses 'corporate capture of water'	DGA Director and officials recognise the need to promote efficiency and improve environmental aspects of water management and to achieve 'a better use of the resource'.
	Lack of data, transparency and enforcement capacity to support changes in the law (factor de uso) that were supposed to increase efficiency of water rights use, but instead have led to increased over exploitation of a number of basins – decreasing resilience of SES in the face of increasing droughts and uncertainty	Evidence of institutional cooperation – (Cooperation convention between CNR and DGA to create a unit to improve transparency and monitoring of water rights – CNR is sending some funds so that they can contract people etc.)

	No long term visions/plan for a number of the most water relevant institutions (e.g. CNR) for how to cope with potential impacts from climate change	Growing number and strength of NGO initiatives which serve to counter balance the pure economic interests in favour for more SES oriented policies; non-formal/public education initiatives also for top decision makers
	Lack of clarity around vision & priority setting for decision making that hampers adaptation decisions around issues such as allocations in increasing uncertain conditions, drought situations; how to prioritise uses under increasing uncertainty – transference of water management to private actors means there is less of an ability to have a unified perspective on how to deal with larger, more complex, issues such as climate change	External experts (World Bank, GTZ etc.) assist in improving understanding and solution finding for water resource management (e.g. GTZ WFD expert hired to work on MMA's National Strategy for River Basin Management)
	The water code differentiates between surface and groundwaters – lack of framework for subterranean waters, from both legal and institutional perspective	Ad hoc conferences from CNR on climate change impacts awareness raising for policy makers, decision makers and agricultural stakeholders (though not integrated systemically into planning, evaluation etc.)
	Previous CNR led programmes for efficiency improvements have now stopped – but replaced by a focus on increasing transparency of rights situation	Judicial Process – at the national level in the supreme court, there is a judge specialised in the water code
Region level	Highly centralised governance structure builds tensions between political and ministerial levels and the technical and operational functions at the basin level; also, technical experts are left frustrated by slow moving and politically led decision making on planning large infrastructural projects	Regional level technical and operational experts' cooperation functions more effectively than political cooperation at the national level; also, technical expertise is considered to be high, including use of modelling for climate change impact information

(continued)

Table 11.1 (continued)

Scale	Major barriers and bridges to adaptation: Chile	
	Barriers	Bridges
	Technical guidelines for declaration of drought are seen as unclear, confusing, and obsolete yet basin managers are limited in the ability to change the legal guidelines; also, there is a mis-match between current hydro-climatic reality and data upon which rights allocation is based; unclear about meaning of certain sections of the provisions (e.g. constructed wells placed at the disposition of irrigators)	Leadership – recognition that technical improvements in monitoring and assessment need to be made to help DGA manage water resource more effectively (i.e. monitoring should be more than 4x per year – but should be permanent, constant and online)
	Transference of water management to private actors (informality of the regulation, e.g., DGAs lack of power to regulate illegal water use) and institutional fragmentation (across DGA, Ministry of Environment, Courts and DOH) means there is less of an ability to generate a unified perspective on planning for longer term large scale issues, such as climate change impacts; mistrust & citizens' perception that often DGA and other authority's intervention, undermine the accuracy of water management further hamper ability to provide coherent basin based guidance for new challenges	At the basin level, there has been cross-sector collaboration between agricultural stakeholders and mining companies, driven by corporate social responsibility (CSR) interests of the mining companies to improve management of water resources in certain basins in Chile (e.g. Valle del Choapa, Limari) – the Aconcagua informal agreements were reported between individual Juntas and mining companies in the upper watershed (mainly seemed to be payoffs for mining impacts on water quality)
	Lack of incentives for water conservation from private water utility company perspective – tariff is renegotiated every 5 years, therefore any profit from efficiency gains are lost in the next 5 year period. Focus on efficiency from pure profit perspective narrows vision and limits ability to prepare for alternative challenges – companies are still focussed on core business of selling more water and increasing consumption – but simultaneously have under invested in infrastructure maintenance (so costs are rising for consumers, while the private companies are experiencing more water losses due to deteriorating infrastructure) – (UN ECLAC)	Emergency drought provisions allow the DGA to potentially quickly call a drought situation and potentially secure fair water allocation and prioritise water uses, to guarantee water allocations to domestic use as well as irrigation, also attempts to reduce the harmful impacts on agriculture by reducing the amount of water for irrigation rather than suspending it altogether;

Unchecked economic development and expansion of irrigation and mining (reportedly impacting the glacier in the upper basin) pressurising increasingly vulnerable water resources Informality of water management approach (unregularised rights and high level of illegal abstraction of groundwater) Pressure from economic actors to open up rights to aquifer, which the DGA calculate to be unsustainably used No regional network or institution responsible for water quality and quantity monitoring – ad hoc studies are outsourced to private companies – leading to a lack of consistency and agreement on status of health and amount of water resources within the basin – no regional evaluation of illegal water usage in the basin – no baseline to make longer term plans/vision for the basin Low institutional capacity and lack of information means that DGA can take up to 5 years to process a water right claim (requests should take no more than 6 months) Institutional fragmentation means that different (but linked) decision affecting water resources are taken either in DOH or DGA also mistrust pits DOH against DGA, in that DOH is perceived to 'collude' with the irrigators, while DGA attempts to police the water rights situation – so that they are not held liable for water losses during drought situations once they have intervened – also causing tensions between the two institutions on the Aconcagua Project, failing to find a resolution to the rights situation due to conflicting studies and opposing views of management of the basin.	DOH led projects which aim to mitigate droughts during scarcity seasons In other basins, Mesa del Agua were created, and a Mesa Tecnica has been established for the Aconcagua Project in order to improve cooperation. Court presence in conflict resolution and water resource management is deemed a welcome necessity, despite frustrations with the process, speed and lack of expertise in water resource issues, since it provides a buffer to potential corruption within governmental institutions (that is perceived as common place in Latin American countries) Institutional potential in the DGA, despite lack of capacity to fulfil role. Aquifer sustainability, surface and ground water sustainability, are part of the DGA mandate. The concept that aquifers are a non-renewable resource, require an analysis of the impact of each request on the sustainability of water resources. Criteria and iterative evaluation for the sustainable use of aquifers are used by the DGA – supposedly linked with permanent/definitive rights assignment – additionally ecological flows do need to be taken into account when new rights are granted (only new rights left in the basin are for groundwater) At the regional level, DGA plans to design and implement special plans of auditing & action; training for users' organisations; registration of authorised abstractions – in order to have certainty about legal and illegal abstraction points; effective systems of abstraction controls, with the obligation to inform the authority; coordination and training with the public attorney, to develop better control and investigation, in cases of infractions and water theft

(continued)

Table 11.1 (continued)

Scale	Major barriers and bridges to adaptation: Chile	
	Barriers	Bridges
	Separation in function between water quality monitoring and setting up water quality rules and standards (e.g. DGA responsible for water quality monitoring, but MMA responsible for water quality)	
	Conflicts and conflicting studies slow down the planning process – Political games played around 'technical' issues – farmers/DOH pressuring DGA to open up the wells (through the drought declaration) so that they can prove third party rights are not effected, and therefore can pave the way for the Aconcagua Project to be agreed upon.	
	Lack of conservation culture or government incentives or programmes to develop conservation culture (not just communication campaign, but the need for water to be more expensive)	
	At the basin level, convergence of disparity in power between economic actors (e.g. mining versus agriculture), lack of institution to protect glaciers and lack of transparency on projects to develop mines in the high Andes, means that mining activates are impacting glaciers and therefore stream flow/storage without accountability or recourse against negative impacts	
	The right of use is the property of the users (Water Code, Art. 5; Chilean Constitution, Art 18 (24); Civil Code, Art 589), but DGA has technical criteria of sustainability, but these are internal directives, not legal provisions and therefore a legal obligation	
	Despite investment at the farm level (irrigation techniques etc.), investment at the river basin level is poor	

Local/User level	Unchecked economic/irrigation growth outpaced water management infrastructure (~ 20,000 ha increase in irrigated land, but no increase in regulating infrastructure 'we are behind by about 50 years') – incremental increase in potential stress situations which cc will aggravate – but no resolution on the horizon from infrastructure point of view. Major perception that the dam will be the silver bullet for irrigators to manage drought (even that they need more than one dam) Lack of conservation mentality (i.e. water efficiency to save water) – incentives for farmers are to increase efficiency for expansion of irrigation, not to decrease water use (and costs); users do not pay volumetric fee for water, but charges for distribution and infrastructure – no associated 'cost of water' to incentivise water efficiency Illegal abstractions are common place (particularly in Section 2 of the basin), impeding fair and proportionate distribution to the rights; additionally, illegal abstraction and use of groundwater (illegal wells) is seen as the solution to droughts (mal-adaptation) Difficulty in enforcement of illegal activities (illegal abstraction; dumping of trash in water ways) at the municipality level (i.e. requesting an inspector to enforce), which further stresses water ways during drought periods; mistrust towards DGA and DOH to enforce regulations and protect water resources (e.g. DOH decisions to allow excessive extraction of gravel from river bed, increasing vulnerability in times of drought)	Water rights framework means the user-based management approach is seen to be flexible for speedy adaptations in times of increased scarcity (ref Tourno, proportional distribution etc.) Local sensitivity and experience to drought periods tends to be high, despite lack of formal training, local farmers know when to restrict the use, and what the necessary measures are when they 'know' it will be a hydrologically scarce growing season (said to be in their 'DNA') Technical solutions to better management of drought are potentially on the table (mix of regulation works (dam) and unified water intakes to avoid water losses and improvement of water distribution so that it is more exact) Potential to mobilise cooperation around specific projects or threats, despite lack of collaboration across different economic sectors in general water resources/river basin management (e.g. mining companies are more interested now in being involved with the Aconcagua Project – earlier they were sceptical, but more recently have become more open to the Junta's interest in coordination and therefore need to understand their current and future potential water use and timing of water use)

(continued)

Table 11.1 (continued)

Scale	Major barriers and bridges to adaptation: Chile	
	Barriers	Bridges
	Institutional informality (i.e. there may or may not be a Junta de Vigilancia, it may or may not function properly/be legalised, there may or may not be cooperation between the different Juntas) heightens upstream – downstream rivalries and can lead to unproportional distribution – power imbalances and lack of cooperation reduce the ability to find user based solutions where there are not major winners and losers (and everyone losing in the end) – reducing resilience of the SES in general Strong adversity against charging for water/economic incentives for increased efficiency and conservation (not just to increase efficiency then exploit more) Privately negotiated agreements and informal collaborations and negotiations are vulnerable to inconsistencies and renegements leading to high mistrust amongst user groups and a lack of confidence in agreements settled on – yet these are one of the key adaptive actions! Lack of investment and resources to pay for infrastructure maintenance at canal user level meaning water losses exacerbate drought impacts; additionally, farmers are expanding plantation areas, without having sufficient rights, aggravated by losses in the canals themselves	Informal legal situation of Juntas can lead to more incentive to cooperate with other Juntas in order to avoid conflict, though in the long run this is not necessarily an option. Individual user groups do have incentives to cooperate (mining, hydropower and agriculture); additionally, actors across the different Junta de Vigilancia see it in their interests to negotiate and cooperate with the Juntas further upstream and downstream – the aim is to avoid conflict since as entrepreneurs, they see it in their interest to invest in efficiency and good relationships

Judicial processes (legal formalisation of Juntas, dispute settlement over water rights etc.) are seen as too formalistic, complicated, time consuming and expensive for water rights holders (court decisions can take up to 10 years for sentences) – lack of a flexible mechanism for conflict resolution is likely to heighten tensions in more severe and recurrent droughts
Lack of transparency and clarity around registered water rights (and supply and prices of water rights transactions) data intensifies challenges for canal associations to administrate rights and water usage, also lack of clarity on rights and ownership of water, reduces the ability to understand how to manage the effects of climate change on these rights, hampering adaptation planning
Rights administration within user associations lack effective and consistent monitoring and assessment and modelling techniques from which to iteratively plan water distribution

Table 11.2 Detailed qualitative analysis of key barriers and bridges to water management solutions for adaptive capacity building across different scales in the Swiss case

Scale	Major barriers and bridges to adaptation: Switzerland	
	Barriers	Bridges
Federal level	Limited authority of the Federal government on environment and water due to Federalism; also, resistance and aversion at lower levels to federal level intervention weakens enforcement powers of federal level legal provisions and policy guidelines for sustainable water management No legislative or regulatory concept of managing water resources at the watershed level, and rules or ordinances for management of conflicts and scarcity are only rudimentary, since there have not been many instances of water conflicts to date Federal legislation across water relevant acts (WPA, FPA, EPA, Hydropower Laws etc.) are neither integrated nor aligned, so while the systems are connected, the laws affecting them are not Enforcement of federal legislation, particularly in relation to water quality impacts from nitrates and pesticides are not very consequent and coherent; additionally, the water quality provisions in WPA are not aligned with the precautionary principle (e.g. as REACH laws are) but mainly focussed on end of pipe solutions	Strong connections with different actors (universities, research institutes, and stakeholder platforms) and levels through research partnerships on climate change and other developing trends Tactical use of financial incentives and subsidies through the Neue Finanz Ausgleich to encourage the implementation of federal legal provisions for sustainable water management practices, water efficiency improvements, along with increased participation and the ecological improvement of water management (infrastructural) projects for the increased resilience of waterways Capacity of non-state actors (environmental NGOs etc.) is strong, with strong legal instrument to influence law and decision making (Verbandbeschwerderecht), which has allowed for the integration of longer term ecological and climate related issues to be taken into account in both legislation and federal policies Integration of ecological considerations into the 1991 WPA moved federal legislation away from a purely technical focus to one based on managing the resilience of the hydrological system (incl environmental flows and revitalization projects)

	Federal administration is perceived sceptically, particularly by the more autonomous and independent cantons (e.g. Valais, Graubunden, which also are mountain cantons) – increasingly the difficulty of consensus building for legislation, policies and plans which push for integration of environmental concerns and longer term challenges from climate change impacts; at the canton level, there can be little interest in a 'national overview' or any suggestions coming from Bern	Despite the sectoral components of different federal acts concerning water (WPA, FPA, EPA, Hydropower Laws), there is a strong 200 year old legal basis from which water is governed, which is supported by corresponding environmental legislation – strong basis which has over time been adapted to face new challenges (i.e. moving from protection against water, to use of water, to the protection of water)
	Strong reliance on financial incentives and subsidies to enforce legal provisions for environmental and social aspects of water management (e.g. in revitalization projects, if the extra money is not needed from the federal government, then there is no obligation for a more participative process that also integrates extra ecological aspects as requested in federal legislation – ref NFA)	Learning networks with federal research institutes and universities, where federal institutions can integrate cantonal institutions to research on water resource related issues (e.g. future irrigation requirements)
	Perceived lack of capacity, know how and preparation in hydropower sector for future challenges (climate change, EU integration etc)	Informal partnerships and knowledge networks (such as WA21) are coordinated from federal level research institutes (e.g. EAWAG, WSL) to foster learning and cooperation as well as cross sector collaboration for solutions to future challenges in water resource management (e.g. climate change impacts, further integration into EU frameworks for energy and water related legislation, rising energy demands etc.) – these institutes attempt to collaborate right down to the commune level (bridging organisations)
	Ecologisation of water resources management and change in federal philosophy towards a more integrated approach has alienated certain groups, in particular the agricultural lobby who perceive the federal government to be in collusion with environmental lobbyists	
Canton level	Institutional fragmentation across different cantonal level authorities and sectors means that there is little common coordination towards water issues; also, sectoral bodies tend to work bilaterally rather than systematically and while there are examples of relationship building and cooperation it is on an ad hoc basis rather than systematic	Crisis of flooding events and environmental degradation of waterways in the past decades opened up opportunity for both canton and federal institutions to shift to longer term more uncertainty based forward planning and awareness of potential future shocks and the implementation of new concepts (e.g., Third Rhône Correction), which encourage flexible and robust approaches that embrace integrated risk management

(continued)

Table 11.2 (continued)

Scale	Major barriers and bridges to adaptation: Switzerland	
	Barriers	Bridges
	Fragmentation of natural hazard related activities are spread across different canton institutions (DSFB and the DWL), reducing possibility for an integrated strategy for long term protection, application of principles of precaution and causality and a uniform concept of safety – duplication of effort increase time and cost inefficiencies	Physical vulnerability and shocks also created an opportunity for non-state actors (e.g. hydropower companies) to formalise their ad hoc adaptations (managing the hydropower reservoirs to absorb the excess flows during high precipitation events) into more permanent and formal institutional strategy in collaboration with the canton authorities
	Lack of leadership in the cantonal administration on water resources management, despite attempts by individuals to better integrate different water related offices in the cantonal administration.	Strong networks between public and private actors integrating research on climate change impacts and other potential future problems into planning decisions, as well as public – private partnerships (e.g., MINERVE) to improve adaptation responses across the canton
	Strength of local autonomy and water sovereignty at the municipal level means that the canton is powerless (politically and legally) to generate canton (and basin) wide plans for water management, generating challenges in the implementation and enforcement of legal provisions at the local level (i.e. delay of implementation of WPA Art. 80 from 2007 to 2012)	Those aspects of the federal law that are priorities are given sufficient resources to be well implemented
	Institutional fragmentation of legislation and incentives create challenges for basin planning and cohesion of plans across different sectors and communes (ref KEV incentives for Trinkwasseranlage Turbinierung Projects)	Projects initiated through national research programmes (Nationalfond & Interegg) to build up the monitoring network for springs/sources (Quellmessestationen)
	Monitoring and assessment on glacial melt is sparse and deemed to be inadequate; as with other areas of Switzerland there is a relative lack of hydrological data for the Valais, especially in relation to precipitation events	Awareness of tertiary climate impacts (precipitation changes impact on likelihood for forest fires in the Valais – creating need for sufficient water storage for protecting against forest fires) leading to the development of plans/concepts by the canton to ensure sufficient water supplies at the local level

Lack of capacity (financial and time constraints) to carry out longer term assessment of data concerning river levels to assess glacial melt impacts on water balance throughout the summer season (i.e. to establish whether the hydrological balance still mirrors provisions in the law for irrigation use); lack of capacity in cantonal administration also provides challenges for the enforcement and implementation of federal laws, and the canton can only define a set of priorities	Self-assessment on whether the institutional structures in place in the Valais are 'fit for purpose' not only for water management challenges now, but it the future (leadership & Centre de Competence de l'Eau)
Focus on the Rhône Corrections meant that other areas of river and water management were not given as close attention and individual projects tended to replace more comprehensive resource planning	Emergency monitoring and reaction units (CERISE) create linkages between canton level and commune levels. While canton level institutions fulfil monitoring and warning roles, liaison with commune level leaders allows responses to be implemented quickly at the local level (evacuation if necessary), but with the coordination and additional informational capacity at the canton level
Enforcement of regulations and legislation within the communes is difficult for the canton	Despite sectoral organisation, and lack of general coordination, attitudes to cooperation are seen as generally good, allowing common solutions to be found to water management challenges and common visions to new challenges to be drawn up at the canton level
Institutional fragmentation of different aspects of water quality monitoring – while the canton is responsible for organising the monitoring of the groundwater (vorkommen) in the Rhône valley, the communes, as owners of the sources/springs (quellen) are themselves responsible for the (Ausscheidung der Quellschutzzonen) monitoring of the quality of the springs (Qualitätsüberwachung)	Plans in place at canton level law makers (Kantonale Gewässersanierungs Plan) to restore and clean up waterways in a poor ecological state due to hydropower activities, to increase ecological resilience. The types of water withdrawals which are affected by the law are clearly defined in the law (WPA Art 80). Details of requirements and status of implementation are available online
Limitations in ability to implement full spectrum of environmental and social considerations in the TRC, due to constraints from past planning decisions (i.e. no space for the river) and opposition to non-traditional flood protection measures (i.e. not just digging deeper into the river bed and strengthening dykes)	Priority projects (TRC) enjoy privileged and direct links to canton leadership (Conseil d'Etat), coordination with the adjoining partner Canton (Vaud) as well as commune representatives, creating both top down and bottom up links

(continued)

Table 11.2 (continued)

Scale	Major barriers and bridges to adaptation: Switzerland	
	Barriers	Bridges
	Length of time needed for application of the law in general can be problematic – e.g. implementation of Art 80 WPA was originally set for a15 year deadline (2007), however, this has now been extended to 2012 – due to lack of capacity at the cantonal administration within its environmental departments and strong autonomy of communes in the Valais Lack of long term holistic planning or investigation concerning multi-uses of the streams in relation to changing hydrological parameters The highly participative process (e.g. TRC) can be seen as tedious and ineffective, particularly when lacking needed leadership to navigate towards a common solution; frustrations also lie in the lack of expert knowledge from participants in certain cases EIA process – lack of knowledge and objectivity concerning EIA process being outsourced to private environmental consultancy companies that may have it in their interest to write up a favourable report for the companies that pay them – this is a regular occurrence with EIA process across many countries	Legal basis of TRC ensures that a participative process has to happen, and helps to ensure that a consensus is built earlier in implementation to avoid the risk of recourse later on – including the involvement of environmental organisations from the start of the project Recognition from private actors (e.g. hydropower operators) of the need to improve the ecological status of waterways, where possible, as required by the law Relative financial dependence of Valais on federal subsidies (one of the poorer cantons in Switzerland) counterbalances their relative autonomy and conservativeness to ensure that federal and cantonal resilience based provisions can still be implemented
Commune level	Strong local autonomy is not always matched by strong institutional capacity and technical expertise. Implementation of federal provisions for environmental flows causes tensions for energy, environmental and agricultural canton departments, because each commune wants to do their own activities and money from water taxes on the energy companies is an important source of income	Local autonomy means that municipalities can act according to their individual needs and geographies, developing local adaptive processes for periods of stress (e.g., municipal regulations, specialised water channels etc.) A sense of ownership over emergency response and land management plans is fostered due to autonomy and responsibility for implementation; also, the highly participative process ensures that over time a consensus is built, requiring considerable coordination and communication

Low accountability if canton perceives communes not to be meeting drinking water standards – challenge of implementing federal provisions such as the Hazard Mapping Concept – ShowMe Maps (federal database to show level and results of implementation) show that majority of communes have maps available, but relatively few have actually been implemented into the spatial planning in the Valais (www.bafu.admin.ch/naturgefahren – ShowMe) Local autonomy also weakens possibility for general coherent planning across small scale political boundaries and general planning 'concepts' are therefore lacking. There is little cross-commune collaboration and the focus is generally only on water payments. Lack of rules and guidelines on how to manage scarcity and rivalries together with the long time frame for hydropower concessions (80–100 years), which creates tensions between lack of long-term basin planning while being fixed into long-term inflexible agreements on water resources use Fragmentation of different levels of projects – e.g. flood management projects for the Rhône from the canton, and for the streams within the communes, means trying to coordinate with separate watercourse management projects that are already running Past planning decisions have enclosed the River Rhône and tributaries and closing off options to improve the ecological resilience of these rivers for integrated flood management – reducing future options to enhance SES resilience In the hydropower sector, future adaptation options are seen to be limited since there is no more 'water to be captured' – so when there is less water, there is less hydropower production	Historical adaptations (Suonen – irrigation system) to the dry conditions in the Valais (from the rain shadow effect of the Alps) has infrastructurally prepared the irrigation sector for a certain range of climate change Commune level planning for extreme events took place in the wake of past precipitation events – The Risiko Planning Hochwasser is a concept for when there are flooding events, including comprehensive information on the measures that need to happen in case of disasters (who they need to contact, population/organisations, evacuations, accommodation in the region/food supplies) so that they are well prepared for a flooding event Local level monitoring and inventories (Trockenweiseinventar; Hochwasserschutzkonzept) Strong support networks across sub-canton and commune levels, particularly for training and maintenance of both institutional and built infrastructure. Despite length and frustrations with participative planning process of TRC, it does allow all voices to be heard at the different local points for project implementation and thereby build a consensus across different local actors in an inclusive manner; lots of coordination is required which means that there has to be a good movement of information and finance between the different levels and that plans do need to be well justified to those that they will impact and affect – does lead to negotiation and input, not just consultation Relative lack of financial capacity at the local level means that deference to cantonal plans for the needed subsidies does assist implementation of ecological resilience based provisions in federal and cantonal law.

(continued)

Table 11.2 (continued)

	Major barriers and bridges to adaptation: Switzerland	
Scale	Barriers	Bridges
	Out-migration and shifts in main economic sector (e.g. from farming to working in industry, i.e. Lonza Factory) means that there are less people (with fewer financial resources as well) to maintain the infrastructure (irrigation system – Suonen) which contributed to the service of evacuating excess flows to the main rivers (e.g. Saasvispa, Mattavispa, Vispa), aggravating the level of damage from the precipitation events, as the water just could wildly flow off the meadows – an issue that is getting worse, as it becomes less and less viable for young people to become farmers in the region (convergence of less farmers and more frequent hazards)	Local knowledge and experience – Strong awareness of changes in local environment and climate impacts, which many have witnessed over their lifetimes
	Water sovereignty at communal level across cantons adds further complexity and challenges to the implementation of the TRC due to the rivalries between different communes (e.g. whether to harvest energy on one river or another), making it difficult to manage the different interests for each canton	Diversification across different communes to extend water connections and utilities beyond the village scale to multi-village scale
	Economic pressures at the local level weaken the implementation of residual flow provisions by the canton, meaning that often only the minimal is done, as leaving more water in the streams would mean that the communes would earn less	Influence of the tourist trade and resurgence of local farming products – organic produce etc., SES resilience and awareness
	Acrimonious disagreements at local level concerning implentation between agricultural groups and canton and ecological groups; high differentiation in participative process depending on commune and interests of local actors (i.e. reports of WWF being excluded from the process in some areas; denied access to planning documents, despite legal right to view them)	

et al. 2000; Yohe 2001; Adger et al. 2007), responses were coded and analysed according to the governance scale at which the problem was dominant.

Tables 11.1 and 11.2 present the different perceptions of stakeholders, as to the elements of the governance system that support or hinder effective water management or its ability to cope with climate variability and impacts. The specifics of the different bridges and barriers are clearly quite different across the two case areas. In Chile, there is a stronger pre-occupation with the help or hindrance that the water market provides at the national and regional level, due to the dominance of the Water Code in the governance system. However, at the local level, trust, enforcement and institutional capacity are great pre-occupations. In Switzerland, stakeholders across all levels concentrated more on the issues of local autonomy, including challenges and strategies related to the decentralised mode of governance, that is a barrier to the integration of water governance. However, synthesis across the cases does allow common themes to be identified that relate to the ability of decision makers (at different scales and sectors) to mobilise and facilitate responses to climate change or variability challenges. Across both cases and levels of governance, inter-jurisdictional issues (with associated challenges in institutional and technical capacity) and information and data are common themes across both bridges and barriers.

11.3 Common Barriers Across the Cases

A common challenge cited by stakeholders related to physical and environmental issues. More precisely problems of old, failing or insufficient infrastructure were common complaints for actors inability to adapt effectively to mounting hydrological challenges (i.e. lack of storage capacity, groundwater exploitation, and poor maintenance of irrigation canals in Chile; challenges in infrastructure maintenance for flood protection and irrigation in the Swiss case). Relatively small scale geographic differentials (i.e. water resources differing from one village to the next) and insurmountable impacts from climate change (i.e. tipping points on snow pack and glacier melt) were also cited commonly as major challenges for adaptation. These barriers relate directly to technical and engineering solutions to climate change adaptation and other water management issues, an area that has traditionally been the prime focus of both adaptation and resource management policies. However, the rest of the discussion centers on the elements of the social infrastructure that determine the decision making environment in which these adaptation decisions are made.

A common challenge across both case areas was the challenge of power imbalances and different levels of capacity across different governance levels. This is an issue that has been cited in other studies as a major barrier to climate change mitigation and adaptation (Burch 2010). In the Chilean case, the main issue at the national and regional water and environment agency levels was the friction between the technical and operational experts and the political sector. Despite high levels of technical knowledge and expertise (e.g. DGA glacier monitoring programme, highly trained engineers, economists etc.), the ability for operational professionals to apply their knowledge and expertise (i.e. setting rules for sustainable groundwater abstraction) is hampered by the inflexible legal framework and political dominance in water governance, since the separation between the political and operational elements of water resource governance is weak. Legally binding decisions at the Supreme Court level (i.e. 2004 Supreme Court decision on 'factor de uso') have been shown to lead to severe consequences at the basin level, for which there are now limited options to improve the situation, since the DGA is powerless to impact property rights.[1] Furthermore, the local courts have a role in water governance as the responsible body for conflict resolution, yet a number of local stakeholders commented on the lack of specific knowledge and proficient resolution that reduced the effectiveness of judicial decisions on water related cases.

In the Swiss case, the challenges of limited authority at federal and canton levels was highlighted across all scales of governance. For example, stakeholders at the regional and national levels pointed to the lack of canton oversight as a key barrier to coordinated planning and directives for water management during critical phases for water provision. Local autonomy and decentralisation means that the canton only plays a supporting role in conflict resolution and water governance. While it is the role of the canton to assist local level actors (i.e. managers of the canal capture points, water utilities etc.) in finding solutions to water provision during critical dry periods, it was noted that they cannot propose solutions, but only help them come to solutions. This was highlighted as both a bridge and a barrier, since while it enables local ownership of planning and issue resolution, it can detract from a coherent and coordinated strategy across the basin, and take a long time. Furthermore, the small scale political arrangement of water management in the Valais, heightens the difficulty for municipalities to coordinate uses and comprehensively plan for longer-term challenges.

In Chile, a pervasive issue was the difficulty of addressing mounting challenges in water resource management through the legal framework, despite the 2005 changes. Key issues were the inability to strike a balance between economic and environmental provisions, so that laws were not weighted to favour exploitation, but to improve incentives for conservation, and to improve the enforcement of environmental provisions and illegal exploitation. Stakeholders also pointed out that the length of time needed to make small changes to the law was incommensurate with

[1] Perhaps the best example of this comes from another Chilean basin, the Copiapo. 'Rio Copiapo, is a symbol for the existing water rights and water laws in Chile in that they are not able to solve this problem. So in the Copiapo, you have a very high level of groundwater extraction and the balance is very negative, up to 18,000,000 m^3 per year of over exploitation – and they cannot find a solution – that is legal extraction'. (Interview, MMA, November, 2010). Another example comes from the negotiation of water rights on the Huasco River between Barrick Good and the Juntas. Negotiations took place with the Junta rather than the individual farmers themselves. Juntas do not have the right to enter into an agreement on water rights, as they do not own them, so it is 'ultra vires'. However, the DGA in this case was unable to stop the agreement, as only the court has the power to do so, by which point individual actors may no longer have the capacity to bring a case forward. (Catholic University of Chile, November, 2010).

the effect, since changes only applied to new rights. In the Aconcagua, all water rights are already assigned. Furthermore, the Water Code establishes a highly informal approach, where 'management' of the resource is devolved to private use right owners. This has created a form of deadlock between this informality of approach, a lack of institutional power and a lack of incentives for actors to collaborate together to find common long term vision of how to prepare for climate change impacts.

Another aspect of this barrier is evident at the Chilean regional level, where the DGA is relatively powerless to manage the issue of un-regularised rights and illegal extraction. The DGA allows for the abstraction of water from wells without water rights, and recognises that this practice is increasing, but that it is not in their power to 'denunciate' illegal usage, only to track use and non-use of legal water rights. A private actor must report the illegal abstraction to another institution (Superintendencia/Judiciary). Additionally, the DGA does not have the ability to limit or reduce water rights during drought periods (although scarce water resources are monitored and portioned by individual user groups).

In the Chilean case, stakeholders noted a lack of consideration for climate change into short and long term evaluation and planning, despite an awareness of the issues that water related climate change impacts posed to different sectors. Despite government training programmes on increasing irrigation efficiency (from the CNR), there is no integration of efficiency with climate impacts on water resources or sustainable management. A major barrier in the Chilean case is the data upon which water allocations and management decisions are based. Despite having a long time series of hydrological data available to calculate water extraction in dry years, the water rights were initially adjusted to a series of wet years, which allowed for a high level of abstraction. A further issue is that there is now very little possibility to update the data upon which these allocations have been granted, since all water rights are allocated and protected by the constitution. Furthermore, while the DGA is responsible for the allocation of water rights, they do not have sufficient information and documentation on registered and unregistered rights to be able to effectively account for climate impacts on hydrological resources in the basin.

In Chile there is strong cultural opposition to a volumetric charge for water usage, particularly in the agricultural sector, where one stakeholder commented that it would be the 'death of agriculture'. The ideological strength of the market system, however, limits the potential for alternative solutions to be considered. One expert pointed to a lack of clarity among irrigators as to the benefits of an increase in water efficiency, such as the sale or lease of water rights, which highlights the absolute focus on maximum use of rights as opposed to any considerations for the long term sustainability of the SES and resilience in the face of climate change impacts.

In the Swiss case, the main barrier related to data and information was seen to be the high complexity of data collection and maintenance, which was a particular problem to the mountain communes. While paucity of data was cited, it is worth noting that the Swiss Alps have a much higher level of data available on water quantity and quality data than the Chilean Andes, or indeed many other areas. The challenge is perhaps more related to the number of different databases and multiple levels at which data is collected and maintained that can create difficulties for water managers who need to use the data in periods of extreme events. In particular, stakeholders pointed to the relative lack of data and transparency on groundwater and water provision in comparison to hydrological hazards.

In Chile, the legislative framework provides for the formation of basin user groups in order to manage the allocation of user rights for surface water. These bodies do not encompass groundwater user rights and tend to be comprised of a single sector, i.e. farmers,[2] which local stakeholders found to be a challenge to resolving allocation and pollution issues, particularly in periods of drought. Furthermore, low levels of trust between different actors (across governance scales, sectors and sections of the river) negatively impacted the consistency in privately negotiated agreements and informal collaborations that govern ad hoc adaptations for coping with extreme drought.

Another issue that is linked to information, but related to behavioural and cultural aspects of that theme is the openness of actors to learning from other experiences or contemplating other solutions. This is a challenge that has been touched upon in other studies. For example Dovers and Hezri (2010) use Smithson's categories of ignorance or irrelevance, which refer to issues that are left un-discussed since they are seen as taboo, or even not requiring verification (Smithson 1989 in Dovers and Hezri 2010) to explain the unwillingness of actors to accept and organise to adapt to potential climate change impacts. In Chile, despite there being an acceptance and awareness of climate change impacts, there is very little reference to comparative experience to guide learning towards developing better governance (including market based) mechanisms to cope with it. There is a strong notion that Chile is too unique to learn from other markets or country experiences. This isolation and strong ideology of one form of water governance hampers openness to innovation to adapt to current and future challenges in relation to increasing pressure on the resource. Related to the openness to learning, is the role that experience and perceptions of change play in developing or hindering mechanisms for coping with change. For instance, in the Swiss case, the lack of experience in coping with intense precipitation events that came as regularly and intensely from 1987 to 2000 has led to a loss of expertise in that fewer farmers were maintaining the infrastructure that could have reduced impacts, thus increasing vulnerability to these events.

Leadership and trust across different scales and institutions was a common barrier in both case areas. In Chile, trust was highlighted as an issue across multiple scales and sectors but the limiting forces on leadership were also seen to be an issue. For example, one stakeholder highlighted how politicians, who wished to strengthen provisions for environmental and societal protection in a more integrated vision of

[2] In other basins, such as the Copiapo, Mesa del Agua were piloted in an attempt to resolve the water management issues. 'The Mesa del Agua originally had no rules, no power, and no standards. It was not a place to take decisions, but just a place where the different interested parties could come to, in order to hear about the plans for different actors, and see presentations. They therefore created a Mesa Tecnica de Agua, and asked the Regional DGA to create a technical group. The members of this group are comprised of 1/3 Government, 1/3 Organised Society, 1/3 Users: Regional DGA, Regional CONAMA, Mining, Water Provision, and Agriculture.' (Interview, MMA, November 2010).

water management, tend to feel impotent to effect change due to the strength and power of stronger economic actors. The ideology of the market has already been mentioned, but it is equally relevant here, since the pure focus on the water market limits the flexibility to change course and open up new options in the face of new challenges (e.g. climate change, increased uncertainty). The focus of research tends to be on the technicalities of the water market, with little interest in the challenges of public policy making and the ability of the legal or governance framework (rather than just the market) to support and absorb issues that will manifest from climate change and other future hydrological challenges.

In Switzerland, stakeholders repeatedly pointed to the lack of leadership across the canton and federal levels, despite the policy guidance provided by the federal administrations for water and environment on integrated risk management, climate impacts and integrated water management amongst other topics. This perhaps highlights the hands off approach that is taken in the decentralised system, where technocrats at the federal level can provide insights into thought leadership on water adaptation, but there is an aversion at lower levels to their leadership or authority to match research innovations with provisions for implementation at local levels.

11.4 Common Bridges Across the Cases

In Chile, the majority of stakeholders across all levels deemed the water rights system to be flexible and adaptive, allowing for re-evaluation and revision, but protecting those with rights and allowing water to be provided for development and economic opportunities. One can argue against both aspects of this statement (see later discussion), but it is included here simply to present the perception of a number of stakeholders interviewed. Notably, more recently, there has been a concerted effort to readjust the balance between economic and environmental priorities, as can be seen through the transition of CONAMA to the MMA, the establishment of environmental courts, combined with a stronger focus on climate change impacts and adaptation. Likewise, the DGA Director recognises the need to promote efficiency and improve environmental aspects of water management and to achieve a better use of the resource, in order to reduce associated vulnerabilities to climate change impacts. The DGA has thus prioritised improving the robustness of the water rights information system (*Cadaster Public de Agua*) as well as transparency by putting it online. Information concerning water availability, hydraulic works and user organisations are also a top priority. While the DGA has detailed plans to direct increasing amounts of resources to update and modernise the water information system, these were nascent plans in late 2010 and early 2011 and so their implementation has not yet been confirmed.

In Switzerland, authorities highlighted the importance of provisions for financial incentives associated with ecological and social benefits, as a vital means of addressing intra-jurisdictional challenges (as was discussed in the background chapters on governance) but also was cited as a being very important for pushing lower levels of governance to take on the more resilience based approach of higher levels of governance. For instance the 'Mehrleistungen in Bereich Schutzbauten: integrales Risikomanagement' (Additional services in protective structures: integrated risk-management) programme from federal environmental office, promotes sustainability and integrated risk management in large projects related to protective measures against forest and water hazards, by providing more funds if multiple criteria (environmental) are integrated and shown to have been implemented 5 years later. Notably, the inclusion of non-state actors in the legislation process is seen as helping to shape and incentivise the uptake of more innovative approaches at the federal and cantonal levels.

Stakeholders across both case areas, though predominantly at regional and federal or national levels, cited the importance of research networks and knowledge partnerships in developing their understanding of the challenges and solutions to climate change impacts. In the Swiss case, these took the form of cross level partnerships, between federal and cantonal agencies that linked with federal universities and research institutes to develop predictive modelling that might inform adaptation and planning decisions. Monitoring partnerships, such as MINERVE and CERISE, co-ordinate information and interpret data concerning extreme events for emergency crisis groups at the canton and commune level.

Regional stakeholders also indicated the strength of support networks across sub-canton and commune levels, particularly for training and the discussion of new challenges on the horizon. One example, that is not directly related to water governance, but does have a relation to climate impacts on the mountain biosphere and linked economy is the integration project in the region of Leukerbad (Upper Valais) known as the 'Vernetzungprojekt nach Oekoqualitätsverordnung' (Interconnected project on Ecological Quality Provision). Organisations such as the 'Landwirtschaftzentrum' (Agricultural Centre), play a coordinating role in these projects across all three sections of the canton (*Oberwallis, Mittelwallis, Unterwallis*), where they foster a direct relationship with the communes and are responsible for assisting with information and advice for farmers at the local level, thus building understanding and capacity to deal with novel challenges. Monitoring projects, such as inventories of the 'Trockenweisen' (Dry Meadows), and integration projects on ecological zones (i.e. projects on the 'steppeflache' supported by the canton and federal government) are also recognised as being a potential opportunity to sensitise farmers about drought, scarcity, and ecological resilience. Lastly, Valais Tourism also plays a role in the development of a body of expertise on climate change adaptation, commissioning a scenario analysis of development paths for tourism in the canton in collaboration with Institute Gottlieb Duttweiler (Girschik et al. 2007).

In Chile, research partnerships were not as predominant a factor as in the Swiss case, but their importance for developing understanding and capacity for problem resolution was noted across all levels of governance. At the national level, international research networks, predominantly with the World Bank and UN-ECLAC, have played a role in developing capacity and learning about the benefits and challenges in the Chilean market approach. Government stakeholders referred to a number of studies with legal experts from leading Chilean universities and other experts (Pena 1997, 2001; Vergara 1998; Vergara-Blanco 2004) that had helped the DGA pinpoint areas of the legislative and institutional framework that needed to be addressed, particularly and especially in light of climate change impacts. This signifies willingness for institutional learning, despite the latitude for change in the Chilean system being seen as limited to market efficiencies and transparency. Similarly, the positive role of training programmes was referred to in the Chilean case, where the CNR supports farmers not only through the legal provision for subsidising efficiency improvements at the farm and canal level, but also through training programmes and workshops.

In both cases areas, actors at the national and regional levels pointed to attempts to improve integration across different scales of governance and sectors in order to provide a more coherent and coordinated response to mountain and novel hydrological challenges. In both case areas, little progress had been made towards this goal. In the Swiss case, the Water Competence Centre proposed in the Canton Valais had not yet been implemented at the time of publishing this research.

References

Adger WN, Agrawala S, Mirza MMQ, Conde C, O'Brien K, Pulhin J, Pulwarty R, Smit B, Takahashi K (2007) Assessment of adaptation practices, options, constraints and capacity. In: Parry ML, Canziani OF, Palutikof JP, van der Linden PJ, Hanson CE (eds) Climate change 2007: impacts, adaptation and vulnerability. Contribution of working group II to the fourth assessment report of the Intergovernmental Panel on Climate Change. Cambridge University Press, Cambridge

Burch S (2009) In pursuit of resilient, low carbon communities: an examination of barriers to action in three Canadian cities. Energy Policy 38(12):7575–7585

Burch S (2010) Transforming barriers into enablers of action on climate change: insights from three municipal case studies in British Columbia, Canada. Glob Environ Chang 20:287–197

Crabbé P, Robin M (2006) Institutional adaptation of water resource infrastructures to climate change in Eastern Ontario. Clim Chang 78(1):103–133

Dovers SR, Hezri AA (2010) Institutions and policy processes: the means to the ends of adaptation. Wiley Interdiscip Rev Clim Chang 1(2):212–231

Folke C, Carpenter S, Elmqvist T, Gunderson L, Holling CS, Walker B (2002) Resilience and sustainable development: building adaptive capacity in a world of transformations. Ambio 31(5):437–440

Folke C, Carpenter SR, Walker B, Scheffer M, Chapin T, Rockström J (2010) Resilience thinking: integrating resilience, adaptability and transformability. Ecol Soc 14(4):20 [online]. URL: http://www.ecologyandsociety.org/vol15/iss24/art20/

Gelcich S, Hughes T, Olsson P, Folke C, Defeo O, Fernández M, Foale S, Gunderson LH, Rodríguez-Sickert C, Scheffer M, Steneck RS, Castilla JC (2010) Navigating transformations in governance of Chilean marine coastal resources. Proc Natl Acad Sci 107(39):16794–16799

Girschik K, Frick K, Bosshart D (2007) Perspectives d'Avenir du Tourism Valaisan. Institute Gottlieb Duttweiler. Available at: http://www.valaistourism.net/Applications/Intranet.nsf/5d3221c185e6e4b1c1256d250061b171/dd1ad92940f2e462c125724b0037afca/$FILE/Valais%20tourisme%20brochure_FR_4.pdf

Herrfahrdt-Pähle E (2010) The transformation towards adaptive water governance regimes in the context of climate change. Universität Osnabrück, Osnabrück

Keskitalo CH, Dannevig H, Hovelsrud G, West JJ, Gerger Swartling Å (2010) Adaptive capacity determinants in developed states: examples from the Nordic countries and Russia. Reg Environ Chang [Online]. doi:10.1007/s10113-010-0182-9

Moser S, Eckstrom J (2010) A framework to diagnose barriers to climate change adaptation. Proc Natl Acad Sci USA. www.pnas.org/cgi/doi/10.1073/pnas.1007887107

Pena H (1997) Modificaciones al Codigo de Aguas y su Aporte a la Gestion del Agua. Direccion General de Aguas. Ministerio de Obras Publicas, Santiao de Chile

Pena H (2001) 20 Anos del Codigo de Aguas: vision Desde la Administracion. Paper presented at the 4th Jornados de Derecho de Aguas, Facultad de Derecho, Pontificia Universidad Catolica de Chile, Santiago de Chile, 19–20 November

Pierson P (2004) Politics in time: history, institutions and social analysis. Princeton University Press, Princeton

Smit B, Burton I, Klein R, Wandel J (2000) An anatomy of adaptation to climate change and variability. Clim Chang 45:223–251

Vergara A (1998) Estatuto Jurídico, tipología y problemas actuales de los derechos de aprovechamientos de agua. Estudios Públicos 69 (verano 1998):163

Vergara-Blanco A (2004) Discrecionalidad administrativa y nuevas limitaciones a los derechos de agua, vol 276. Centro de Estudios Publicos, Santiago

Yohe GW (2001) Mitigative capacity: the mirror image of adaptive capacity on the emissions side. Clim Chang 49:247–262

Chapter 12
Operationalising Adaptive Capacity

Abstract Drawing on the three earlier analytical steps, presented in Chaps. 10 and 11, this chapter explicates the process of developing more nuanced indicators of adaptive capacity from the original determinants presented in Part I. Thus, drawing on the original determinants discussed in Part I, together with both outcome assessments and the emergent themes in the bridges and barriers analysis, the indicator section that follows will elucidate how the regime, knowledge and network based indicators could provide a framework to address the emergent issues from this set of analysis. The indicators and their operationalised criteria are presented, and contextual sensitivities across the cases are discussed. Finally commonalities and linkages across the different indicators are explicated.

Keywords Rhône, Canton Valais, Switzerland • Aconcagua, Region V, Chile • Operationalising adaptive capacity • Nuanced governance indicators of adaptive capacity • Challenges across scales

12.1 Triangulating Towards a More Nuanced and Empirically Based Set of Adaptive Capacity Indicators

To recap, the initial set of determinants, drawn from the literature, were used to explore adaptive capacity through semi-structured interviews in order to gather information that could be used to assess the forms of adaptive responses in relation (but not exclusively) to climate related stresses and operationalise the determinants into more nuanced indicators and their criteria. The first chapter (Chap. 10) in Part III presented the analysis and characterisation of the adaptive mechanisms identified across the difference governance scales according to the categories of transformation, persistent adaptation and passive responses. The correlating governance mechanisms were discussed in relation to the different categories of adaptive action to

M. Hill, *Climate Change and Water Governance: Adaptive Capacity in Chile and Switzerland*, Advances in Global Change Research 54,
DOI 10.1007/978-94-007-5796-7_12, © Springer Science+Business Media Dordrecht 2013

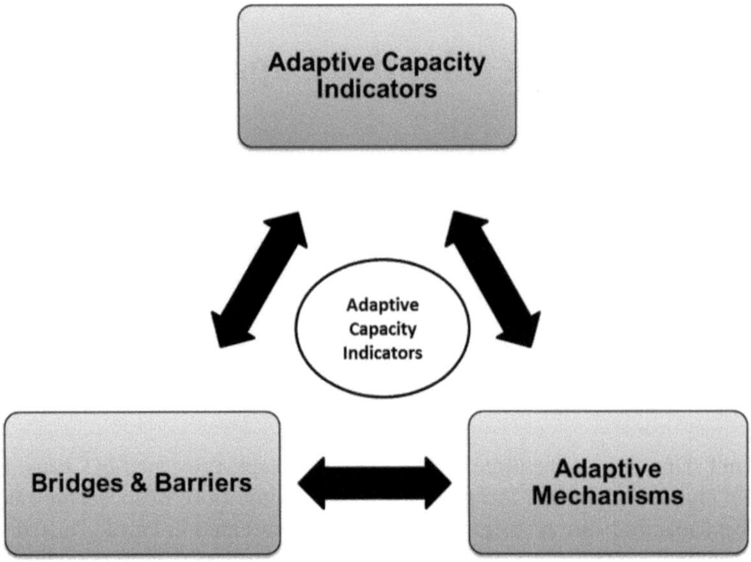

Fig. 12.1 Triangulating towards a more empirically grounded understanding of governance related adaptive capacity indicators

generate a better understanding of the governance actions associated with more transformational adaptation.

Next, analysis of bridges and barriers to adaptation in relation to extreme events allows a better understanding of adaptive capacity to be built up, primarily through the identification of a set of favourable conditions for fostering adaptive capacity. The process of analysis of these sets of conditions that help or hinder adaptability to extreme events, combined with the analysis against different forms of adaptive action were both important steps, combining both inductive and deductive techniques, to identify common themes across the case studies that allowed for the development and operationalisation of a set of indicators from the broad determinants of adaptive capacity (Fig. 12.1).

12.2 Regime

Regime refers to the sets of rules, legal and property rights framework, which determine what water stakeholders can and cannot do. It relates to the ownership and use of water resources as well as the rules and regulations that determine the management of the water resources, which water rights owners must follow. It encompasses the legislative and regulatory framework as well as the property rights system and policy framework. It also comprises the dynamics and power relations between different political and administrative levels from federal or national levels, to regional governments and administrations, to local level authorities or user associations (Table 12.1).

12.3 Knowledge

Knowledge refers to the informational inputs into a governance system that determine what actors know. These indicators encompass elements in the system, which allow for holistic and balanced management decisions that incorporate not only economic, but social and environmental information. It also establishes how uncertainty, and potential heightened levels of uncertainty, is accounted for in decision making. It refers to the timeframe in which plans and management techniques are evaluated and implemented, and the scale at which climate change or adaptive capacity is considered as relevant to the effectiveness of evaluation and planning under changing climatic conditions.

Perceptions of change also influence planning and management decisions. Available and accessible information and knowledge is vital for informing adaptation decision making, including the use of climate and hydrological information systems, effective deployment of 'objective' scientific information across different networks or levels of decision making, and the integration of different kinds of knowledge into decision making (e.g. traditional knowledge and experience). Monitoring and assessment frameworks are of course a primary requisite to ensure that adequate levels of information about water resources are available (Table 12.2).

12.4 Networks

Networks refers to the way in which actors interact and cooperate. It encapsulates the connectivity between groups and stakeholders that allows knowledge to be shared and common solutions to be negotiated for integrated solving of complex challenges. However, connectivity alone may not imply a willingness to cooperate during extreme climate stress, which demands accessible, expedient and effective conflict resolution mechanisms. The level and type of interaction between different stakeholders within the basin or sector also influences the motivations for connectivity between groups to cooperate and find common ground for building, as does the mode of coordination and delegation across different political and administrative layers.

Mismatches between authority in rule setting and lack of agency in management can deadlock actors on water management issues, diminishing the actor system's ability to resolve increasingly complex problems. One important aspect of networks is that the means of integrating scientific information is integrated into decision

Table 12.1 Operationalised indicators of adaptive capacity relating to regime components of the governance system with case examples

Regime indicators	Operationalisation	Case example: Chile	Case example: Switzerland
Ownership	*Consistency & Certainty*: To avoid recurring fundamental shifts in rights and governance frameworks from government to government. Legal certainty around ownerships and use rights is required for water owners and investments in water rights.	Multiple changes in legal framework (Agrarian Reform, Water Code) over the past 50 years has led to a situation where the rights ownership framework is opaque and unclear – yet inflexible in adapting less water (legal extraction can overrun actual availability). Water protected as a property right in the constitution providing high security and certainty to rights holders (but many of these are now seen as derechos de papel/paper rights.	Shifting relevance of traditional private rights, shifting institutions for their management. 80–100 year concessions periods for hydropower usage rights.
	Coverage: Coverage of all water rights/uses (removing blind spots from rights framework – e.g. glaciers, precipitation) CC means a closer eye needs to be paid to these unregulated areas.	Glaciers and groundwater have a weaker institutional framework than surface waters.	Increasing volumes of water for artificial snow production is currently unregulated, negotiated mainly through private company agreements; certain groundwater uses (agriculture) have no oversight, concessions or quotas.
	Clarity: In meaning of water rights (where they are fixed; how much the flow rate is; who owns them) and in application of the law and translation of legal framework to responsible parties/water owners at watershed, local to national levels.	Border problem with ESVAL & S2/S3; farmers exchanging rights, but no record; selling on rights, but still using them; process of legalising/institutionalising user groups can be impeded because of complex judicial and legal procedures; self-organisation - responsibility at the user level not happened for groundwater in Aconcagua.	Environmental flows are only to be taken into account in new concessions; EPA parameters for environmental performance lack clarity and precision; grundeignturm rights are very complicated, and therefore difficult for lawyers to really understand.

Responsibility	*Coordination*: Designated institutions need to have an overview to tackle the larger problems (i.e. unified perspective on cross sector issues, longer term issues).	Transference of water management to private sphere limits coordination and collaboration on trickier, longer term issues; projects are presented to MIDEPLAN only from their ministry's core perspective; no priority setting between sectors and uses; groundwater framework weak compared to surface water	Lack of cohesive and coordinated energy/water/environment policy across different administrative bodies.
	Rule setting: Finding a balance between decision making and political drivers of water resources management and the technical expertise required to build policy across the different levels and layers of responsibility: Ministerial Targets/Regulation; Judicial Decisions; Regional Regulations & Agreements; Local Directives, Agreements; Company/Association Agreements.	Water Code rules on the management of water rights, and its application is interpreted by the courts in individual cases. The individual water rights owners have the mandate to manage, and intervention/enforcement from public authorities has to be requested by a rights holder. Centrally driven water policy, determined by presidential priorities. Water provision regulated and prices set by a government regulator.	Designated levels of responsibility from federal policy to cantonal legislation and ordnances then to local directives (Bewilligung, Ordnung) and private agreements, creating linkages for enforcement and monitoring across uses and jurisdictions. Regulations differ per commune; canton legislation takes time to be in step with federal legislation and direction; private sector actors are responsible for implementation of certain restoration/re-naturalisation provisions.
	Expediency: Affordable and accessible access to informed judgements on water conflicts and issues within a basin.	Associations or the courts are responsible for resolving conflicts between users. Court process is lengthy, expensive and gridlock in conflict resolution prevents the legal institutionalisation of certain user groups, which erodes their ability to manage flows within their section.	Few court cases concerning water resources; administrative route provided for aggrieved parties to denounce planned projects at the relevant administrative level; participation in planning aims to pre-empt conflicts, but in TRC led to drawn out acrimonious negotiations in pre-implementation phase.

(continued)

Table 12.1 (continued)

		Case example	
Regime indicators	Operationalisation	Chile	Switzerland
Preparedness	*Pre-emptive provisioning*: Emergency provisions for hydrological extremes that can be initiated quickly for effective management of water scarcity or periods of flooding	Drought provision in Water Code provides for institutional response for drought management through a presidential decree according to technical parameters of Resolution 39 (1984). Informal institutional responses intra and inter JdV are invoked.	Policy frameworks for longer term flood management complemented by canton and local directives and management plans for pro-active and reactive flood response. No rules on scarcity management.
	Flexibility: Proportional Reduction of water rights; prioritisation of certain uses.	No prioritisation of uses in 'normal' periods; Tourno allows for proportional reduction of water rights within the separate sections of the basin; DGA prioritises domestic water, then irrigation during decreed drought periods; flexible reduction in rights distribution according to % of flow.	Priorities for set periods in canton legal framework; domestic water is prioritised during scarcity situations.
Integration	*Environmental Protection*: Provisions for the protection of aquatic ecosystems, including both quality and quantity provisions.	Water code differentiates between surface and ground waters; drought declarations allow provisional rights for surface- and groundwater and remove limits of ecological flows.	WPA provides for quality and quantity protection of waterways and aquatic ecosystems; environmental protection responsibility of Canton.

	Systemic Integration: Whether environmental provisions are just a 'tack on' that should be fulfilled if possible – or if sustainability of the SES is taken as a goal of the legislative framework.	Environmental protection framework is weak, with little monitoring and enforcement and weak EIA process; environmental flows unaccounted for in existing rights, only new rights; DGA's (responsible for surface and groundwater sustainability) technical criteria of sustainability have no legal basis, but are internal guidelines. MMA separately sets and monitors water quality (Norma secondarias de cualidad de agua).	Environmental flows to be accounted for in new hydropower concession period (but challenges in enforcement); social and ecological sustainability, and 'room for waterways' integrated into Cantonal Water Law (2008).
Accountability	*Neutrality*: Appropriate checks and balances to avoid agency capture in resource related ministries (e.g. National Commission for Energy, Mines) and to enable fairness across economically strong and weak actors.	Economic priority across ministries gear water management for exploitation; register of trades and rights happens with the Conservador, which is private but they have a monopoly for each region, and is under the control of the judiciary (with accusations of nepotism).	Provisions concerning project and planning proposals and EIA provide for multi-stakeholder input (VBR); environmental department responsible for evaluating EIAs from energy and other departments.
	Enforcement: Designated and clear lines of responsibility for monitoring and enforcing the application of legal provisions and regulatory requirements concerning water use and protection.	Self-enforcement amongst users for denouncing illegal abstraction creates challenges in the denouncement and punishment of water crimes, which exacerbates the impacts of drought periods on water courses; enforcement in the hands of the courts rather than DGA; Superintendencia is responsible for quality monitoring and ensuring water user does not exceed the utilities property rights.	Canton is responsible for enforcing legislation and regulations at local and canton level, but cannot impact on water sovereignty at local level; special 'water police' monitor/charge for illegal extractions and use.

(continued)

Table 12.1 (continued)

Regime indicators	Operationalisation	Case example: Chile	Case example: Switzerland
	Formality/Informality: Ability of regulators or water managers to enforce; nature of the relationship between the rule setters/governors and those being governed (i.e. private sector); how hands off the approach is.	High autonomy of individual water rights owners; JdV does not have mandate to manage, but to enable just allocation of water rights; state has a subsidiary role, with public authorities lacking agency and capacity to manage water resources to incentivise cooperation across watersheds and attain more sustainable level of water management as well as limit level of illegal extraction; high number of unregistered rights and illegal abstraction.	Use of water and management of waterways is most strongly informed by cantonal law, which is required to remain in step with federal legislation. Federal state takes a back seat in water management. Communes are encouraged to manage the resource in a way that respects the legal principles only through the subsidy incentives. Hydropower operations are regulated through cantonal level laws and commune level contracts; levels of commune autonomy are particularly high in the Valais, especially in the high mountain communes, challenging enforcement.
Effectiveness	*Implement-ability*: Translation of de Jure provisions into de Facto outcomes - effectiveness of protection and use provisions for sustainable water management.	Payment for non-use to reduce hoarding has not worked in reality as farmers prefer to protect rights, and fees are negligible for hydropower and mining companies; aims of market distribution ineffective because farmers value security over profit under conditions of scarcity. Technical expertise hampered in implementation by political and legal constraints.	Complex interplay between federal/cantonal legal provisions for sustainable and integrated water management and local level sovereignty; environmental flow provision is a point of tension between the energy, agriculture and environmental protection departments, in terms of implementation rather than legal provisions; conflicting view on implementation of TRC concept, pitting economy (agricultural land owners) versus ecology (political/environmental stakeholders).

Holistic: Provisions in legal and rights framework of incentive to use water efficiently and effectively across multiple uses, including incentives for conservation of water resources.	Water market led to highly connected supply and sanitation but no price for water (nor differential pricing according to use/sector) removes incentives for demand side or water balance management; Law 18.450 incentivises irrigation efficiency; inequities (upstream : downstream) and over-use (no environmental flows; full use of water rights volumes).	Focus on end of pipe solutions; lack of incentives/provisions for water conservation.
Capacity: Matching up resources at the level of enforcement	Lack of capacity in the DGA to administer water rights, transactions and hydrological investigations left to private actors; funds from MOP for drought response are low in years with other disasters (e.g. earthquake of 2010), limiting ability to implement projects to minimise economic impacts of drought.	Challenges in enforcement at canton and local level due to lack of capacity

Table 12.2 Operationalised indicators of adaptive capacity relating to knowledge components of the governance system with case examples

Knowledge indicators	Operationalisation	Case examples: Chile	Case examples: Switzerland
Evaluation and planning	*Neutrality*: Limiting scale of subjectivity in research and evaluation inputs into planning decisions, i.e. opposing information and studies being used in the management of water resources to justify certain decisions on infrastructure, water rights or management pathways. A healthy level of diversity is in different techniques and researchers can contribute to the scientific robustness of decision making, but neutral quality assessment can remove bias.	Inter-linkage between political and technical aspects of water management disrupts, slows down and politicises technical assessments and their implementation, leading to contentions over the technical feasibility studies (pitching private consultancy assessments against the DGA assessments) of groundwater sustainability, the Puntilla de Viento Dam and related groundwater exploitation studies. Ability to create long term plans beyond the term of one president/administration is a challenge (e.g. IWRM plans of Bachelet government).	Sectoral motives infiltrate studies commissioned for the TRC; technical studies are used by both sides to justify 'their' proposed technique for flood management. Private engineering offices play a large role in technical support to municipalities in conjunction with public administrative departments.
	Reactivity/Longevity: Development of both short term coping strategies for ad hoc climate shocks and extremes (inter-annual droughts and flooding) and longer term adaptation plans for larger shifts from climate change impacts (changes in glacier melt, snow pack, seasonality).	Reliance mainly on traditional/observation of rainfall amounts to predict preparation needs for a drought the coming summer at farm/canal level. Uncertainty calculations and climate impacts not taken specifically into account in planning of large projects (e.g. Aconcagua Project).	Proactive and prophylactic approach to planning (10 year cycle of planning in hydropower sector; canton and municipal ordinances on disaster management procedures; local risk hazard mapping) flood management plans are cantonal and federal responsibility to be implemented at local level (warning times are about 6–8 hours). MINERVE generates short to medium term precipitation prognosis to forecast bad weather events and expected water volumes to be managed. For water supply, water quality safety systems indicate if spring levels are reaching critically low levels from 1 year to the next.

Inclusion: Integration of uncertainty and climate change into plans, through increased accounting for complexity and uncertainty in managing water resources. Integration of social-ecological-economic criteria (including a valid weighting to each component) as a means of better accounting for complexity and building resilience. Integrating conservation with efficiency, rather than the promotion of efficiency for increased exploitation of water resources.	Cost benefit analyses of water policy decisions in normal times (social discount factor=6 %) and in extreme periods, e.g. of the losses due to the drought compared to the value of investing in pumps to tackle the drought (e.g. assessment of loss of VAT taxes through decreased production);physical consequences of the market are not accounted for in sector specific technical focus of adaptation options (i.e. desalination, national dams policy, irrigation efficiency, etc.) under consideration to meet rising demand under decreasing supply. Studies, modelling and seminars on climate change impacts, but approach is fragmented, lack of holistic and coherent cross sector/ministry strategy for managing and adapting to climate change impacts; climate change projections not included in forward modelling for Aconcagua Project.	TRC diagnosed the security of the Rhône, the quality of the dikes, deficits in terms of the size of hydro in order to meet security objectives, and improve the quality of the environment and the socio-economic aspects along the river, while taking climate risks into account through residual risk calculations. TRC implementation plan will not remain static and fixed, but 10 year evaluation period. Most vulnerable areas deemed priority measures, iterative approaches aim to balance sectoral priorities in implementation. Tendency for water supply on the other hand (irrigation, domestic supply) to take climate change into account is dependent on the municipality. Universities and private engineering consultancies conducting sectoral based studies; water protection office is taking part in an INTEREGG project with an Italian partner to observe springs in linkage with climate change. Climate change taken into account in planning and research but not at the operational level for hydropower. NFA as an incentive for inclusion of multi-criteria for integrated risk management.

(continued)

Table 12.2 (continued)

Knowledge indicators	Operationalisation	Case examples: Chile	Case examples: Switzerland
	Applicability: Appropriateness of the data sets (time period, data points etc.) to the decision making process How data/findings are applied to decision making water resources – being able to apply information to decisions because they are in the hands of technical experts not just politicians.	Issue of wet years being used to assess water availability from which to make rights allocation; data set from which Resolution 39 is constructed is out of date against current hydrological parameters. Final decisions on projects rest at presidential level (on a 4 year rotation); 80 % water security measure over 30 year period to inform permanent water rights allocation, but in Copiapo, one wet year was chosen so abstraction is adjusted to a wet season as droughts increase. Difficult to find and apply data on environmental and social costs	Politically, every 4–5 years there is a new communal authority, requiring knowledge to be transferred and re-assimilated. Challenges to cantonal oversight from local level sovereignty of water and implementation. Technical and hydrological data informs the management of different sources of water (e.g. spring monitoring for domestic supply), with real time data and annual averages, as well as hazard response systems (MINERVE & CERISE). Application of data from measuring stations (e.g. expected flows) to flood protection plans for those communes that have completed implementation.
Monitoring and assessment	*Consistency*: Consistency in different data sets and information; consistency and coordination in collation of data.	Central perception in Santiago on functionality of water resource management differs significantly from that in the regions, where impacts of governance approach are experienced. Wide range of actors (public and private) conduct mapping investigations, but lack of clarity concerning decision making process on contrasting data.	MINERVE requires consistent monitoring of precipitation for increased accuracy in forecasting and prognosis (canton & hydropower); observations and evaluation responsibility presides at canton level, maintaining the overview over the municipalities for ad hoc extremes and longer term planning project such as the TRC. Hydrological data network is extensive but not target-oriented, and cross-over/discrepancies between different levels and sector involved in monitoring, and the different demarcations of responsibility.

	Diversity: Diversity of inputs into the decision making system, e.g. Early Warning Systems, Hazard Mapping, Water Quality Monitoring, Snow and Glacier Monitoring etc.	Technical capacity on monitoring is comparable to developed contexts, but disparity between technical expertise and the monitoring inputs available to administrative departments and water managers.	Trockenweiseninventar, Suoneninventar, Inventories on irrigation in fields and meadow, high collation of different data sets. Less data available on glacier melt contributions to run-off.
	Coverage: Extensiveness and accuracy of the monitoring network, including functionality of equipment and usability of data for accurate decision making.	State monitors snow and precipitation, but irrigators also rely on private meteorological stations (e.g. Mina Andina) for weather information. CONAMA responsible for water quality monitoring, DGA for water quantity monitoring (4× per year), ONEMI for enforcement of quality failures. Monitoring difficult to coordinate and then implement controls. No national network and no consistent and coherent annual/monthly/regular monitoring of quality issues. DGA recently initiated a Chile wide glacier monitoring programme.	Hourly quantity monitoring across public and private sectors; canton quality monitoring as part of a national network co-ordinated through the federal level. Private sector collaboration on monitoring network (e.g. Universities, Canton, Hydropower companies; Engineering consultancies establishing stream sediment monitoring). In mountain areas, monitoring network is less advanced than elsewhere, but more monitoring stations are being implemented to improve understanding of spring levels under climate change.
Transparency	*Availability/Coverage*: Availability of information on water resources and ease of access to that information across all water stakeholders, across both the public and private sectors.	Lack of available, systematised and accessible information on water rights, water judgments, water market and prices, and the health and availability of water resources. Public water registry is out of date,	Online publication and access of canton and federal data, plans (e.g. Kantonale Gewässersanierungs Plan, CERISE) across different platforms (MeteoSuisse, BAFU, vs.ch, planat.ch). Data on snow-production, water use, hydropower use are difficult to access, and spread across multiple companies and communes (though available

(continued)

Table 12.2 (continued)

Knowledge indicators	Operationalisation	Case examples: Chile	Case examples: Switzerland
	Maintenance of water resources data to ensure relevance of the information to water management decision making.	and incomplete and non-electronic. Odepa.cl – available online data; DGA has no oversight on the trading and transaction of rights, market is seen as 'dark'; information on impacts in upper basin/mining companies is inaccessible. Multiple institutions manage water rights data (Conservador de Bienes Raises; DGA, Junta) – non-systematised and chaotic management.	online at many commune websites). MINERVE data sharing is still paper based in contrast to the online publication of plans and data concerning TRC. MINERVE implements a convention (signed between the owners of the hydro-power installations and the Valais) for an exchange of information during crisis periods.
	Communication: Resources and networks for communicating relevant information prior to and during periods of extremes. Communication for capacity building and education for building understanding on water and environmental conservation, including awareness on local climate change impacts and adaptation options.	Lack of education and communication on water conservation; personal and public communication (e.g. newspapers) on drought preparation, advice provision from JdV presidents on managing irrigation and crop production in drought periods.	Detailed communication of TRC through online communications, newsletters and participative fora to explain the justifications for why it is necessary to intervene on the watercourses, and to secure the plain. Attempts at canton level to be more pre-emptive in their communication with municipal level stakeholders. Local press and media used to communicate water supply provisions during scarcity periods and warnings during emergency flooding events. MINERVE convention provides for public communication during extreme events. Canton is legally required to ensure communication on habitation in hazardous areas and where houses can be built (flood and avalanche protection).

Perceptions	*Awareness*: Level of awareness amongst water managers and rights holders and government bodies of hydrological change and increasing uncertainty/ unreliability. Perceptive and sensitive to change but not reduced to apathy by it.	Observational awareness from irrigators and water managers that climate is changing, with reduced snow pack, melting glaciers and precipitation changes, but lack of popular awareness on water conservation; Acknowledgement that climate impacts will heighten and hydrological resources will be increasingly diminished; DGA/ Presidential acceptance of climate challenge.	Observations of change in the local climate (temperature, snow pack, timing of snow melt, glacier retreat, isotherm, permafrost) heighten the awareness of rate of climate change. In the region, engagement on climate is higher in the winter tourism and hydropower sector, where increasing glacier melt heightens hydropower production but increases material flows, in the reservoirs than in water provision, where communes mainly use non-concession spring water and there is a strong perception of Valais as the 'water tower' of Europe. Preparations for one scale of change (dry summers, glacier melt) but apathy towards more drastic ones (glacier disappearance, water scarcity).
	Openness: Openness to learning lessons from external experts and experiences, beyond their own perceptions of uniqueness. Ability to cope with change and willingness to bend to new management solutions or paradigms, while not retaining ideological rigidity.	Strong perception of the uniqueness of the Chilean model, therefore limiting ability to learn from other experiences. Old guard in some of the JdVs may be seen unwilling to move with the times; perceptions of success of the Chilean water market. MMA – other water/environmental authorities are looking very strongly to the experience in Europe of implementation the WFD; DGA/ presidential acceptance of weaknesses in the system/institutions – but not of the approach. Because of the uniqueness of the Chilean model, they do not think that it is possible for Chile to set up the same structure as basin management may follow in other countries.	Belief that historical adaptations to drier climate positions Valais agriculture to be better prepared than rest of Switzerland; research partnerships and knowledge exchange associations within and outside of canton suggest openness to learning from external experts and other regions.

(continued)

Table 12.2 (continued)

Knowledge indicators	Operationalisation	Case examples	
		Chile	Switzerland
Experience and expertise	*Experience*: Formation of experience for managing water resources in extreme periods through years of professional experience, social memory of past extreme events and droughts and training programmes. Different forms of experience; mix of technical competence and traditional knowledge for management of complex problems and novel threats.	Lack of professional training/education in water resources management for those private actors 'managing' water rights, for normal or drought periods. JdV/Canalista positions tend to be non-professional and part time. Strong role of tradition ('culture of water') and history in management of the canals. Experience of managing water scarcity differs from section to section and region to region. CNR/CORFO seminars and training programmes aim to improve management, techniques and efficiency and knowledge transfer in visits abroad. In drought management DGA relies on provision of technical information/parameters, while irrigators' intuition guides the management process. Lack of local knowledge/capacity in DGA drought interveners.	Disparities across municipalities and sectors; post-event debriefings occur at company level, allowing lessons to be applied to management of and preparation for future events; training courses on water management specialised for water management within different sectors through commune and cantonal administrations (e.g. L'association Valaisaine de distributer d'eau'; Societe Suisse de l'industrie du gaz et d'eau; Landwirtschaftzentrum; seminars on extreme events in Dienstelle für Naturgefahr). Long history of experience of rain shadow effect and annual training exercises on disaster response. Mix of experts in TRC (engineers, hydrologists, ecologists, territorial management etc.)
	Expertise: Conflict resolution processes need to be informed by both in depth legal knowledge, but also by the hydro-climatic and environmental impacts in which the judgement would result.	Court responsibility for resolving water conflicts result in judgements that steer the course of water management (ref Supreme Court decision on Factor de Uso). Local/regional judges may lack water specific expertise, leading to bad decisions.	Scope to improve technification in agricultural sector, farmers are increasingly part time, impacting traditional associations for management of irrigation right; at local level, commune presidents may assume responsibility for water management decisions, presiding over the post for short periods and on part time basis; certain water management responsibilities at local level (gefahrenkarte,.

Multiple forms and disciplines of expertise contributing to solution building for complex inter-connected problems, as faced by water managers under climate change.	and causing actors to circumvent the official process where possible. Level of technical expertise is high in water science, but ability to inform decision making and management from technical perspective is obstructed. Technification of water management in irrigation in the basin is mainly traditional (e.g. Bocatoma; irrigation efficiency; measuring/monitoring of canals; reliance on 'culture of water' rather than professional water management training). At government level, technical expertise is limited to hydrology and economics, no focus on public policy.	sanierung etc.) and canton level (NAQUA monitoring) tend to be outsourced to private engineering consultancies (though water suppliers have in house expertise as well) with the requisite expertise to work on the project in conjunction with the specialist cantonal department.
Secularism/Independence: Separation of political and technical management of water provision and management to ensure neutrality and continuity in water management beyond short term election cycles and political dogma (i.e. retaining knowledge). Distinction between political and operational pools of resources; separation of political process and technical expertise on water management.	Lack of independence and political secularism of scholars and experts informing water governance (collusions of World Bank, Neo-liberal politicians, economic agenda, financial contributions etc.). Operational/technical expertise at the regional level more functional than at the central political level.	Administrative departments involved with water management at canton and commune are not political appointees, but professional positions that remain in posts despite political outcomes. In commune councils elected officials constitute the political group and technical staff the service group for both watercourse management and provision; changes in legal framework are subject to or initiated by a broader range of stakeholder due to the direct democratic process.

making, i.e. the networks through which information is disseminated and shared. Collaboration and information sharing across different actors and levels elucidates the extensive and pervasive challenge of getting stakeholders to cooperate and collaborate either formally or informally (Table 12.3).

12.5 Contextual Sensitivities

Despite the cases being highly varied from both a physical and institutional perspective, common elements could be operationalised according to each indicator. Similar underlying challenges in developing and mobilising responses played out in different contexts and ways as the tables above indicate and the following section shall discuss them in more depth. Moreover, despite these different contexts, it is important to attempt to generate common lessons from contrasting systems since one of the challenges in climate change adaptation is scaling up local lessons to be applicable across different contexts (Herrfahrdt-Pähle 2010; Smit and Wandel 2006).

One point that should be addressed again at this point is the recognition that lessons have been drawn from both the flood and drought context. Despite the differences in the frameworks for responding and managing these hydrological situations, both provide important lessons into the governance mechanisms that influence adaptive processes. Notably, other studies that have compared the two contexts have also found that case areas confronted with recent flooding have more advanced strategies (according to AIM approaches), in comparison to drought response and adaptation (Huntjens et al. 2011).

Some studies have suggested the integration of a resilience based approach to be easier for flood protection than drought resilience due to the different kind of risk perception associated with each form of stress; moreover, the varying availability of solutions to flood and drought (Huntjens et al. 2011) make it easier to bring stakeholders together to find more innovative solutions in the window of opportunity after flooding events, where safety is the primary concern. Interestingly, the window of opportunity for those innovations may be very short, since it was recognised that a few years beyond the last flooding issues, urgency and awareness on the need for a more novel and long lasting approach to flooding already is fading. This means that the TRC is entering similar emotive territory as issues surrounding water stress, where stakeholders protect rights and ownership of resources (be it water or land) that are being threatened.

In addition, in developing and operationalising the indicators, a core tension of adaptive capacity emerged. The challenge of balancing flexible adaptive solutions and mechanisms at local and user levels with the policy guidance and legal certainty required from higher administrative levels for longer term processes, transpired to be a common issue across the different sectors in both cases. The following section discusses the operationalised determinants of adaptive capacity and details case examples that illuminate their role and importance in adapting to water management challenges in the case areas, as they relate to climate change and variability. It also highlights how the issue of flexibility or predictability relates to each indicator.

Table 12.3 Operationalised indicators of adaptive capacity relating to network components of the governance system with case examples

Network indicators	Operationalisation	Case example: Chile	Case example: Switzerland
Cooperation (Collaboration)	*Negotiation*: How actors negotiate amongst themselves to resolve water management issues (sectors, actors, government, levels) How actors resolve conflicts, reach agreements on water distribution, security and pollution.	Lack of a formal flexible mechanism of conflict resolution; court costs require resolutions to be financed by the JdV rather than Association of Canalistas, who are financially weaker; user to user negotiation on water issues, and agreements between JdV and mining and hydroelectric companies in the basin; utilities negotiate water prices nodes every 5 years with Superintencia; individualism and autonomy amongst rights holders challenges ability to agree a canal or JdV level strategy or solution and lack of interest to cooperate.	Consultation between private sector (hydropower companies) and commune level; Balance between federal provisions (legal base for financing), individual or group interested stakeholders, and fulfillment of functions concerning security, but also the ecology of the watercourses; lengthy and complex participative settlement of project implementation (leads to weakening of more radical innovation in TRC); conflict resolution expected at commune, with mediation role played by semi-administrative bodes such as Landwirtschaftzentrum; community resource management deemed as an important factor for solidarity and conflict resolution (attempting to foster it despite advances in modern irrigation infrastructure).
	Modes of organisation: How actors collaborate and cooperate across different sectors (public/private), uses (private/private – i.e. hydropower, energy, agriculture, utilities, industry/mining) and scales (local, regional, national, international) – .	Formation of JdV are provided for in the Water Code; Self-organised cooperation between JdV and canals during periods of drought allows flexible agreements and private compensation on water releases from one section to another that helps spread the risk and impacts more evenly and reduce social conflict – based on variable personal relationships and trust (i.e. functions much better between certain sections than others). Drought provision in Water Code instructs	Cross-sector collaboration through the Krisenstab for extreme periods, but little cross-sector co-ordination or contact on water resource management during 'normal' periods; individual company to company or commune to company agreements on water releases for artificial snow production; hydropower operators, universities cooperate and share information and flood management responsibility with canton and commune administrations through the convention for MINERVE;

(continued)

Table 12.3 (continued)

		Case example	
Network indicators	Operationalisation	Chile	Switzerland
	e.g. self-organisation versus legal frameworks that require cooperation.	DGA on compensation and liability, while CNR is guided by Law 18.450 for promotions and financial support of irrigation and efficiency.	water optimisation between hydropower operators (that are reliant on same resource); increasing levels of collaboration between communes (e.g. joint planning and training); sector, regional and local focus to partnerships and platforms for cooperation and joint learning.
	Incentivisation: Mechanisms to incentivise cooperation amongst water stakeholders within a basin (co-ordinating institutions, common visions). Government Subsidies and financing mechanisms inform how higher levels of government interact with and enforce actions at lower levels. Methods of inducing conformity to national, regional or basin priorities: federal/national subsidies, financial support, incentives etc.	Bi-lateral agreements between different sets of rights holders; increasing cooperation between mining sector and farmers (e.g. 10 million peso donation from Mina Andina to JdV, administered by the CNR as the executive and technical body for utilisation of the funds); private financing through water rights owners for maintenance and upkeep of distribution system - users pay for the operation and maintenance of infrastructure, while the state bears cost of the capital investment (Water Law 1981/1123); Government is liable to pay compensation for affectation to 3rd party rights from DGA drought management decisions (i.e. if new wells affect the rights of other owners), and compensate for the costs of extra infrastructure required for water supply.	NFA provides subsidies for increased participation in the implementation of water management projects (including meeting ecological criteria); reliance on financial and technical capacity at federal and canton levels provides for higher levels of cooperation at commune levels (where autonomy and sovereignty is strong); rivalries persist between different communes and the cantons on decision making on watercourse management – but attempts to reconcile them through participative consultation in TRC and final technical implementation signed off by Valais (and Vaud) Council; participative project planning/approval process is resource and time intensive, but aims to foster consensus and cohesion (Hydropower, TRC etc).

	Collaboration: Individual power relations, levels of trust, relative strength of social fibres holding groups together, levels of trust/distrust between parties. Power balances/imbalances affecting cooperation or collaboration. *Nature of support structures*: Compensation, remediation assistance, advice and technical support, financial assistance.	Broken agreements between JdV leads to a lack of trust across upstream and downstream sections of the basin and diminishes the ability to informally manage water sharing, thereby requiring intervention from the DGA; weaker economic actors (including smaller farmers, or those last in the canal) struggle to protect their water rights from quality or quantity damage by stronger economic actors (e.g. mining); negotiations between irrigators and farmers concern increased groundwater exploitation or compensation from drought declaration to pay for constructing more wells; political battle to amend the Water Code was ideological, acrimonious and lengthy for minor changes; general lack of trust between water rights owners community (including DOH) and DGA over Aconcagua Project and well abstraction.	Canton provides financial support for remediation when the communes cannot cover the costs (e.g. 1993); Investment in irrigation infrastructure through both canton and federal subsidies and technical assistance (Landwirtschaftszentrum) and programmes supported by charities at the national level, with the aim of maintaining both built and social infrastructure of the Suonen for social cohesion and conflict resolution; Financial and technical/training assistance provided to communes for water course management and implementation of cantonal provisions (hazard mapping, zoning); collaborative adaptation research projects (Regionprojekt, Verwerktung Markt am Handel) between tourism and agriculture; distrust of canton/ecological sectors and agricultural lobby over land issues in TRC.
Participation	Participation (not just consultation) in the political process, providing a voice in decision-making either directly or through legitimate intermediate institutions that represent their interests.	Water management is effectively in the hands of water rights owners, providing for strength of participation according to the amounts of rights that are owned; civil society participation (as per the IWRM model) is weakly provided for through consultation in the EIA process (or in other basins the Mesa del Agua).	Participative process in TRC implementation (COREPIL) attempted to reconcile different stakeholder positions, but is time and resource intensive and reduces the innovative elements of the project; participation in development and building planning open to organisations (Verbandbeschwerderecht; EIA) and individuals, though in reality may not be implemented.

(continued)

Table 12.3 (continued)

Network indicators	Operationalisation	Case example: Chile	Case example: Switzerland
Knowledge partnerships	*Integration/Dissemination*: Means of integrating scientific information into decision making, e.g. partnerships with academia and research organisations at national or international level.	Canvassing by means of submission of proposals to DGA to allow greater exploitation of the aquifer; relationship between academic institutions (both major universities and research bodies) is heavily financed by the strongest economic actors; regional and national government agencies rely on technical studies, data and research from major universities and private consultancies (e.g. CEASA water quality studies).	Convention was signed between the owners of the installations, a state university and the Valais for provision of meteorological and climate data for crisis management as part of MINERVE (including informing the public); hydrological data for water course management deemed behind other hazards (e.g. avalanche, rock falls etc); collaborative research between hydrological institutes and hydropower companies to generate short to medium term precipitation prognosis/forecast.
	Exchange & Support: Informal or formal networks to share and exchange best practice, lessons learnt, and technical solutions. Development of understanding, expertise, knowledge sharing, best practise through government support and advisory services.	DOH, CODELCO, CRA, CNR constitute the executive Mesa Tecnica de Agua in Aconcagua, which works with different factors such as regional DGA, MMA, ESVAL and other State institutions to resolve issues relating to the Aconcagua Project; International experts (EU, World Bank; UN-ECLAC) consulted with and employed by government ministries (MMA, DGA); Influence of foreign scholars (Milton Freedman and the Chicago School[a]) infamous in development of 1980 Constitution; Access to consultants for irrigation projects can be limited for poorer farmers (for CNR proposals); CORFO (corporation of promotion) organises international expeditions to enhance technical capacity of producers (agriculture, industrial etc).	Research partnerships between private companies (hydropower, manufacturing firms) and academia (Technical Institutions, e.g. Luzern, EPFL, WSL, EAWAG) to improve management of surface flows; MINERVE not yet formalised, and paper-based; TRC relies on partnerships with academia for climate change projections; Collaborative sectoral associations (Association of Valais Water Distributers; SDOC/SDRC; Societe Suisse de l'industrie du gaz et d'eau, Swiss Mountain Water) ; WA21 provides an inter-sectoral national level platform for knowledge sharing and dissemination on IWRM across Switzerland; Federal offices are tasked with support and information provision to lower administrative levels; collaborative research projects aim to inform (but not prescribe) decision making (ACQWA; Interegg 3).

Institutional integration (Co-ordination)	*Co-ordination/Clarity*: Level of integration between different water related institutions; Co-ordinating institutions; Relationship between natural and institutional landscape. Clearly defined roles and means of coordination between institutions (e.g. ministries) at different levels of government and different sectors.	Quality (MMA) and quantity (DGA) are managed separately; Rivers are divided up into different independent sections (Aconcagua has 4 sections with 3 JdV) so that delivery of rights are autonomously managed per section; different responsibilities for water spread across the different ministries (mining and hydropower have the money and the power); drought commission was created by the president in conjunction with the drought decree to integrate Interior Ministry, MOP, and Ministry of Agriculture, presided over by the Interior Minister; narrow cooperation between the ministries during 'normal' times; Mesa Tecnica includes JdVs, DOH and CODELCO fully, ESVAL participates intermittently; surface and groundwater managed through different instruments, but the new law improves integration of the two resources through an instrument to improve the infiltration of the aquifers to recover water table levels; land and water rights also separated in law.	Duplication of effort, trying to better coordinate on natural hazard management as challenges mount; Water management at canton level is sector based working group, who are conducting preparatory meetings to establish a Wasserkompetenz Zentrum within the cantonal administration; lack of co-ordination across different sources/uses (springs, groundwater, glacier, surface waters, lateral streams, Rhône); examples of integration of different water management challenges into one institutional/ infrastructural response in adaptive actions (e.g. MINERVE, TRC); increasing connection and integration between communes on water supply; Coordination within and across Valais (and also Vaud) for TRC, but not to external stakeholders in the rest of the Basin (Geneva, France); CIPEL provides platform for coordination and collaboration across the different communes and countries sharing the Lac Leman; cross sector coordination through CERISE in water crisis situations; Hydropower teams at commune levels have close relationships with communal organisations.

(continued)

Table 12.3 (continued)

Network indicators	Operationalisation	Case example	
		Chile	Switzerland
	Power & Balance: Institutional power relationships, complexity of institutional relationships; power imbalances/relationships between different sectors/economic groups/ministries.	Media and press articles used to influence decision making on rights allocation and the Aconcagua project; rights administration disagreements between DOH & irrigators and the DGA; power struggles between CNR and MOP, DOH and DGA, DGA and MMA reduces efficacy of coordination and roles – relationship changes from one administration to the next; strength and independence of powerful economic actors, and their infiltration into political, academic and judicial decision making.	Shifting roles and responsibilities across private and public spaces (in agriculture to public and in flood defence to private) according to diminishing and rising capacity; competing policy priorities across sectors (e.g. micro-hydropower versus environmental flows provisions) with no overview or integration into TRC; agricultural and hydropower stakeholders at odds against the ecologicalisation of water management at canton and federal levels; communes are independent and autonomous from canton and federal administrations, but are subject to canton and federal laws, and reliant on subsidies and support.
Levels of decision making	*Administrative Authority*: Relative authority at different administrative levels; level of centralisation or decentralisation.	Centralist government with high degree of Presidential powers and priority setting (e.g. Presidential decision required on projects, drought declaration etc.); Presidential interventions on project approval; regional branches (operational level) of ministries handcuffed by planning and policy decisions by national bodies (political level) and planning offices (MIDEPLAN) in Santiago (seen as 'King'), with lack of effective feedback mechanism between central and regional authorities; Courts have authority to define precedents in the implementation of the Water Code, thereby retain a powerful role in water 'management'; canal and JdV actors express 'powerlessness' against government institutions.	Decentralised system of federalism, with role of implementation at canton level, and in the Valais, the communes are particularly autonomous; subsidies for following ecological and security priorities of the 2001 Federal Directive for watercourse management is key for balancing the autonomy of decision making at the commune level (subsidised up to 95 % for Gefahrenkarte, ZonenPlan, Bauzone etc) enabling canton and federal environmental agencies to ensuring the implementation of priority policies and concepts; canton administrations are responsible for sectoral coordination and support, but communes for implementation, with technical and financial support from canton; Valais, as a relatively poor canton is reliant on federal subsidies.

Agency & Autonomy: Subsidiarity of government institutions; role of private solutions and self-organisation.	Limits to public authorities' ability to actively manage or regulate water resources; state has a subsidiary role in water management, which is in the hands of the private water rights holders; water users have independence and autonomy and the role of denouncing illegal extraction or pollution to the DGA; centralist bureaucracy is sidestepped in rights registration and bypassed by private actor negotiations, agreements and adaptive actions to cope with periods of drought; DGA unable to affect property rights, therefore emphasis is on users to self-organise to resolve drought induced conflicts; in extreme drought periods, government assistance and intervention is at the request of the irrigators.	Federal and cantonal governments have a subsidiary role in water management at the commune level, limiting their authority (right to redress) on water management across the country and region; water decisions reside at the commune public level, but are influenced by canton level coordination and support, particularly in communes where finances or capacity is low; private user groups are gradually morphing into public supported institutions; Increasingly responsibility for maintenance of irrigation infrastructure is shifting to public hand from the private, while canton is taking more oversight over energy concessions; provisions for participation in decision making at local, canton and federal levels take time, but build consensus; Constitutional Right to Petition allows for citizens to self-organise and influence the direction of water management at a federal level.

[a]Klein (2008) and Valdes (1995)

12.6 Synthesis: Commonalities and Linkages Across Indicators

12.6.1 Regime

In the Chilean case, the Water Code is the prime determinant of the administration and management of water, in its incarnation as a water right, rather than a holistic water resource. Furthermore, the Water Code cannot be seen in isolation from the Constitution, in which the private property of water rights are provided for (Article 19.24), guaranteeing the security of these water rights, reducing the ability of legislators to significantly reform the water rights situation and remedy the over-allocation and over-exploitation of water resources in the central and northern basins of Chile, without impinging on constitutionally protected property rights.

Moreover, *regime* is a particularly complicated issue in Chile because of the great influence that the neo-liberal doctrine implemented by the Pinochet regime had and still has on resource management across the country. The rules and regulations concerning water resources should therefore not be analysed in isolation from the changes implemented by the Pinochet regime, which continue to influence resource management in Chile in general. Under the neo-liberal doctrine, water rights were created into a commodity, separate from land, which could be bought and sold as any other commodity. After the new Constitution was crafted and passed in 1980 (the coup took place in 1973), in quick succession, the Water Code was then passed in 1981, the Electricity Act in 1982 and the Mining Code in 1983, each transferring power to the private sector, and de-regulating the governance of natural resources (consumptive water use, non-consumptive water use for energy and minerals for mining). These framework laws have a pervasive effect on almost every aspect of water governance, and every aspect of water resources governance impacts each of the indicators listed in the tables above. The balancing of the exploitative focus of these new laws and tendency to resolve conflicts in favour of stronger economic parties has received attention from other scholars researching the challenges facing water management in Chile (Bauer 2004; Budds 2004; Carruthers 2001). However, this issue has limited relevance in the Aconcagua region, and is more prevalent in the northern and southern areas of Chile, and therefore will not be tackled in this chapter.

The results of the *Regime* indicators, point to a tension between the lack of formality in regulating water resources and the highly formalistic centralised mode of rule setting and policy formation.

The ability of the DGA to have a management role in a basin is extremely limited. The DGA is not allowed to interfere with any third party water rights, once they have already been granted. During times of drought, this is expressly provided for under Art. 314 (Water Code) that states that if *a third party is affected due to declaration of hydrological scarcity, the state is responsible for compensation.* This provision refers to the protection granted to rights holders from negative impacts of decisions taken by the DGA during a declared drought period, specifically the permission for water users to access groundwater to which they do not have permanent rights. Similarly the weak position of the DGA in enforcing against illegal abstraction highlights the limited ability that the public authorities have to regulate and manage water resources.

Yet, this informality of approach takes place within strict codified rules of water governance. For example in the Junta de Vigilancia for Sect. 2, the process of formally legalising it through the official procedures has been an issue for over a decade due to the inability to resolve the issue of its borders. The mismatch between the institutional structures and the natural structure is the division of river basins into different sections, with weak connection between them. Only in periods of declared drought, have these institutional borders been dismantled (DGA taking over the river, and ignoring the sections of Juntas), as a last resort. Any other form of intervention on the river by the DGA also has to be demanded by water rights owners, in relation to economic, financial or allocation issues. However the DGA's lack of enforcement capability entails a regulation void, in which the certainty water owners cherish according to the Water Code has become meaningless in the basin's reality of over-exploitation and increasing drought periods. The lack of a regulating hand and management capacity from the government side is thus seen by some experts to have detrimental effect on the effectiveness of the water market (Dourojeanni and Jouravlev 1999). The lack of price, and clarity over prices, hampers the market system functioning to meet its goals.

The relationships between the judiciary, legal registries and the political and economic elite in Chile has also received broad attention in other studies (Alvarez 2005; Bauer 2004; Budds 2004; Carruthers 2001) and therefore was not a major focus in the research. However, the relevance of a lack of neutrality and a politicisation of water management decision making does impact the adaptation choices that are presented to the technical bodies responsible for more operational elements of water management in Chile. For example, an interview account detailed the pressure that the Bachelet government had placed on the *Superintendencia de Servicios Sanititarios* to force certain utilities in the north to move to desalination, as there was concern that there was not enough water in these northern basins to meet both domestic and mining demand. Technical managers within the *Superintendencia de Servicios Sanititarios* resisted government pressure, as it was deemed unacceptable to transfer the costs of desalination onto domestic consumers, who would have had to pay 3–4 times more for their water delivery, while the mining companies retained the water rights in the upper watershed, with no guarantees of covering the difference in pricing.

Switzerland has a complex, broad mix of different forms of ownership, responsibility, enforcement and rule setting at different levels of administrative government that provides autonomy at the local level in the Valais to set rules and regulations according to the needs and particularities of the commune. However, there are examples of ordnances and directives, which have yet to be implemented in the Valais, either at canton or commune level. For example, the Federal Ordinance for Drinking Water Provision in Emergency Periods (*Bundesverordnung für Trinkwasserversorgung in Notzeiten*), which proposes an organisational structure to deal with drinking water in any kind of crisis, has yet to be implemented by the canton. Separate from the hazard maps, the canton is required to develop a water map (*wasserkarte/wasseratlas*), which is yet to be achieved. Viewed in relation to the challenges of implementation discussion in Chap. 5, the challenge of implementing prophylactic planning tools to manage indeterminate hazards is hampered by the lack of enforcement power that higher levels of administration have to enforce the precautions implemented at lower levels (either canton or commune).

In the Valais, the rules and regulations which guide water pricing, provision and use tend to be set in commune or canton level regulations (*reglemente*), conventions, concessions and agreements, which allow for some flexibility in revising rules to adapt to newer challenges. However, the length of hydropower concessions means that windows of opportunity for revision seldom appear. This long term fixed nature of the concessions becomes more critical during periods of higher water scarcity, where concession water may be required to replenish reservoir stocks for domestic consumers (e.g. in times of scarcity, SIB may request EOS (Energie Ouest Suisse) to replenish Lac de Louvie, yet until the concession is renegotiated in 2040, no fixed emergency plan for periods of scarcity can be implemented).

The decentralised form of governance in Switzerland is nevertheless still segregated along sectoral divisions, with coordination across the different sector-specific institutions intermittent and irregular, particularly across the energy, water and environment policy process. Both micro-hydropower and the TRC are policy priorities currently, each with their own relationship to climate change mitigation and adaptation. The volume of micro-hydro planned at the commune level in the Valais, is potentially in conflict with the attempts to enhance the social-ecological features of the Rhône floodplain through the TRC. While implementation for both projects resides at the commune level, the canton has oversight for the TRC while the communes have responsibility for micro-hydro, meaning that there is a lack of oversight or integration across the canton and at federal levels.

12.6.2 Knowledge

The situation in Chile is particularly interesting with regards to knowledge, and highlights many of the contradictions and challenges that characterise the Chilean case. Gaps in data and information in the water market are a key issue across water resources management. As the mantra goes, what you measure you manage, and in the case of the Chilean market, the focus is on data for the market (which itself is lacking), while the void of data on ecosystem impacts of the water market signals the lack of concern for managing the watershed system holistically. The DGA (Desmadryl 2010) has expressed their prioritisation of improving and updating the water information system, not only improving the quality and coverage of data in the system, but creating a more accessible online platform to improve transparency. The CNR has also been tasked to assist in the improvement of rights data, due to the lack of capacity within the DGA. Furthermore, more information on hydrogeological and geophysical studies was also deemed to be necessary to establish and assess quantity and quality aspects of watersheds. According to the DGA Director, the following issues are priorities: designing and implementing special plans of auditing; action plans at the regional level; training for users' organisations; registration of authorised abstractions (to be able to clarify legal and illegal abstraction points); effective systems of abstraction controls, with the obligation to inform the authority; coordination and training with the public attorney, to develop better control and investigation, in cases of infractions and water theft.

While these policy priorities point to a broadening of the informational focus from within the DGA, further investigation would be needed to ensure that the goals have been translated from intention to implementation. Interviews suggested that across water experts in Chile, technical ability is high (though concerns were raised about technical capacity of the MMA at the regional level) but the challenge resides in the inclusion and matching of the technical capacity with where decision making capacity lies. The focus on improving the state of the water market in Chile rests on improving information and transparency of the market to improve its ability to achieve efficient allocation, rather than improve the range of information and regulations that inform market allocations that would account for a broader set of objectives, including increasing the resilience of the social-ecological systems that are impacted by the water market to adapt to changing hydrology.

Another example of this issue comes from the example of MIDEPLAN, the ministry in which projects are evaluated. MIDEPLAN is unable to evaluate projects from a perspective other than the core mandate and purpose of the institution from which it is being proposed. For example, the DOH can only present projects from an irrigation perspective, the Ministry of Mines from a mining perspective, negating any potential integration of benefits or indeed impacts. The challenge is not the level of expertise in Chile, where hydrologists and engineers have the capacity to calculate ecosystem demands and impacts according to water availability, but the paucity of this information and its linkage to the actual rules of water management are detrimental to the holistic resilience of water management. In certain areas, namely groundwater and ecosystem health, the irregularity of monitoring and absence of a co-ordinated monitoring and observation network has led to a lack of data that erodes the DGA's ability to manage the related water rights, especially to be able to manage the groundwater rights during the declared drought periods, when groundwater is more heavily exploited in the Aconcagua.

In the Chilean case, despite legal provisions to enforce obligations on monitoring and abstraction controls, internal DGA guidelines on declaring restricted areas and provisions that require user associations to be established for both groundwater and surface water (requiring these associations to model and monitor abstraction levels) (see Chap. 8 and Chap. 10), major challenges exist in their enforcement. The lack of measurement of non-market based data, and the gaps in information concerning water rights, are major impediments to the DGA's ability to effectively take control of management during periods of declared drought, when they are expected to be able to do so. In order to be able to mitigate or manage the effects of increased droughts on water rights under climate change conditions, the inability to account for the rights and usage that presently exists is a major limiting factor in the capacity to adapt. This incompleteness of knowledge is mirrored in the relative weighting of economic and environmental issues in the legislative process. Quality and water management rules are set by analysing the economic implication of the proposed rule on the relevant sectors (agriculture, mining and hydro-electricity). It is the responsibility of the department of economic analysis in the MMA to assess proposed rules (according to the planning standards imposed by the Ministry of Planning – MIDEPLAN) with a social discount rate of 6 %.[1] If the economic costs are measured to be too high (according to the equation used) then the new rule will not be passed by through the political route.

Another facet to the knowledge related challenges in the Chilean case, are the tensions between the legal and technical spheres of knowledge and agency and the political or administrative. One manifestation of this is the impotence of the DGA to mediate issues between rights holders in a basin, and the resulting role of the judiciary in conflict resolution. The judicialisation of environmental management has meant that judges, who lack expertise on hydrological or environmental matters, dictate precedents in water resources management at the watershed or national level. There is a chasm between the level of expertise in political and technical decision making, yet there is an ineffective separation of political from technical matters, that handcuffs and frustrates the operational level, weakening their ability to provide workable solutions to mounting challenges at the basin level.

The positive emergence of a plan to implement a set of environmental courts (proposed in conjunction with the emergence of the MMA out of CONAMA) is also likely to be hampered by capacity constraints in expertise. Such courts could be a vital tool to develop more effective and expedient conflict resolution, particularly in periods of drought. However, a long term effort would be required in training and capacity building within the judiciary for such tribunals to approach water conflicts with a more holistic knowledge of the system characteristics, rather than a shallow formal interpretation of the Water Code, which is overbearingly influenced by economics.

Climate change is being observed across Chile, mainly through the reduction in glaciers and snow pack, which is matched by almost all interviewees recognising that decreasing water availability in the Aconcagua and most regions in central and northern Chile will require improvements in the organisation of water management, the information that informs it, and the ability to settle disputes, in order to avoid mounting conflicts. The DGA recently initiated a glacier monitoring programme; there is a general lack of data that would be required to manage climate change impacts, including water rights information, water availability, riparian ecosystem health; but the integration of climate scenarios into water resources planning, both infrastructure and rights management, within the Aconcagua is currently not taking place. However, across the specific sectors, climate change impact studies have taken place (either by the ministries themselves or academia) and there is an

[1] For comparison the Stern Report used 2%, while Nordhaus used 3% (OECD 1997).

advanced level of information on climate change scenarios at global and regional levels, with associated adaptation options per sector (agriculture, mining, energy etc.). The challenge is to integrate climate change impacts data not only into policy and decision making at basin levels, but also developing adaptation options that move beyond sector specific technological fixes to hydrological changes. Furthermore, develop plans at the basin level that would enable adaptation across the different sectors that could minimise further degradation of the social-ecological system.

The sectoral focus towards climate change impact studies means that the subsidiarity of the environment in climate change adaptation is further reinforced. MMA reports on climate change impacts present technical solutions towards mitigation and adaptation to climate change for other water users (irrigation, energy and water supply sectors), but rarely presents the environmental perspective in terms of how to reduce the negative impacts of climate change on environmental degradation and the potential role that enhancing ecological resilience may have in adaptation choices.

The data sets and monitoring networks available for the Alps are in general far more developed and extensive than in the Andes. Despite criticism that the monitoring network is not as historical, widespread or well-maintained as in other areas of Switzerland, the observational data, perception and awareness of climate change is high. Climate modelling information in the cantonal administration and larger hydropower companies' information is integrated into planning for the TRC and the development of larger hydropower management decisions. However, this is contrasted by lack of integration of climate information across other sectors and at lower administrative levels that provides data on water provision as opposed to information on intense precipitation events.

Where climate information is included at the operational commune level, it is mainly related to increases in intense precipitation and natural hazards, particularly as they relate to the ability to employ water turbines for energy production. As part of the Emergency Plan, required at the commune level, there is a requirement to integrate data on stream flow levels, and their response to precipitation. At present, most of the instruments to enable this level of monitoring are not yet implemented, but it is planned to set up a central database, so that whoever requires the information for decision making in extreme events can easily access it. At present, managers rely on maps that estimate the correlation between precipitation volumes and run off.

The current lack of possibility for cantonal oversight over planning and water related developments (refer to the deeper discussion in Part II) at the commune level is one challenge in developing an integrated and coordinated response to more complex and novel challenges, such as those posed by climate change. The separation of responsibility, data bases and information products across sectoral, administrative[2]

[2] E.g. The canton the groundwater monitoring in the Rhône Valley (Monitoring der Grudnwasservorkommen), but not the monitoring of the springs (Quellen), for which the communes are responsible.

and resource (i.e. surface, groundwater, quality, quantity) lines has developed over many decades, but is seen as a challenge to those trying to prepare the response system to better cope with increase precipitation extremes and related natural hazards. While in policy and management planning, the canton's authority is weak and subsidiary, it does have a supportive role in the organisation of the various monitoring networks (Spring Protection, Quality Monitoring, Groundwater levels) when the communes themselves are unable to fulfil their duty.

One of the main challenges discussed in Part II are those concerning the implementation of legal provisions at local levels, where capacity and expertise are variable. Developing the requisite professional overview and forward looking coordinated response to broader challenges (and related investment decisions) that climate change entails at the local political level, where politicians are often in part time posts, is an on-going challenge that institutes, such as EAWAG, are focussing on. On the other hand, at the technical level, support from federal and cantonal authorities is more prevalent. In the case of the hazard maps, which the communes must generate for their tributaries, the federal government acts as a technical support and the canton acts as a connector between the federal government and the communes. In terms of the Geoplans (*Notfallplanung Hochwasser Kanton Wallis*), the communes are also responsible for implementation of the plan, but rely heavily on informational inputs from special engineering consultancies.

A common thread through the Swiss case was the perception of climate change as a problem for the next generation, something to worry about in 10–20 years' time, but not an issue that needed to be dealt with now. Similarly to the Chilean area, stakeholders live at close proximity to glaciated and snow covered areas, and therefore have observed glacier recession, changes in snow coverage, increasing instability of permafrost and changes in precipitation volumes. Local water managers are well aware that as these changes intensify, discharge into springs will affect volumes available for water supply, but deem it to be a problem that will need to be resolved in either 5–10 or 10–20 years' time. More pressing problems relate to the challenges in balancing rivalries between hydroelectricity and social-ecological demands on waterways, with increasing regulation for ecological flows in competition with the development of micro-hydropower and new pumped storage (particularly as the energy landscape is changed in the aftermath of the Fukushima Daiichi nuclear disaster).

Increased storage capacity and man-made springs are technical options proposed for future management of more extreme climate change, when water supply may no longer meet demand. However, stakeholders, such as farmers and water suppliers, who have always dealt with the relative scarcity in the Valais, suggest that their past experiences and tactics in managing the extremes of low to high precipitation means that they are relatively well prepared for the measures that need to be implemented to manage such extremes. Therefore, while climate change may not explicitly be factored into the planning process across local levels, stakeholders suggest that it subconsciously is part of decision making. Furthermore, heightened awareness of climate related risks in relation to extreme events have a narrow window of opportunity. Managers of the TRC and DSFB note the difficulties in developing the level of scientific knowledge and understanding of factors that should be taken into account in water resources planning, as the memory of the earlier flooding experiences fade.

At the national and federal level, climate change is explicitly taken into account by the federal administrative bodies for environmental and water issues (FOEN, Water Office) as well as national collaborative platforms such as WA21, leading to attempts to foster cross-sector collaboration (MOUs[3]) to share learning and generate integrated solutions to future challenges in hydropower, watershed management and water infrastructure management. Despite the generally high awareness that climate change will lead to a multitude of heightened challenges, a means of finding holistic inter-linked solutions to the challenges of climate change remains elusive at federal and canton levels.

12.6.3 Networks

Within the Chilean case, the lack of unified management across the basin that limits the potential for integration and coordination across the different sectors was detailed in Chap. 5. The DGA has presented its aim of transitioning towards more unified basin management in order to mitigate the escalation of water resource conflicts, by strengthening organisations at the basin level for increased integration and participation of water stakeholders in watershed management. The uniqueness of the Chilean legal and economic framework for water is deemed to be the major impediment to a more integrated approach, but despite this, there have been attempts at strengthening inter-sectoral and basin cooperation through different user organisations, such as the Junta de Vigilancia (which is legally open to all water owners within a basin, although usually only includes irrigators) and the Mesa del Agua to better coordinate and generate information for enhanced decision making and dispute resolution.

The IWRM basin institutions, the Mesa del Agua, were only piloted in three basins in Chile, with limited effect and agency. But in the Aconcagua, a *Mesa Tecnica de Aconcagua* had been set up to better coordinate the stakeholder groups in favour of the Aconcagua Project. At present it includes the DOH, CODELCO and the CRA (*Confederacion de Regantes de Aconcagua* – Aconcagua Irrigators Confederation). The *Mesa Tecnica* also maintains a dialogue with the DGA, CONAMA, ESVAL, CNR and other state institutions, which are important for disputes relating to the project, but are not regular members of the body.

The coordination of stakeholders with a common interest for the realisation of the Aconcagua Project indicates that increased collaboration across the basin can be realised around a specific project. A more challenging, yet potentially more

[3] MOU between the different partners to create a new form of collaboration rather than another federal department.

rewarding, step would be to develop this platform as neutral territory to negotiate and resolve the 10 year long stand-off surrounding the project, by making it more inclusive and providing an arena for dissemination of scientific studies on the basin. This could potentially overcome the current impasse on the scientific basis for groundwater reserves, where the DGA and DOH (together with the irrigators themselves) both have contradictory studies behind their arguments for exploitation or protection of the groundwater in the basin. The impasse over groundwater availability and the Aconcagua project reflects the levels of distrust between the different administrative bodies, the different sectors and the different upstream and downstream actors within the case area. Irrigators view the DGA as blocking their private adaptation ability by closing the groundwater reserves for new rights, since it has been declared to be beyond sustainable extraction. The lack of trust and agreement on scientific evidence blocks the actors' ability to build opinions from a common basis, leading eventually to a further depletion of increasingly vulnerable scarce resources.

The division of the Aconcagua River into four different sections, with three functioning Junta de Vigilancia (and only two that are legalised) means that during periods of scarcity and drought, upstream rights owners are better positioned to control flows to the rest of the Juntas in the basin. While proportional reduction is negotiated across the basin, in reality the upper rights owners may not follow through on promises. Some irrigators felt an overarching basin organisation would allow a more efficient, expedient and less costly resolution in cases of such power imbalances, not only between different rights holders, but in holding the government agencies to account as well.

Although formal routes of negotiation and conflict resolution can be costly and lengthy, the autonomy of water rights holders means that private negotiation between different actors and actor groups (e.g. Utilities, Mining companies and Junta de Vigilancia, and regional officials) often can replace the more time and resource consuming official routes. However, other studies have also shown that this private bargaining and negotiation can also lead to injustices for the weaker political and economic actors (Alvarez 2005). It also closes the door for other stakeholders to participate in decision making over water resources, to which they may or may not have rights, but an interest in its equitable and sustainable management (i.e. other users of the ecosystem services provided from the watershed, e.g. coastal fishermen, domestic water users, environmental bodies).

Money is an amazing motivator. In both case areas, financial subsidies play an important role in incentivising and enhancing levels of cooperation between different actor groups and levels. In the Chilean case, drought declarations come with levels of both enhanced coordination through the DGA, but also increasing availability of financing for adaptation from the Ministry of Interior (through the DGA). Drought declarations also imply increased liability of the government if third party rights are affected by DGA approval of provisional groundwater abstraction. As climate impacts mount, DGA intervention, according to current rules, would mount. Yet while water rights owners are adverse to increasing levels of government involvement in water management, there is less of an aversion to government money to finance infrastructure for groundwater exploitation (by the DGA) to cope with lower surface flows and subsidies for water infrastructure including both large scale dams (by the DOH) and smaller farm scale irrigation efficiency improvements (by the CNR).

While subsidies exist for the construction of water infrastructure (Water Code, Art. 1123), operation and maintenance is left in the hands of the irrigators. The state invests in the construction of infrastructure, transferring it to the private rights owners once complete. When the title passes to the farmers on a newly built irrigation project, a financing agreement is put in place so that over 25 years, for example, farmers pay for the infrastructure, and the water rights belong to the farmers themselves. According to interviews at the DOH, a new law is in preparation with respect to enhancing sustainability in such projects. However, in terms of operation and maintenance, many of the water canals in the Aconcagua Basin (particularly some of the longer ones such as the Waddington that stretches over 100 km) have high leakage rates, with irrigators at the end of the line often not receiving any portion of their rights allocation due to evapotranspiration and leakages along the way, reinforcing and heightening the impacts experienced during drought periods.

It is not just financial assistance that can be fruitful in fixing some of the underlying challenges to sustainable water management, but capacity building programmes also can play an important role, as learning networks and knowledge exchange can open up the possibilities of applying lessons learnt from one area for innovative approaches to challenges in another. Capacity programmes run by CORFO provide farmers (as well as other sectors) with the opportunity for foreign travel to learn more about techniques and technologies for specific areas of interest. Within the case areas, one stakeholder referred to such a sponsored trip to France as highlighting for him the value of basin organisations for conflict resolution. This suggests that while national institutions see the value that knowledge exchange can have and the importance of developing learning networks for capacity building, the mechanisms to share and integrate accumulated knowledge are lacking. This leads to a failure to translate new insights from external cases to complex challenges within the basin.

The DGA has committed to strengthening the level of expertise and the range of knowledge in the water user associations and their empowerment for using it in dispute settlement and eventually become stronger partners to the DGA in its own mandate to efficiently manage the distribution of water resources. However, clashes between the authority of different government organisations and the agency and autonomy of rights owners at different levels and sectors are one of the defining challenges in the Chilean case. The public authorities in Chile at the regional level are limited in their ability to actively engage in the management of the resource. The DGA, effectively, can administer the allocation of water rights (according to availability) and record the transference of rights in the market. Water management is thereby transferred to the private sector, and the independence and autonomy that this grants water users is not matched in incentives for enhancing levels of cooperation between them.

Despite the lack of capacity within the DGA for actual water management, it is still expected to take on the management of the basin in the most challenging and contentious periods of extreme drought, to effectively control disputes and finance coping strategies. At the other end of the scale, Chile is not only highly centralised but there is also a low rate of delegation from the Presidential and Ministerial level on water resource related decisions, decisions that might otherwise be taken in the operational rather than political sphere. Infrastructural projects, drought decrees, ministerial committees on irrigation development, changes in quality rules are all taken at minister level if not higher. Information gathered by technocrats inform these, but a number of interviewees at regional or operational levels (in DGA, MMA, DOH) expressed their impotence when informing highly politicised and sector specific decisions on water management issues. This suggests that the informational quality of water related decision making is hampered by the lack of functioning cross scale networks, limiting the ability of the authorities at multiple levels to resolve interlinked and complex problems related to increasing stress on a dwindling resource.

The Swiss case represents an interesting contrast to the Chilean in terms of networks. While it is a highly decentralised governance system, it wrestles with similar challenges in terms of subsidiarity of government role and the autonomy of the communes (rather than private rights owners) that challenge building cohesive, integrative and cooperative solutions for adapting to climate change. Levels of cooperation and collaboration differ depending on the scale and sector. Despite the sectoral and small scale arrangement of water resources management that tests the ability to plan for and implement integrated solutions to more complex problems (i.e. challenges of implementing TRC, addressing scarcity issues across neighbouring communes), there are many examples of partnerships and networks across the region engaging in climate challenges. Often, these partnerships are still sector specific, but at least extend out beyond the commune and canton, for the purposes of sharing best practices, technologies, and learning from the experiences of other areas.

While each commune in the Valais does have autonomy over their own water resource, the independent water suppliers have more recently set up the 'Association valaisanne des distributeurs d'eau (AVDE)' which conducts meetings twice per year (a technical day and a general assembly). One stakeholder highlighted how 'each year we choose a different issue, for example protection zones. We get different specialists from inside and outside the region to come and speak about it to inform all of us who are practically implicated in the issue. There are about 40–50 members, and it is a good place to exchange ideas, and get a better understanding for different issues, come up with solutions'. While the association is voluntary, cantonal representatives also take part. The association also runs training courses specialised in drinking water provision. A similar association exists for utilities (*SSIGE: Societe Suisse de l'industrie du gaz et d'eau*[4]), which meet for seminars once per

[4] http://www.svgw.ch/francais/pagesnav/PO.htm

year, while federal research institutes such as EAWAG collaborate on studies and dissemination with the communes.

The canton agricultural administration (Amt für Strukturverbesserung) also plays a role in encouraging (and subsidising through federal assistance) the maintenance of traditional water management structures at local levels, as a stakeholder explained: 'They are trying to support/encourage the maintenance of these organisations because they assist in the upkeep of the infrastructure/minimise costs at the local level. In the old system, the whole village was implicated in the management of the water and the canals. Whether you took water from the top of the canal or bottom of the canal, you still needed to work the same to maintain it – it fostered a unique solidarity. Grundeigenturm still stay with the land. The canton is providing subsidies to maintain the canals and the geteilschaft'. The canton reduces the increasing financial and capacity burden upon them by fostering flexible conflict resolution mechanisms that are less costly than judicial or administrative routes, while also maintaining responsibilities for the upkeep of traditional infrastructure (that plays an important role during extreme precipitation events as well as the summer irrigation season). The investment projects on the Suonen, however, did not have any impact on reducing water demand.

Despite the examples of cooperation and collaboration, Chap. 5 discussed the challenges of coordination and integration across the Valais for the different facets and scales of water resources management. In direct relation to hazards (climate related or no) the crisis management groups, functioning at the commune level, indicate strong coordination across the different sectors and stakeholders, to ensure that emergency responses to hazards are well-prepared and structured. The TRC represents an attempt to better coordinate horizontally across stakeholder interests and vertically from the federal level to the commune level of implementation. The federal administration is eager to create coherence in hydraulic engineering projects across the country under the 'Loi des Course d'Eau', so that the provisions relating to ecosystem health and integrated risk management may be better adhered to.

The same approach is being taken at the canton level for natural hazards, where a stakeholder explained that 'the idea is to create important synergies across each domain with a relation to natural hazards and create a more uniform approach (technically and financially) to the domain. Currently NH tasks are spread across the DSFB and the DWL. Reorganization would foster an integrated strategy for long term protection, the application of principles of precaution and causality and a uniform concept of safety. It would remove duplication of effort and therefore reduce time and cost inefficiencies. The briefing note is asking the Conseil d'Etat to rethink how the section is organised'. The pressure of more frequent and intense hydrological events has pushed federal and canton levels to better coordinate across institutional boundaries, to differing degrees of success.

The financial incentives provided by the federal government aim to encourage an implementation of the TRC that reflects the legal guidelines in canton and federal legislation. The commune's reliance on subsidies is a key tool of authority for the higher levels of administration, enabling the canton and federal government to craft responses and projects that integrate the more progressive elements of the law that accounts for social-ecological resilience. Despite the financial incentives, challenges in negotiating the different frames of reference and priorities which stakeholders bring with them to the TRC planning committees, has proved highly contentious and time intensive so far, often to the final detriment of resilience based aspects of the project. Moreover, as in Chile, changing climate is making the government potentially more financially liable, as communes are hit by extreme events, the costs of which they cannot cover. The event in 1993 caused damage that most of the communes were unable to cover, requiring the canton to foot the bill for most of the damage. The Swiss parliament has already started discussing the need for a considerable increase in financial resources for protection against flooding and other natural hazards, likely to increase with climate change[5] (FOEN 2011).

The challenges of coordination across policy frameworks and plans has already been discussed under the *Knowledge* indicators, specifically in relation to the micro-hydropower and environmental protection agendas. But this specific issue also points to the issue of harmonising the competing interests through balanced negotiations and participation. In the case of the TRC, stakeholder participation has not diminished the generation of two competing fronts in the discourse. The field is divided between the agricultural front who are unimpressed with the potential loss of land (losers in the ecologicalisation of water management), and the politico-technical and environmental front who favour heightened protection of the natural environment as a means of boosting social-ecological resilience. While in the past corrections, the technical engineering approach followed a harder path, the fact that softer solutions are being sought in the TRC, means that a new equilibrium has been reached, which is little consolation to those actors losing productive land. Participation provides an arena for these voices to be heard and negotiate, but not necessarily an efficient means of resolving such complex issues in an effective time frame.[6]

Interestingly, these issues suggest a very different form of mismatched authority and agency in the Swiss case than in the Chilean. As a decentralised federalised country, the federal administration may set the framework policies and rules, which water management in the cantons and communes should adhere to, but this is limited to strategic guidance, direction and subsidies, while the communes (with less technical and financial capacity) must maintain priority uses and provide solutions in times of water scarcity. Furthermore, in the Valais, the strong autonomy of the

[5] http://www.news.admin.ch/message/index.html?lang=de&msg-id=41748 (aftermath of 2011 flooding event); http://www.parlament.ch/d/suche/seiten/geschaefte.aspx?gesch_id=20083752 (Parliamentary discussions for financial period 2008–2011).

[6] Refer to the comments from an engineer in the TRC: 'It is difficult because they are defending their interests, but my feeling is that they are not really entering into the dialogue, nor are working towards a compromise. We have proposed compensations to these people, but they don't really want to talk, they just stay defending their alternative proposed solution. The process of trying to reconcile these two different views in the participative process of the TRC takes up a significant amount of time. We are working at the communal level as well to help the process along.'

communes can limit the ability for the canton to give advice, negotiate and coordinate; nevertheless, plans such as the *Plan Generaux d'evacuation des Eaux* are implemented, and the canton provides an important link (e.g. launching studies) between the communes and the federal level.

Examples of shifting responsibilities between the public and private sectors and the different levels of administration point to the limitations of the traditional approach to water management at the lowest possible level in the face of increasingly more complex problems. At the commune level, certain private activities have been transferred to the public realm. For example the management and remediation of extreme weather damage is no longer managed privately, but was seen to not only require the commune structure but even canton level commitment, since the level of remediation work and investment began to exceed the ability of the traditional structures in place to cope with it (e.g. in Baltschieder, which was heavily impacted in 2000, the commune had to take over the clean-up operation, rebuilding the streams and water infrastructure, in conjunction with the canton). This shift in the aftermath of such extreme events has tended to remain post event, and although the traditional associations have stayed in place, their roles are diminished, but fostered and supported by the commune or canton (see above).

In other areas, there has been shift of responsibility from commune to canton. For example, the new hydropower concession will no longer be purely in the commune's authority, but administered through both the communal and cantonal levels. Under the new concession period, not only can communes become shareholders in order to take part of the installation under the commune's ownership, but the concession is reviewed and agreed to by the canton (e.g. to ensure environmental flow requirements, refer to Table A7 and Sect. 5.2). In general, the canton has tended only to intervene in issues pertaining to water provision and hydropower at the commune level when there has been a problem in which the cantonal authority needs to intervene. The increase in oversight at the canton level allows an increase in oversight and implementation of ecosystem provisions, while respecting the strength of communal sovereignty over its resources. It also provides a separate body that can negotiate between rights holders and users in the case of conflicts in the complex negotiation of concession agreements as has been shown to be important in other studies of climate change adaptation (Huntjens et al. 2010).

The value of trust in building cooperation and collaboration for resolving complex water management problems is also relevant in the negotiation of energy concessions. The concessions represent a long-term, multi-generational use rights based relationship between state and private actors, and are the basis of significant investment. Concerns about climate impacts are an important factor that must be taken into account in this relationship. The polarisation of the different interests in the TRC has not necessarily been reflected in the multi-stakeholder negotiation process of concession renegotiation in other areas of Switzerland. For example in Glarus, the inclusion of environmental organisations in the concession negotiation has ensured that ecological factors are considered in the agreement, thereby less recourse from environmental stakeholders in the aftermath of the agreement and a smoother passage through the approval process. However, this level of negotiation is only possible when all stakeholders are afforded an equal place at the table.

Leadership is an aspect of networks that have been shown in other studies to play a role in driving water management systems to innovate and test out novel approaches (Cook et al. 2011). In Chile, the challenges of trust building across different public and private administrative bodies are in part linked to the issue of the informality and lack of accountability or responsibility for water resources management (as opposed to rights distribution). In the Swiss case, the role of environmental organisations in challenging and changing the debate around water resources management has had an impact in shifting water resources policy and legislation to a more integrative approach.

The organisational leadership prescribed to environmental groups has been reinforced by the linkages between them and prominent administrative figures and technical experts in the cantonal administration in the development of the TRC plan and other areas of innovation in water management at the canton level. However, these networks tend to be weaker and once local user level stakeholders are included (e.g. actor participation in the COREPIL) less knowledge exchange based; and it is at this level that cohesion and collaboration in adaptation approach breaks down. Perhaps, investing in the level and quality of knowledge sharing networks to the user level, instead of bringing the plans to them for consultation, would be a means of establishing more functional networks across the policy-implementation gap.

Further investigation into the role of leadership in developing new techniques and innovation for climate relevant problems in water systems could be well served by techniques such as social network analysis. A better understanding of the role of leadership could also improve our understanding of the area managers and decision makers, might concentrate on to better navigate the bridges and barriers affecting adaptive capacity of water governance. Other studies have investigated the role of policy entrepreneurs[7] (NeWater) in terms of their ability to utilise windows of opportunity to translate novel strategies proposed within shadow networks into more mainstream approaches considered within formal policy arenas. These studies suggest that policy entrepreneurs provide a vital disruptive function in change-resistant institutions, allowing policy change to incrementally lead to institutional change.

[7] http://www.newater.info/index.php?pid=1056 – define policy entrepreneurs as (1) they anticipate windows of opportunity by developing and testing attractive policy alternatives and demonstrating their feasibility; (2) they employ strategies of venue manipulation, venue shopping and/or create new venues to be able to insert new ideas, which have been developed in shadow networks, into formal decision making forums, and (3) they use narratives or other discursive strategies to frame an issue strategically, and by that to attract supporters and justify change.

References

Alvarez P (2005) Un itinéraire de l'eau, Approche géographique et agronomique d'une gestion de l'irrigation en zone aride du Chili. Université de Orléans, Orléans

Bauer C (2004) Siren song: Chilean water law as a model for international reform. Resources for the Future, Washington, DC

Budds J (2004) Power, nature and neoliberalism: the political ecology of water in Chile. Singap J Trop Geogr 25(3):322–342

Carruthers D (2001) Environmental politics in Chile: legacies of dictatorship and democracy. Third World Q 22(3):343–358

Cook J, Hill M, Freeman S, Levine E (2011) Shifting course: climate adaptation for water management institutions. WWF US, Washington, DC

Desmadryl M (2010) La visión del gobierno chileno sobre el futuro del agua en Chile. Paper presented at the Seminario Internacional de Fundación Copec-Universidad Católica /Agua: Desafíos de su escasez, Santiago de Chile, 16 November 2010

Dourojeanni A, Jouravlev AS (1999) El Código de Aguas de Chile: entre la ideología y la realidad, vol LC/R.1897. Comisión Económica para América Latina y el Caribe, Santiago de Chile

FOEN (2011) Leben mit Naturgefahren: Ziele und Handlungsschwerpunkte des Bundesamts für Umwelt (BAFU) im Umgang mit Naturgefahren. Federal Office for the Environment: Natural Hazards Prevention Division, Bern

Herrfahrdt-Pähle E (2010) The transformation towards adaptive water governance regimes in the context of climate change. Universität Osnabrück, Osnabrück

Huntjens P, Pahl-Wostl C, Grin J (2010) Climate change adaptation in European river basins. Reg Environ Chang 10:263–284

Huntjens P, Pahl-Wostl C, Rihoux B, Schlüter M, Flachner Z, Neto S, Koskova R, Dickens C, Nabide Kiti I (2011) Adaptive water management and policy learning in a changing climate: a formal comparative analysis of eight water management regimes in Europe, Africa and Asia. Environ Policy Gov 21:145–163

Klein N (2008) The Shock Doctrine: The Rise of Disaster Capitalism. Penguin Press, London, U.K.

OECD (1997) Better understanding our cities: the role of urban indicators. Organisation for Economic Co-operation and Development, Paris

Smit B, Wandel J (2006) Adaptation, adaptive capacity and vulnerability. Glob Environ Chang 16(3):282–292

Valdes JG (1995) Pinochet's Economists: The Chicago School of Economics in Chile (Historical Perspectives on Modern Economics). Cambridge University Press.

Chapter 13
Assessing Adaptive Capacity

Abstract In the final chapter of Part III the case evidence is reviewed according to the adaptive capacity indicators and analysed for positive fulfilment of the operationalised criteria. Furthermore, a synthesised review of the correlation between adaptive responses and indicators of adaptive capacity is presented. Results indicate a correlation between more transformative and persistent adaptive actions and the decentralised governance context of the Swiss case. The centralised and yet neoliberal market model of the Chilean case is dominated by a number of passive actions, which can be seen to correlate with potential longer term degradation of the resilience of the social-ecological system. However, both cases are correlated with a number of persistent adaptive actions.

Keywords Rhône, Canton Valais, Switzerland • Aconcagua, Region V, Chile • Degradation of social-ecological system resilience • Persistent adaptive actions • Operationalising adaptive capacity indicators

13.1 Analysing the Case Evidence: Indicator Coding

The following section presents the analysis of the case evidence according to the indicators developed in the previous chapter and operationalised criteria. Green indicates positive fulfilment, while purple indicates mixed fulfilment and orange indicates negative fulfilment.

M. Hill, *Climate Change and Water Governance: Adaptive Capacity in Chile and Switzerland*, Advances in Global Change Research 54,
DOI 10.1007/978-94-007-5796-7_13,

13.1.1 *Regime*

Table 13.1 Operationalised indicators of adaptive capacity relating to regime components of the governance system with case examples

		Case area	
Regime indicators	Operationalisation	Chile	Switzerland
Ownership	*Consistency & Certainty*: Legal certainty around ownerships and use rights	Multiple changes in legal framework (Agrarian Reform, Water Code) over the past 50 years has led to a situation where the rights ownership framework is opaque and unclear – yet inflexible in adapting less water (legal extraction can overrun actual availability). Water protected as a property right in the constitution providing high security and certainty to rights holders (but many of these are now seen as *derechos de papel*/paper rights).	Shifting relevance of traditional private rights, shifting institutions for their management. 80–100 year concessions periods for hydropower usage rights
	Coverage: Coverage of all water rights/uses	Glaciers and groundwater have a weaker institutional framework than surface waters.	Increasing volumes of water for artificial snow production is currently unregulated, negotiated mainly through private company agreements; certain groundwater uses (agriculture) have no oversight, concessions or quotas.
	Clarity: In application of the law demarcation of rights, and translation of legal framework at watershed, local to national levels.	Border problem with ESVAL & S2/S3; farmers exchanging rights, but no record; selling on rights, but still using them; process of legalising/institutionalising user groups can be impeded because of complex judicial and legal procedures; self-organisation – responsibility at the user level not happened for groundwater in Aconcagua.	Environmental flows are only to be taken into account in new concessions; EPA parameters for environmental performance lack clarity and precision; *grundeignturm* rights are very complicated, and therefore difficult for lawyers to really understand.

Responsibility	*Coordination*: Designated institutions need to have an overview to tackle the larger problems.	Transference of water management to private sphere limits coordination and collaboration on trickier, longer term issues; projects are presented to MIDEPLAN only from their ministry's core perspective; no priority setting between sectors and uses; groundwater framework weak compared to surface water.	Lack of cohesive and coordinated energy/water/environment policy across different administrative bodies.
	Rule Setting: Balance between technical and political inputs into water governance and across levels.	Water Code rules on the management of water rights, and its application is interpreted by the courts in individual cases. The individual water rights owners have the mandate to manage, and intervention/enforcement from public authorities has to be requested by a rights holder. Centrally driven water policy, determined by presidential priorities. Water provision regulated and prices set by a government regulator.	Designated levels of responsibility from federal policy to cantonal legislation and ordnances then to local directives (*Bewilligung, Ordnung*) and private agreements, creating linkages for enforcement and monitoring across uses and jurisdictions. Regulations differ per commune; canton legislation takes time to be in step with federal legislation and direction; private sector actors are responsible for implementation of certain restoration/re-naturalisation provisions.
	Expediency: Affordable and accessible access to informed judgements on water conflicts and issues within a basin.	Associations or the courts are responsible for resolving conflicts between users. Court process is lengthy, expensive and gridlock in conflict resolution prevents the legal institutionalisation of certain user groups, which erodes their ability to manage flows within their section.	Few court cases concerning water resources; administrative route provided for aggrieved parties to denounce planned projects at the relevant administrative level; participation in planning aims to pre-empt conflicts, but in TRC led to drawn out acrimonious negotiations in pre-implementation phase.
Preparedness	*Pre-emptive provisioning*: Emergency provisions for hydrological extremes.	Drought provision in Water Code provides for institutional response for drought management through a presidential decree according to technical parameters of Resolution 39 (1984). Informal institutional responses intra and inter JdV are invoked.	Policy frameworks for longer term flood management complemented by canton and local directives and management plans for pro-active and reactive flood response. No rules on scarcity management.

(continued)

Table 13.1 (continued)

Regime indicators	Operationalisation	Case area Chile	 Switzerland
	Flexibility: Proportional reduction or prioritisation of water rights/uses.	No prioritisation of uses in 'normal' periods; *Tourno* allows for proportional reduction of water rights within the separate sections of the basin; DGA prioritises domestic water, then irrigation during decreed drought periods; flexible reduction in rights distribution according to % of flow.	Priorities for set periods in canton legal framework; domestic water is prioritised during scarcity situations.
Integration	*Environmental Protection*: Provisions for the protection of aquatic ecosystems.	Water code differentiates between surface and ground waters; drought declarations allow provisional rights for surface- and groundwater and remove limits of ecological flows.	WPA provides for quality and quantity protection of waterways and aquatic ecosystems; environmental protection responsibility of Canton.
	Systemic Integration: Sustainability of the SES is taken as a goal not just tacked on.	Environmental protection framework is weak, with little monitoring and enforcement and weak EIA process; environmental flows unaccounted for in existing rights, only new rights; DGA's (responsible for surface and groundwater sustainability) technical criteria of sustainability have no legal basis, but are internal guidelines. MMA separately sets and monitors water quality (*Norma secondarias de cualidad de agua*).	Environmental flows to be accounted for in new hydropower concession period (but challenges in enforcement); social and ecological sustainability, and 'room for waterways' integrated into Cantonal Water Law (2008).
Accountability	*Neutrality*: Appropriate checks and balances; fairness across economically diverse actors.	Economic priority across ministries gear water management for exploitation; register of trades and rights happens with the *Conservador*, which is private but they have a monopoly for each region, and is under the control of the judiciary (with accusations of nepotism).	Provisions concerning project and planning proposals and EIA provide for multi-stakeholder input (VBR); environmental department responsible for evaluating EIAs from energy and other departments..

Enforcement: Clear responsibilities for monitoring and enforcement.	Self-enforcement amongst users for denouncing illegal abstraction creates challenges in the denouncement and punishment of water crimes, which exacerbates the impacts of drought periods on water courses; enforcement in the hands of the courts rather than DGA; *Superintendencia* is responsible for quality monitoring and ensuring water user does not exceed the utilities property rights.	Canton is responsible for enforcing legislation and regulations at local and canton level, but cannot impact on water sovereignty at local level; special 'water police' monitor/charge for illegal extractions and use.
Formality: Ability of regulators or water managers to enforce regulations.	High autonomy of individual water rights owners; JdV does not have mandate to manage, but to enable just allocation of water rights; state has a subsidiary role, with public authorities lacking agency and capacity to manage water resources to incentivise cooperation across watersheds and attain more sustainable level of water management as well as limit level of illegal extraction; high number of unregistered rights and illegal abstraction.	Use of water and management of waterways is most strongly informed by cantonal law, which is required to remain in step with federal legislation. Federal state takes a back seat in water management. Communes are encouraged to manage the resource in a way that respects the legal principles only through the subsidy incentives. Hydropower operations are regulated through cantonal level laws and commune level contracts; levels of commune autonomy are particularly high in the Valais, especially in the high mountain communes, challenging enforcement.

(continued)

Table 13.1 (continued)

Regime indicators	Operationalisation	Case area Chile	Switzerland
Effectiveness	*Implement-ability*: Translation of de Jure provisions into de Facto outcomes.	Payment for non-use to reduce hoarding has not worked in reality as farmers prefer to protect rights, and fees are negligible for hydropower and mining companies; aims of market distribution ineffective because farmers value security over profit under conditions of scarcity. Technical expertise hampered in implementation by political and legal constraints.	Complex interplay between federal/cantonal legal provisions for sustainable and integrated water management and local level sovereignty; environmental flow provision is a point of tension between the energy, agriculture and environmental protection departments, in terms of implementation rather than legal provisions; conflicting view on implementation of TRC concept, pitting economy (agricultural land owners) versus ecology (political/environmental stakeholders).
	Holistic: Incentives to use water efficiently and effectively across multiple uses (incl. conservation).	Water market led to highly connected supply and sanitation but no price for water (nor differential pricing according to use/sector) removes incentives for demand side or water balance management; Law 18.450 incentivises irrigation efficiency; inequities (upstream : downstream) and over-use (no environmental flows; full use of water rights volumes).	Focus on end of pipe solutions; lack of incentives/provisions for water conservation.
	Capacity: Matching up resources at the level of enforcement.	Lack of capacity in the DGA to administer water rights, transactions and hydrological investigations left to private actors; funds from MOP for drought response are low in years with other disasters (e.g. earthquake of 2010), limiting ability to implement projects to minimise economic impacts of drought.	Challenges in enforcement at canton and local level due to lack of capacity; but means of balancing this through federal assistance

13.1.2 *Knowledge*

Table 13.2 Operationalised indicators of adaptive capacity relating to knowledge components of the governance system with case examples

Knowledge indicators	Operationalisation	Case examples	
		Chile	Switzerland
Evaluation and Planning	*Neutrality*: Objective research and evaluation inputs into planning decisions.	Inter-linkage between political and technical aspects of water management disrupts, slows down and politicises technical assessments and their implementation, leading to contentions over the technical feasibility studies (pitching private consultancy assessments against the DGA assessments) of groundwater sustainability, the *Puntilla de Viento* Dam and related groundwater exploitation studies. Ability to create long term plans beyond the term of one president/ administration is a challenge (e.g. IWRM plans of Bachelet government).	Sectoral motives infiltrate studies commissioned for the TRC; technical studies are used by both sides to justify 'their' proposed technique for flood management. Private engineering offices play a large role in technical support to municipalities in conjunction with public administrative departments.
	Reactivity/Longevity: Development of both short term coping plans and longer term adaptation plans.	Reliance mainly on traditional/observation of rainfall amounts to predict preparation needs for a drought the coming summer at farm/canal level. Uncertainty calculations and climate impacts not taken specifically into account in planning of large projects (e.g. Aconcagua Project).	Proactive and prophylactic approach to planning | (10 year cycle of planning in hydropower sector; canton and municipal ordinances on disaster management procedures; local risk hazard mapping) flood management plans are cantonal and federal responsibility to be implemented at local level (warning times are about 6–8 h). MINERVE generates short to medium term precipitation prognosis to forecast bad weather events and expected water volumes to be managed. For water supply, water quality safety systems indicate if spring levels are reaching critically low levels from 1 year to the next.

(continued)

Table 13.2 (continued)

Knowledge indicators	Operationalisation	Case examples	
		Chile	Switzerland
	Inclusion: Integration of climate change and environmental impacts into planning process.	Cost benefit analyses of water policy decisions in normal times (social discount factor=6 %) and in extreme periods, e.g. of the losses due to the drought compared to the value of investing in pumps to tackle the drought (e.g. assessment of loss of VAT taxes through decreased production);physical consequences of the market are not accounted for in sector specific technical focus of adaptation options (i.e. desalination, national dams policy, irrigation efficiency, etc.) under consideration to meet rising demand under decreasing supply. Studies, modelling and seminars on climate change *impacts*, but approach is fragmented, lack of holistic and coherent cross sector/ministry strategy for managing and *adapting* to climate change impacts; climate change projections not included in forward modelling for Aconcagua Project.	TRC diagnosed the security of the Rhône, the quality of the dikes, deficits in terms of the size of hydro in order to meet security objectives, and improve the quality of the environment and the socio-economic aspects along the river, while taking climate risks into account through residual risk calculations. TRC implementation plan will not remain static and fixed, but 10 year evaluation period. Most vulnerable areas deemed priority measures, iterative approaches aim to balance sectoral priorities in implementation. Tendency for water supply on the other hand (irrigation, domestic supply) to take climate change into account is dependent on the municipality. Universities and private engineering consultancies conducting sectoral based studies; water protection office is taking part in an INTEREGG project with an Italian partner to observe springs in linkage with climate change. Climate change taken into account in planning and research but not at the operational level for hydropower. NFA as an incentive for inclusion of multi-criteria for integrated risk management.

	Applicability: Appropriateness and application of the data to the decision making process.	Issue of wet years being used to assess water availability from which to make rights allocation; data set from which Resolution 39 is constructed is out of date against current hydrological parameters. Final decisions on projects rest at presidential level (on a 4 year rotation); 80 % water security measure over 30 year period to inform permanent water rights allocation, but in Copiapo, one wet year was chosen so abstraction is adjusted to a wet season as droughts increase. Difficult to find and apply data on environmental and social costs.	Every 4–5 years there is a new communal authority, requiring knowledge to be transferred and re-assimilated. Challenges to cantonal oversight from local level sovereignty of water and implementation. Technical and hydrological data informs the management of different sources of water (e.g. spring monitoring for domestic supply), with real time data and annual averages, as well as hazard response systems (MINERVE & CERISE). Application of data from measuring stations (e.g. expected flows) to flood protection plans for those communes that have completed implementation.
Monitoring and assessment	*Consistency*: Consistency across data sets, coordination in collation.	Central perception in Santiago on functionality of water resource management differs significantly from that in the regions, where impacts of governance approach are experienced. Wide range of actors (public and private) conduct mapping investigations, but lack of clarity concerning decision making process on contrasting data.	MINERVE requires consistent monitoring of precipitation for increased accuracy in forecasting and prognosis (canton & hydropower); observations and evaluation responsibility presides at canton level, maintaining the overview over the municipalities for ad hoc extremes and longer term planning project such as the TRC. Hydrological data network is extensive but not target-oriented, and cross-over/discrepancies between different levels and sector involved in monitoring, and the different demarcations of responsibility.
	Diversity: Diversity of inputs into the decision making system.	Disparity between technical expertise and the monitoring inputs available to administrative departments and water managers.	Inventories on irrigation and fields and meadow, high collation of different data sets; comparatively less data available on glacier melt contributions to run-off, but research programmes to fill gap.

(continued)

Table 13.2 (continued)

Knowledge indicators	Operationalisation	Case examples	
		Chile	Switzerland
	Coverage: Extensiveness and accuracy of the monitoring network.	State monitors snow and precipitation, but irrigators also rely on private meteorological stations (e.g. Mina Andina) for weather information. CONAMA responsible for water quality monitoring, DGA for water quantity monitoring (4× per year), ONEMI for enforcement of quality failures. Monitoring difficult to coordinate and then implement controls. No national network and no consistent and coherent annual/monthly/regular monitoring of quality issues. DGA recently initiated a Chile wide glacier monitoring programme.	Hourly quantity monitoring across public and private sectors; canton quality monitoring as part of a national network co-ordinated through the federal level. Private sector collaboration on monitoring network (e.g. Universities, Canton, Hydropower companies; Engineering consultancies establishing stream sediment monitoring). In mountain areas, monitoring network is less advanced than elsewhere, but more monitoring stations are being implemented to improve understanding of spring levels under climate change.
Transparency	*Availability/Coverage*: Availability, relevance of and access to information on water resources.	Lack of available, systematised and accessible information on water rights, water judgments, water market and prices, and the health and availability of water resources. Public water registry is out of date, and incomplete and non-electronic. Odepa.cl – available online data; DGA has no oversight on the trading and transaction of rights, market is seen as 'dark'; information on impacts in upper basin/mining companies is inaccessible. Multiple institutions manage water rights data (*Conservador de Bienes Raises*; DGA, Junta) – non-systematised and chaotic management.	Online publication and access of canton and federal data, plans (e.g. *Kantonale Gewässersanierungs* Plan, CERISE) across different platforms (MeteoSuisse, BAFU, vs.ch, planat.ch). Data on snow-production, water use, hydropower use are difficult to access, and spread across multiple companies and communes (though available online at many commune websites). MINERVE data sharing is still paper based in contrast to the online publication of plans and data concerning TRC. MINERVE implements a convention (signed between the owners of the hydropower installations and the Valais) for an exchange of information during crisis periods.

	Communication: Communicating relevant information for extreme periods and capacity building.	Lack of education and communication on water conservation; personal and public communication on drought preparation, advice provision from JdV presidents on managing irrigation and crop production in drought periods.	Detailed communication of TRC through online communications, newsletters and participative fora to explain the justifications for why it is necessary to intervene on the watercourses, and to secure the plain. Attempts at canton level to be more pre-emptive in their communication with municipal level stakeholders. Local press and media used to communicate water supply provisions during scarcity periods and warnings during emergency flooding events. MINERVE convention provides for public communication during extreme events. Canton is legally required to ensure communication on habitation in hazardous areas and where houses can be built (flood and avalanche protection).
Perceptions	*Awareness*: Awareness of climate change impacts.	Observational awareness from irrigators and water managers that climate is changing, with reduced snow pack, melting glaciers and precipitation changes, but lack of popular awareness on water conservation; Acknowledgement that climate impacts will heighten and hydrological resources will be increasingly diminished; DGA/Presidential acceptance of climate challenge.	Observations of change in the local climate (temperature, snow pack, timing of snow melt, glacier retreat, isotherm, permafrost) heighten the awareness of rate of climate change. In the region, engagement on climate is higher in the winter tourism and hydropower sector, where increasing glacier melt heightens hydropower production but increases material flows, in the reservoirs than in water provision, where communes mainly use non-concession spring water and there is a strong perception of Valais as the 'water tower' of Europe. Preparations for one scale of change (dry summers, glacier melt) but apathy towards more drastic ones (glacier disappearance, water scarcity).

(continued)

Table 13.2 (continued)

Knowledge indicators	Operationalisation	Case examples	
		Chile	Switzerland
	Openness: Openness to learning and willingness to adopt new solutions or paradigms.	Perception of the uniqueness and success of the Chilean model, therefore reticence in ability to learn from other experiences; older water users seen to be unwilling to move with the times; water and environmental authorities look to European WFD; acceptance of challenges in institutions, but not of the approach.	Belief that historical adaptations positions Valais better than rest of Switzerland; research partnerships and knowledge exchange associations within and outside of canton suggest openness to learning from external experts and other regions.
Experience and expertise	*Experience*: Mix of technical competence, traditional knowledge, social memory, training and years of experience.	Lack of professional training/education in water resources management for those private actors 'managing' water rights, for normal or drought periods. JdV/Canalista positions tend to be non-professional and part time. Strong role of tradition ('culture of water') and history in management of the canals. Experience of managing water scarcity differs from section to section and region to region. CNR/CORFO seminars and training programmes aim to improve management, techniques and efficiency and knowledge transfer in visits abroad. In drought management DGA relies on provision of technical information/parameters, while irrigators' intuition guides the management process. Lack of local knowledge/capacity in DGA drought interveners.	Disparities across municipalities and sectors; post-event debriefings occur at company level, allowing lessons to be applied to management of and preparation for future events; training courses on water management specialised for water management within different sectors through commune and cantonal administrations (e.g. *L'association Valaisaine de distributer d'eau'*; *Societe Suisse de l'industrie du gaz et d'eau*; *Landwirtschaftzentrum*; seminars on extreme events in *Dienstelle für Naturgefahr*). Long history of experience of rain shadow effect and annual training exercises on disaster response. Mix of experts in TRC (engineers, hydrologists, ecologists, territorial management etc.)

Expertise: Multiple forms and disciplines of expertise contributing to water related decisions.	Court responsibility for resolving water conflicts result in judgements that steer the course of water management (ref Supreme Court decision on Factor de Uso). Local/regional judges may lack water specific expertise, leading to bad decisions and causing actors to circumvent the official process where possible. Level of technical expertise is high in water science, but ability to inform decision making and management from technical perspective is obstructed. Technification of water management in irrigation in the basin is mainly traditional (e.g. Bocatoma; irrigation efficiency; measuring/monitoring of canals; reliance on 'culture of water' rather than professional water management training). At government level, technical expertise is limited to hydrology and economics, no focus on public policy.	Scope to improve technification in agricultural sector, farmers are increasingly part time, impacting traditional associations for management of irrigation right; at local level, commune presidents may assume responsibility for water management decisions, presiding over the post for short periods and on part time basis; certain water management responsibilities at local level (*gefahrenkarte*, *sanierung* etc.) and canton level (NAQUA monitoring) tend to be outsourced to private engineering consultancies (though water suppliers have in house expertise as well) with the requisite expertise to work on the project in conjunction with the specialist cantonal department.
Secularism/Independence: Separation of political and technical facets for neutrality and continuity in water governance.	Lack of independence and political secularism of scholars and experts informing water governance (collusions of World Bank, Neo-liberal politicians, economic agenda, financial contributions etc.). Operational/technical expertise at the regional level more functional than at the central political level.	Administrative departments involved with water management at canton and commune are not political appointees, but professional positions that remain in posts despite political outcomes. In commune councils elected officials constitute the political group and technical staff the service group for both watercourse management and provision; changes in legal framework are subject to or initiated by a broader range of stakeholder due to the direct democratic process.

(continued)

13.1.3 Networks

Table 13.3 Operationalised indicators of adaptive capacity relating to network components of the governance system with case examples

		Case example	
Network indicators	Operationalisation	Chile	Switzerland
Cooperation (Collaboration)	*Negotiation*: Ability to negotiate and resolve conflicts, reach agreements on water distribution, security and pollution.	Lack of a formal flexible mechanism of conflict resolution; court costs require resolutions to be financed by the JdV rather than Association of Canalistas, who are financially weaker; user to user negotiation on water issues, and agreements between JdV and mining and hydroelectric companies in the basin; utilities negotiate water prices nodes every 5 years with *Superintencia*; individualism and autonomy amongst rights holders challenges ability to agree a canal or JdV level strategy or solution and lack of interest to cooperate.	Consultation between private sector (hydropower companies) and commune level; balance between federal provisions (legal base for financing), individual or group interested stakeholders, and fulfilment of functions concerning security, but also the ecology of the watercourses; lengthy and complex participative settlement of project implementation (leads to weakening of more radical innovation in TRC); conflict resolution expected at commune, with mediation role played by semi-administrative bodes such as *Landwirtschaftzentrum*; community resource management deemed as an important factor for solidarity and conflict resolution (attempting to foster it despite advances in modern irrigation infrastructure).
	Modes of organisation: Institutional platforms for actors to collaborate and cooperate across different sectors, uses and scales.	Formation of JdV are provided for in the Water Code; Self-organised cooperation between JdV and canals during periods of drought allows flexible agreements and private compensation on water releases from one section to another that helps spread the risk and impacts more evenly and reduce social conflict – based on variable personal relationships and trust (i.e. functions much better between certain sections than others). Drought provision in Water Code instructs DGA on compensation and liability, while CNR is guided by Law 18.450 for promotions and financial support of irrigation and efficiency.	Cross-sector collaboration through the Krisenstab for extreme periods, but little cross-sector co-ordination or contact on water resource management during 'normal' periods; individual company to company or commune to company agreements on water releases for artificial snow production; hydropower operators, universities cooperate and share information and flood management responsibility with canton and commune administrations through the convention for MINERVE; water optimisation between hydropower operators (that are reliant on same resource); increasing levels of collaboration between communes (e.g. joint planning and training); sector, regional and local focus to partnerships and platforms for cooperation and joint learning.

Incentivisation: Mechanisms to incentivise cooperation amongst water stakeholders within a basin.	Bi-lateral agreements between different sets of rights holders; increasing cooperation between mining sector and farmers (e.g. 10 million peso donation from Mina Andina to JdV, administered by the CNR as the executive and technical body for utilisation of the funds); private financing through water rights owners for maintenance and upkeep of distribution system - users pay for the operation and maintenance of infrastructure, while the state bears cost of the capital investment (Water Law 1981/1123); Government is liable to pay compensation for affectation to 3rd party rights from DGA drought management decisions (i.e. if new wells affect the rights of other owners), and compensate for the costs of extra infrastructure required for water supply.	NFA provides subsidies for increased participation in the implementation of water management projects (including meeting ecological criteria); reliance on financial and technical capacity at federal and canton levels provides for higher levels of cooperation at commune levels (where autonomy and sovereignty is strong); rivalries persist between different communes and the cantons on decision making on watercourse management – but attempts to reconcile them through participative consultation in TRC and final technical implementation signed off by Valais (and Vaud) Council; participative project planning/approval process is resource and time intensive, but aims to foster consensus and cohesion (Hydropower, TRC etc.).

(continued)

Table 13.3 (continued)

Network indicators	Operationalisation	Case example: Chile	Case example: Switzerland
	Collaboration: Trust, power balances, mixed nature of support structures.	Broken agreements between JdV leads to a lack of trust across upstream and downstream sections of the basin and diminishes the ability to informally manage water sharing, thereby requiring intervention from the DGA; weaker economic actors (including smaller farmers, or those last in the canal) struggle to protect their water rights from quality or quantity damage by stronger economic actors (e.g. mining); negotiations between irrigators and farmers concern increased groundwater exploitation or compensation from drought declaration to pay for constructing more wells; political battle to amend the Water Code was ideological, acrimonious and lengthy for minor changes; general lack of trust between water rights owners community (including DOH) and DGA over Aconcagua Project and well abstraction.	Canton provides financial support for remediation when the communes cannot cover the costs (e.g. 1993); Investment in irrigation infrastructure through both canton and federal subsidies and technical assistance (*Landwirtschaftszentrum*) and programmes supported by charities at the national level, with the aim of maintaining both built and social infrastructure of the *Suonen* for social cohesion and conflict resolution; Financial and technical/training assistance provided to communes for water course management and implementation of cantonal provisions (hazard mapping, zoning); collaborative adaptation research projects (*Regionprojekt, Verwerktung Markt am Handel*) between tourism and agriculture; distrust of canton/ecological sectors and agricultural lobby over land issues in TRC.
Participation (Ref GA)	*Participation*: Providing a voice in decision-making across water stakeholders.	Water management is effectively in the hands of water rights owners, providing for strength of participation according to the amounts of rights that are owned; civil society participation (as per the IWRM model) is weakly provided for through consultation in the EIA process (or in other basins the Mesa del Agua).	Participative process in TRC implementation (COREPIL) attempted to reconcile different stakeholder positions, but is time and resource intensive and reduces the innovative elements of the project; participation in development and building planning open to organisations (*Verbandbeschwerderecht;* EIA) and individuals, though in reality may not be implemented.

Knowledge Partnerships	*Integration/ Dissemination*: Partnerships and clear processes to integrating scientific information into decision making.	Canvassing by means of submission of proposals to DGA to allow greater exploitation of the aquifer; relationship between academic institutions (both major universities and research bodies) is heavily financed by the strongest economic actors; regional and national government agencies rely on technical studies, data and research from major universities and private consultancies (e.g. CEASA water quality studies).	Convention was signed between the owners of the installations, a state university and the Valais for provision of meteorological and climate data for crisis management as part of MINERVE (including informing the public); hydrological data for water course management deemed behind other hazards (e.g. avalanche, rock falls etc.); collaborative research between hydrological institutes and hydropower companies to generate short to medium term precipitation prognosis/forecast.
	Exchange & Support: Informal or formal networks to share and exchange best practice, lessons learnt, and technical solutions.	DOH, CODELCO, CRA, CNR constitute the executive Mesa Tecnica de Agua in Aconcagua, which works with different actors such as regional DGA, MMA, ESVAL, and other State institutions to resolve issues relating to the Aconcagua Project; International experts (EU, World Bank; UN-ECLAC) consulted with and employed by government ministries (MMA, DGA); Influence of foreign scholars (Milton Freedman and the Chicago School[a]) infamous in development of 1980 Constitution; Access to consultants for irrigation projects can be limited for poorer farmers (for CNR proposals); CORFO (corporation of promotion) organises international expeditions to enhance technical capacity of producers (agriculture, industrial etc.).	Research partnerships between private companies (hydropower, manufacturing firms) and academia (Technical Institutions, e.g. Luzern, EPFL, WSL, EAWAG) to improve management of surface flows; MINERVE not yet formalised, and paper-based; TRC relies on partnerships with academia for climate change projections; Collaborative sectoral associations (Association of Valais Water Distributers; SDOC/SDRC; *Societe Suisse de l'industrie du gaz et d'eau, Swiss Mountain Water*); WA21 provides an inter-sectoral national level platform for knowledge sharing and dissemination on IWRM across Switzerland; Federal offices are tasked with support and information provision to lower administrative levels; collaborative research projects aim to inform (but not prescribe) decision making (ACQWA; Interegg 3).

(continued)

Table 13.3 (continued)

Network indicators	Operationalisation	Case example	
		Chile	Switzerland
Institutional integration (Co-ordination)	*Co-ordination/ Clarity*: Clearly defined roles and means of coordination between institutions, levels and relationship to the physical boundaries.	Quality (MMA) and quantity (DGA) are managed separately; Rivers are divided up into different independent sections (Aconcagua has 4 sections with 3 JdV) so that delivery of rights are autonomously managed per section; different responsibilities for water spread across the different ministries (mining and hydropower have the money and the power); drought commission was created by the president in conjunction with the drought decree to integrate Interior Ministry, MOP, and Ministry of Agriculture, presided over by the Interior Minister; narrow cooperation between the ministries during 'normal' times; Mesa Tecnica includes JdVs, DOH and CODELCO fully, ESVAL participates intermittently; surface and groundwater managed through different instruments, but the new law improves integration of the two resources through an instrument to improve the infiltration of the aquifers to recover water table levels; land and water rights also separated in law.	Duplication of effort, trying to better coordinate on natural hazard management as challenges mount; Water management at canton level is sector based working group, who are conducting preparatory meetings to establish a Wasserkompetenz Zentrum within the cantonal administration; lack of co-ordination across different sources/uses (springs, groundwater, glacier, surface waters, lateral streams, Rhône); examples of integration of different water management challenges into one institutional/infrastructural response in adaptive actions (e.g. MINERVE, TRC); increasing connection and integration between communes on water supply; Coordination within and across Valais (and also Vaud) for TRC, but not to external stakeholders in the rest of the Basin (Geneva, France); CIPEL provides platform for coordination and collaboration across the different communes and countries sharing the Lac Leman; cross sector coordination through CERISE in water crisis situations; Hydropower teams at commune levels have close relationships with communal organisations.

	Power & Balance: Balance across institutional relationships.	Media and press articles used to influence decision making on rights allocation and the Aconcagua project; rights administration disagreements between DOH & irrigators and the DGA; power struggles between CNR and MOP, DOH and DGA, DGA and MMA reduces efficacy of coordination and roles – relationship changes from one administration to the next; strength and independence of powerful economic actors, and their infiltration into political, academic and judicial decision making.	Shifting roles and responsibilities across private and public spaces (in agriculture to public and in flood defence to private) according to diminishing and rising capacity; competing policy priorities across sectors (e.g. micro-hydropower versus environmental flows provisions) with no overview or integration into TRC; agricultural and hydropower stakeholders at odds against the ecologicalisation of water management at canton and federal levels; communes are independent and autonomous from canton and federal administrations, but are subject to canton and federal laws, and reliant on subsidies and support.
Levels of decision making	*Administrative Authority*: Relative authority at different administrative levels enables challenges to be resolved.	Centralist government with high degree of Presidential powers and priority setting (e.g. Presidential decision required on projects, drought declaration etc.); Presidential interventions on project approval; regional branches (operational level) of ministries handcuffed by planning and policy decisions by national bodies (political level) and planning offices (MIDEPLAN) in Santiago (seen as'King'), with lack of effective feedback mechanism between central and regional authorities; Courts have authority to define precedents in the implementation of the Water Code, thereby retain a powerful role in water 'management'; canal and JdV actors express 'powerlessness' against government institutions.	Decentralised system of federalism, with role of implementation at canton level, and in the Valais, the communes are particularly autonomous; subsidies for following ecological and security priorities of the 2001 Federal Directive for watercourse management is key for balancing the autonomy of decision making at the commune level (subsidised up to 95 % for *Gefahrenkarte, ZonenPlan, Bauzone etc.*) enabling canton and federal environmental agencies to ensuring the implementation of priority policies and concepts; canton administrations are responsible for sectoral coordination and support, but communes for implementation, with technical and financial support from canton; Valais, as a relatively poor canton is reliant on federal subsidies.

(continued)

Table 13.3 (continued)

Network indicators	Operationalisation	Case example Chile	 Switzerland
	Agency & Autonomy: Relative agency and autonomy of sectors and levels enables challenges to be resolved.	Limits to public authorities' ability to actively manage or regulate water resources; state has a subsidiary role in water management, which is in the hands of the private water rights holders; water users have independence and autonomy and the role of denouncing illegal extraction or pollution to the DGA; centralist bureaucracy is sidestepped in rights registration and bypassed by private actor negotiations, agreements and adaptive actions to cope with periods of drought; DGA unable to affect property rights, therefore emphasis is on users to self-organise to resolve drought induced conflicts; in extreme drought periods, government assistance and interventions is at the request of the irrigators.	Federal and cantonal governments have a subsidiary role in water management at the commune level, limiting their authority (right to redress) on water management across the country and region; water decisions reside at the commune public level, but are influenced by canton level coordination and support, particularly in communes where finances or capacity is low; private user groups are gradually morphing into public supported institutions; Increasingly responsibility for maintenance of irrigation infrastructure is shifting to public hand from the private, while canton is taking more oversight over energy concessions; provisions for participation in decision making at local, canton and federal levels take time, but build consensus; Constitutional Right to Petition allows for citizens to self-organise and influence the direction of water management at a federal level.

[a]Klein (2008) and Valdes (1995)

13.2 Synthesis

Figures 13.1 and 13.2 present a synthesis of the results so that the linkages between governance context and adaptive capacity can be more clearly shown. Drawing on Tables 10.3 and 10.4, Fig. 13.1 highlights the concentration of categories of adaptive outcome per case area. The Swiss case has a higher concentration of transformative and persistent adaptation outcomes, while the Chilean case has a higher concentration of passive and persistent cases. In the framing of adaptive capacity as the enhanced ability to transform or adapt to new challenges or states, the inference follows that the higher the adaptive capacity, the more transformative the adaptive outcomes should be.

As hypothesised in Part I, the adaptive outcomes from the Swiss cases correspond with more transformative and adaptive actions and management approaches as well as a more positive correlation with the adaptive capacity indicators (Fig. 13.2). On the other hand, the adaptive outcomes in the Chilean case correspond with less transformative outcomes as well as a less positive correlation to the adaptive capacity indicators (Fig. 13.2).

In characterising the governance context, it is worth repeating that the Chilean governance mode is represented by a centralist, neo-liberal market model while the Swiss case is representative of a decentralised multi-level governance model. As has been discussed in the sections above, despite the different governance models, both cases share common challenges in the development and mobilisation of proactive and reactive adaptive capacity, perhaps partly since both models ascribe a similar level if not type of autonomy to the local level (in Chile to the user rights holders and in Switzerland at the local level). Figure 13.2 provides a more nuanced characterisation of the governance context by presenting a synthesis of the governance related indicators of adaptive capacity as operationalised in Chap. 12 and assessed in the previous section above.

Results indicate a correlation between more transformative and persistent adaptive actions and the decentralised governance context of the Swiss case. The centralised and yet neo-liberal market model of the Chilean case is dominated by a number of passive actions, which can be seen to correlate with potential longer term degradation of the resilience of the social-ecological system (e.g. Water Code, Art 314 allowing for more intensive exploitation of groundwater resources). However, both cases are correlated with a number of persistent adaptive actions.

Figure 13.2 clearly highlights the tension between the rigid and inflexible legislative context in the Chilean case, with the higher levels of autonomy at the user levels, which frustrates and constricts the ability of water managers and the owners of use rights to adapt in a more proactive manner to hydrological changes and stresses in the basin. While reactive coping techniques can be quickly called upon through the networks and traditions that exist, more long term preperations and transformative approaches for meeting the mounting challenges are blocked by lack of trust and cooperation, lack of agency at regional operational levels and lack of accessible and appropriate information on water resources.

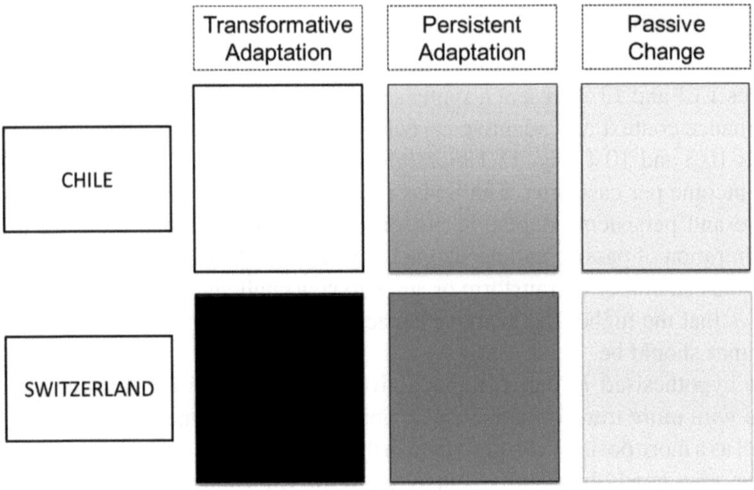

Fig. 13.1 Overview of adaptation context: categorisation of policy and adaptive action response

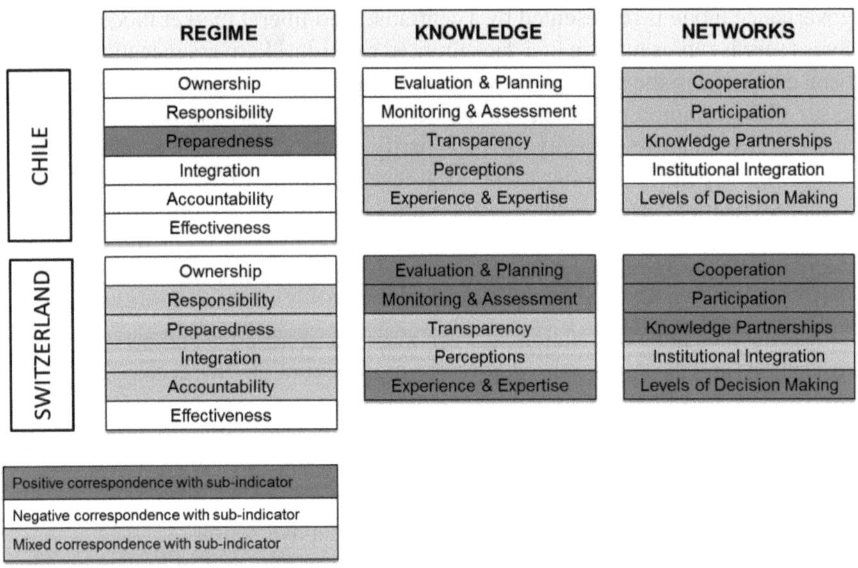

Fig. 13.2 Synthesis of results for case area evidence and positive or negative correlation with operationalised indicators of adaptive capacity

In contrast the Swiss indictator summary points to highly networked layers of governance that allow knowledge and learning to be transferred vertically across different levels of capacity (across federal university partnerships to local level support institutions) but has greater challenges with horozontal integration. While the

Swiss case is defined by more transformative and persistent adaptive outcomes, Fig. 13.2 highlights the areas of rule making and division of responsibility that remain a challenge.

For some of the indicators highighted as mixed in purple, there was a wide range of fulfillments. In terms of Transparency, the indicator has very mixed examples in both cases, but the Swiss case was more positive than the Chile. Levels of Decision Making was also mixed with negative criteria in the Chilean case, as was Institutional Integration in the Swiss case. The synthesis presented could also be a useful tool to help guide decision makers on where resources would be best used to address elements of the governance system that hinder adaptive capacity and where to foster elements that enable adaptive capacity.

Pinpointing the different indicators that are problematic (orange shaded indicators), and specifying the sub-indicators, as presented in Tables 13.1, 13.2, and 13.3, could help direct resources to repair problematic points of the governance framework and better utilise enabling elements. The following section of discussion will elaborate the particular tensions that emerged from the analysis of adaptive capacity and highlight how the operationalised indicators can help address the implicit trade-offs. The discussion of the trade-offs across different scales of governance and change will be followed by the further development of a framework or tool to guide decision makers on how to address both proactive and reactive elements of adaptive capacity. By better understanding the governance components of both sets of responses, the aim is to allow water managers and decision makers not only to generate more transformative approaches that would not limit future choices, but also to foster reactive parts of the system necessary for smaller and faster transformations in extreme events.

13.3 Part III Conclusion

Investigating adaptive capacity through the exploration of the governance related adaptive responses in relation to extreme events, allowed greater focus on the tensions and trade-offs implicit in not only anticipatively developing the capacity to manage and cope with environmental and climate variability, but also in mobilising that capacity in reaction to stresses as they manifest. Categorising these responses in terms of their transformative, persistent or passive adaptation characteristics enables linkages to be assigned between certain governance approaches and mechanisms with greater transformative potential.

In Chap. 10, it was found that more transformative responses were associated with the following governance mechanisms:

Regime

- Mix of federal legislation and cantonal legislation setting the framework for the most transformational elements of the TRC (integration of uncertainty and climate information, integrated risk management based on social-ecological resilience).

- Linkage of environmental provisions within these laws and environmental goals with subsidy programmes and economic incentives.

Knowledge

- Elongated and iterative planning time horizons, enabled through a diverse range of impact studies and multi-stakeholder investigations to allow for compromise and balance in the project.
- Integration of climate change adjusted risk and uncertainty into planning, acknowledging that current levels of flows may be surpassed in the future.
- Flexibility allowed through the implementation plan so that the technical experts, rather than politicians, can define the planning process.
- Scientific and technical monitoring and modelling are relied upon to diagnose vulnerabilities, and communication programmes intended to translate the outcome studies into justifications for the project with local level stakeholders.

Networks

- Reliance on federal financial support allows the federal level (more transformative approach) to have some power, but regional particularities and needs are accounted for through the decentralised implementation structure.
- Each scale has its own source of power and agency leading to an extenuated impasse in passing the implementation plan, but the potential to negotiate a common, integrated solution.

While more passive approaches were associated with the following governance mechanisms:

Regime

- Drought declaration supports coping but allows further exploitation of 'vulnerable' ground water sources.
- Informality of water governance in 'normal' periods associated with a lack of capacity and knowledge in institutions that are called upon for drought management.
- Legal guidelines exist for the management of increasing flooding issues (governmental policy guidance), but there is a void of guidance and rules on scarcity or stress.

Knowledge

- No requirement to account for uncertainty through climate change impacts or inter-annual variability.
- Loss of knowledge during drought interventions, since government actors lack the capacity and familiarity of the basin.
- Lack of coherence across different evaluations and assessments of the hydrological resources available limits the ability of both public agencies and private actors to agree on basin planning.
- Strong awareness amongst water owners that hydrological patterns are shifting has not translated to enhanced use of technology, monitoring, modelling or

integration of uncertainty into the management and planning of water resources in the basin.

- Ideological rigidity limits the ability to change the framework rules which govern the water system and constricts actors' views of how to resolve the complex emerging problems.
- Lacks preparedness and planning for possible scarcity situations due to the perception of climate change as an issue to be taken into account for long term horizon planning but not yet for operational day to day management.

Networks

- Lack of trust between actors impedes integrated solutions to common problems.
- Power imbalances between different ministries and government institutions are associated with the environment and weaker economic actors being side-lined in water resource management.

Chapter 11 then presented and discussed the major bridges and barriers to adaptive capacity. Results were drawn from the coding and analysing of interviews, which identified points made about the challenges and ability of the systems to respond to climate related stresses, to identify aspects of the governance system that can stop, delay, hinder, or help actors during the process of adaptation. Results from the bridges and barriers analysis revealed common challenges across both case areas and scales of governance relating to conflicts between inter-jurisdictional agency and autonomy, institutional and technical capacity, as well as information and data availability and accessibility.

In Chile, despite stakeholders at the national level suggesting the flexibility of the Water Market was a major bridge to mobilising flexible solutions to hydrological variation, water users at the local level did not point to market transactions themselves as a mechanism that enabled coping during drought periods. Instead, a lack of information and trust, as well as enforcement issues and institutional capacity were the main preoccupations at the local level that detracted from the potential positive effects of flexible management at the user level. In Switzerland, stakeholders across all levels concentrated more on the issues of local autonomy, including challenges and strategies related to the decentralised mode of governance, as a potential barrier to integrated long term planning of water resources that could develop proactive capacity. Across both case areas, stakeholders pointed to the importance of research networks and knowledge partnerships in developing their understanding of the challenges and solutions to climate change impacts.

Drawing on these analytical steps, Chap. 12 presented a set of more nuanced indicators of adaptive capacity that had been developed from the original determinants presented in Part I. Finally, this chapter has presented an analysis of the fulfilment of the adaptive capacity indicators and a synthesis review of the correlation between adaptive mechanisms and the operationalised indicators of adaptive capacity. Assessment of the indicators showed challenges across the regime categories in the Chilean case, mainly in relation to the dichotomy of a rigid and inflexible legislative context with a high level of autonomy at the user levels, constricting the ability of the networks of water manager and rights owners to adapt in a more proactive manner to hydrological changes and stresses in the basin (hampering the generally positive correlations within network and knowledge indicators).

While reactive coping techniques can be quickly called on through the networks and traditions that exist, more long term preparations and transformative approaches for meeting the mounting challenges are blocked by lack of trust and cooperation, lack of agency at regional operational levels and lack of accessible and appropriate information on water resources. In contrast, the Swiss indicator summary points to the highly networked layers of governance that allow knowledge and learning to be transferred vertically across different levels of capacity, but that has greater challenges with horizontal integration, which mirror the challenges associated with the implementation of IWRM. Despite being characterised by more transformative and persistent adaptation, Fig. 13.2 highlights that the areas of effective enforcement, division of responsibilities and 'locking in' certain uses remain a challenge.

References

Klein N (2008) The Shock Doctrine: The Rise of Disaster Capitalism. Penguin Press, London, U.K.

Valdes JG (1995) Pinochet's Economists: The Chicago School of Economics in Chile (Historical Perspectives on Modern Economics). Cambridge University Press.

Part IV
Challenges in Developing and Mobilising Adaptive Capacity

Chapter 14
Balancing Structural Conflicts Across Scales to Develop and Mobilise Adaptive Capacity

Abstract Building on the three stages of analysis presented in Part III, this first chapter of Part IV discusses the challenges of developing and mobilising adaptive capacity across the complex spatial and temporal scales that emerged as key themes in earlier analysis. Across the spatial scale, there is a challenge in balancing guidance and certainty from higher levels of governance with flexibility of autonomous actors to respond quickly to challenges at the local scale. Furthermore, adaptation to certain stress conditions within one scale or magnitude of change was found to not necessarily imply long-term adaptability to conditions whose persistence and impacts will be more pervasive.

Keywords Rhône, Canton Valais, Switzerland • Aconcagua, Region V, Chile • Long term adaptability • Short term reactive capacity • Tensions across governance scales • Adaptation to climate change and variability • Balancing flexibility and predictability

14.1 The Spatial Scale

A common thread that emerged in analysis of adaptive actions and related governance mechanisms was the underlying tension of balancing guidance and certainty from higher levels of governance with flexibility and autonomy of users and rights holders at lower scales. It is a challenge that is further heightened in times of stress in the case areas, which instigate a heightened involvement of central or regional government agencies, whether from a financial or organisational capacity. The results presented in the previous chapters in Part III elucidate the empirical evidence related to the sub indicators of adaptive capacity and thus allow for trade-offs to be identified in the relationship between the requirement for clear rules and certainty to guide the development of adaptive behaviour and rules to mobilise adaptive actions

M. Hill, *Climate Change and Water Governance: Adaptive Capacity in Chile and Switzerland*, Advances in Global Change Research 54,
DOI 10.1007/978-94-007-5796-7_14, © Springer Science+Business Media Dordrecht 2013

in extreme events with the requisite flexibility for local actors to react and plan according to their individual needs.

While clarity in rules and legal certainty is fundamental for accountability in water governance (see Part II), it can also lock ownership and use rights into codified norms that are based on out of date hydrological data and patterns. In terms of evidence in 'responsibility', coordinating and organising institutions are needed for inter-connected water policy and management particularly in the face of complex and uncertain challenges. But there is a need to recognise local individualities and needs, which can go unconsidered at higher levels of administration. While participative processes can address this dichotomy, they can also stall agreements on projects and frustrate multiple stakeholders (especially if not matched with requisite knowledge and information assets). Other studies have discussed similar challenges in relation to balancing legitimacy and accountability through IWRM based approaches with adaptive management criteria of flexibility, experimentation and self-organisation (Engle et al. 2011). Indeed, this trade-off is elucidated in their comparison of IWRM and adaptive management criteria in the case of Brazilian water governance. Engle et al. (2011) found that 'centralization of decisions in the hands of the technical agency may facilitate the implementation of experiments as well as afford a level of flexibility that may be incompatible with more decentralized systems'.

Evidence from the 'preparedness' indicators suggests that the rules at higher levels that guide stakeholders at lower levels for managing extreme hydrological situations need not only to take the local reality into account, but also be matched with capacity at local levels so that provisions can be effectively interpreted and implemented. Finally, all three indicators reveal evidence for the struggle to find a balance between autonomy and strength of user rights for managing their resource, while holding disparate actors together through a formalised set of enforceable provisions that allow for the sustainable management of the resource and bring actors together to resolve common problems. High levels of informality may devolve agency to lower levels, but if this is not matched with guidance, incentives and the requisite knowledge to cooperate on complex challenges, it is associated with policies that lead to the passive degradation of the SES.

The challenge through both these preparatory and reaction periods represent a balancing of the trade off between flexibility and predictability to optimise adaptive capacity. It may be described as the search for juggling structure, guidance, and policy certainty at higher administrative scales, in a manner that also facilitates and supports autonomous adaptations at local levels. Succesfully balancing this trade-off could help to maintain the ability of a governance approach to allow for both reactive and proactive adaptive capacity to be built and mobilised. To reiterate from earlier discussion, while reactive and autonomous adaptation is the ability to change and adapt to new threats or realities that have manifested (Tompkins and Adger 2005), proactive adaptation can in turn be categorised as longer term preparations for different scales of change.

Flexibility can be seen as short term transformation potential, i.e. the ability to change course, reorganise, and mobilise quickly if the SES is on an unsustainable and dangerous trajectory, or faced with a sudden shock, by mobilising its ability to reactively adapt. On the other hand, predictability is linked to the need for legal certainty and guidance for building longer term transformational potential. It also refers to long term policy planning that enables a system to become proactive in its adaptation to a particular type of extreme. In order to not only develop adaptive capacity, but mobilise it to both variability and larger scale changes, the cases elucidate the importance of building both reactive and proactive adaptive capacity.

While proactive adaptive capacity can be associated with predictability and guidance at higher levels, reactive capacity is enabled through flexibility and autonomy at lower governance levels. One of the major challenges in climate change adaptation is therefore navigating this balance between fostering the flexibility needed to deal with an increase in the likelihood of complex and unexpected changes from climate change (Ebbesson 2010) while maintaining the certainty and guidance for longer term preparedness through legislative, regulatory and policy frameworks.

Other studies have focussed on the high level trade-offs that policy and decision makers face in any democratic system when considering climate change risks and adaptation, through socio-political and economic factors (Tompkins and Adger 2005). Short term political cycles, limited public attention on longer term challenges and judgements on risks and costs of climate change dilute the urgent context in which climate mitigation and adaptation should take place. Tompkins and Adger (2005) refer to the trade-offs between cost, risks and socio-political factors as being 'encompassed in the shape of the indifference curve between reactive and anticipatory management' (p 565), which are navigated by the institutional landscape made up of government and civil society actors, as well as individual agents. While in their article, both mitigation and adaptation are the unit of analysis, similar trade-offs are identified within the focus on adaptation alone. Trade offs are also present in decision making on investments at different governance levels for adaptation. Decision makers must decide at what level, and in what form (social, technical, financial) to invest limited resources (temporal, financial, educational).

Figure 14.1 suggests a representation of how this adaptation trade-off manifests across the suggested core tension in adaptive capacity. The figure purposefully does not represent this as a linear regression, from highly predictive enabling proactive adaptation while highly flexible facilitating reactive adaptation. It is not suggesting a linear relationship between the two elements of adaptive capacity, but more an intersecting connection, with elements of proactive adaptive capacity enabling succesful reactive adaptive capacity (e.g. TRC). Likewise, increasing numbers of reactions to extreme events may have the potential to impact longer term preparedness for climate change, by taking advantage of windows of opportunity to push through plans relating to adaptation. The proceeding section further discusses how this tension manifests across the different indicators, while the following section will propose a multi-scale framework to address the tension in the process of developing both reactive and proactive adaptive capacity.

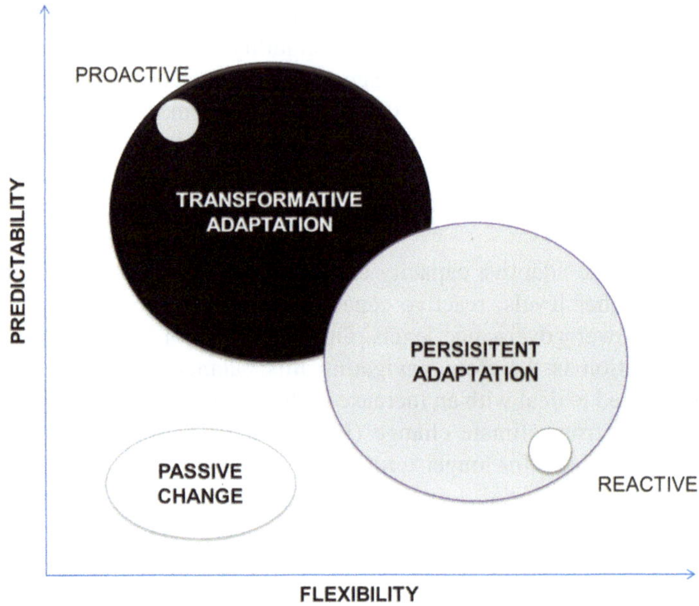

Fig. 14.1 Balancing out the core tension (flexibility and predictability) in order to generate both proactive (longer term preparedness to climate change impacts) and reactive (flexibility and quick reactions to climate events) adaptive capacity

14.1.1 Regime

Certainty and the rule of law are fundamental in a governance system to ensure governments are subject to the law, providing trust in their rule, predictability in planning decisions and security in longer term investment (Cosens 2010; Craig 2009; Ebbesson 2010; Ruhl 2009). *Regime* indicators represent the elements of the system that provide this level of predictability to guide the actions of institutions and individual actors in managing water resources. The legal and property rights framework is crucial for specifying ownership and use of water resources as well as the rules and regulations that determine the management of the water resources, which water rights owners must follow. *Regime* indicators of *ownership* and *responsibility* are important in clarifying the rules that denote rights, duties, privileges, power and responsibility (Ebbesson 2010) that impact how an SES is managed.

Within the Swiss case, federal and cantonal legislative provisions that provide the duties of ecological integration in spatial planning and integrated flood management are driving forces behind the transformative elements of the TRC (Fig. 10.2, Box 10.1). Furthermore, the subsidy mechanisms in the NFA for beneficial ecoligical outcomes and participatory approaches to commune level projects are also associated with more transformative approaches. On the other hand, in the Swiss case, the length of legal certainty bestowed upon the hydropower concessions as well as legislated priorities for irrigation within cantonal legislation, locks in set water

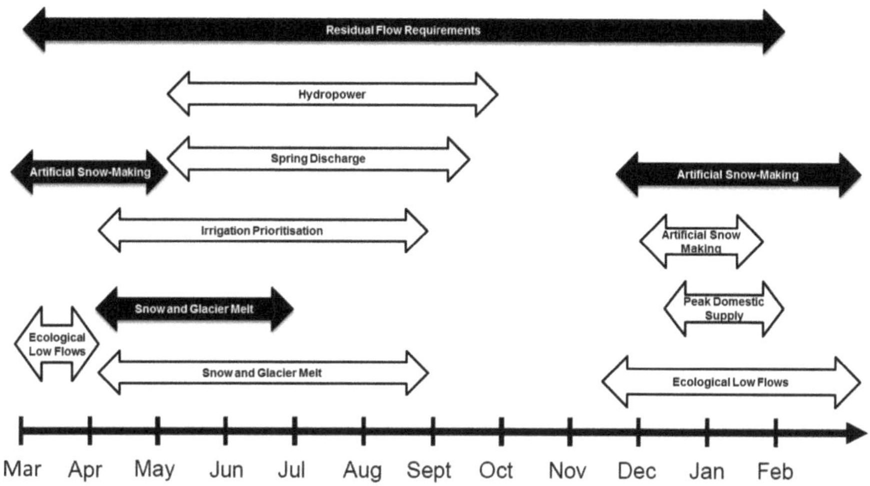

Fig. 14.2 Seasonal timeline of water uses (*White arrows* show the range of traditional water uses, and *black arrows* showing how these rivalries are developing with shifting seasonality and increasing exploitation)

allocations and priorities over timescales during which the hydro-climatic environment is projected to undergo significant change.

Figure 14.2 represents the complexities that shifting seasonality might imply for such long term codified rights, guaranteeing or prioritising allocation to certain users over or during set time periods (Hydropower Concessions, Law on the Use of Hydropower (Art. 42), WPA (Arts. 31, 33, 34 and 36)). It shows that there is genuine cause for concern, in that the multiple rivalries of the streams are gradually being subjected to either physical changes in seasonality or through increased variability, demand and new legislative requirements that are at present not being investigated in an integrated or holistic manner. While some stakeholders express concern that these increasing rivalries will be a challenge for management of water resources in the canton, it is presently still under the radar of cantonal legislators and sectoral policy makers.

Furthermore, any diminishment of spring water (for domestic supply) in relation to glacier melt is likely to be supplemented with increased groundwater exploitation that may have repercussions on surface water recharge. In the Valais, these challenges are in some way bridged by the complex balance of strong local autonomy and sovereignty of water rights with an increasing reliance on federal and cantonal subsidies that aim to encourage implementation of federal and cantonal ecosystem based provisions at the local level to foster a resilience based approach to increasing hydrological extremes. However, the residual flow requirements themselves are a more recent addition to the WPA, and so despite the importance of their role in the protection of the riparian ecosystem and health of the waterways, do have challenging repercussions for socio-economic based rivalries.

It is the common perception of the Chilean system that the unique level of autonomy of the water rights owners allows the system to be highly flexible and adaptive, since they are not constrained by the inefficiencies of government and thus can self-organise to manage solutions at their own level, amongst themselves. However, in practice, the water rights and legal situation in Chile are based on principles that neither promote conservation, preservation of currently scarce resources (though efficiency is an aim) nor protect vulnerable riparian ecosystems. The role of the water rights holders themselves, whether part of a Junta or not, is seen to be one of documentation and distribution rather than any responsibility for management of the resource. Yet in Chile, the subsidiary role of government within the neo-liberal model delegates as many responsibilities as possible to the private actor, leaving a gap between resource use and resource management, that currently no one within the basin is really filling. Even more problematic, the rights structure and information upon which the rights allocation has been based, has allowed for the legal over allocation of the basin, which due to the certainty of the rights themselves (guaranteed by the Constitution and Water Code), is inflexible and non-adaptive to decreasing availability of water.

14.1.2 Knowledge

Knowledge indicators encompass the long term development and integration of climate information as well as the perceptions of environmental issues; whether or not climate change is taken into account in planning and decision making timeframes. Often, stakeholders elucidated how climate change impacts seemed too distant, insurmountable or uncertain to incorporate into current evaluation and planning. While data may be at hand to adequately assist coping strategies with drought or flooding events, in depth studies, monitoring and climate projections may not be accessible for informing longer term planning strategies.

In the Swiss case, monitoring and assessment networks are maintained and used across multiple levels and sectors and there are a number of federal and regional studies and collaborations on long term climate change projections. While the MINERVE and TRC provide examples of climate change integration into longer term planning, at other levels (i.e. local) or in other areas of water management (water provision) long-term effects from climate change (e.g., shifting seasonality of hydrological regime; glacial melt tipping points) seem too far away or too daunting to incorporate into local water management planning. The examples of collaborative and iterative science driven projects can be found in the hydropower sector and the TRC project that integrates climate projections in an iterative and integrative manner for sustainable watercourse management for both short- and long-term coping. So, while the series of flooding events were seen to serve as a wakeup call for political and policy action on developing a longer-term integrative and uncertainty based approach to watercourse management, in most areas of the Valais, alterations in water availability from changes in glacier and snow-melt have been more variable across the canton.

In the Chilean case, there are greater challenges in developing the baseline of adequate data to effectively manage water quality challenges and administer the allocation of water rights. However, interestingly, this is not linked to a low level of expertise, but rather to capacity challenges in the designated institutions for water management, as opposed to the other ministries or sectors that have a stake in water resources. It is the application of water and climate information to both short and long term water management decisions that is the challenge. Therefore, while there is evidence of climate change relevant studies and evaluations being present across sector-specific institutions, there is a struggle to apply this information thematically to water challenges and to holistic water management planning (e.g. reservoir and groundwater planning for Aconcagua Project).

Furthermore, the lack of relevance of water data and calculations for drought management and the historically short time periods used for water allocations, suggests a lack of applicability and appropriateness of information for both short and long term management. Other studies have noted the challenges of overcoming institutional complexity and inertia to ensure that models and data are not maintained after they have been rendered useless (Peters 1987 in Tompkins and Adger 2005). Furthermore, national level studies on climate change impacts and adaptation tend to be sector specific. There are also few mechanisms to objectively evaluate the ability of the local water system (physical and institutional) to cope with increased drought situations or integrate climate impacts into basin level water resource planning.

In both cases, observational awareness of climate change impacts do not automatically translate into an integration of climate change relevant adaptation strategies for coping with the longer term impacts of the change that is being observed. Additionally, the massive implications of greater magnitudes of change induce a level of apathy across different sectors (hydropower, domestic water provision) that reinforce the notion that planning for larger scales of change is pointless. In the Swiss water provision context, the acknowledgement that larger scale changes are likely to occur is tempered by the understanding that drastic impacts from glacier reduction will not manifest over the next generation, and therefore there is no need to include preparations for such impacts at present. However, it is the laws, contracts and infrastructural projects that are being planned now that will need to be relevant and adequate in 10–20 years, just as climate impacts heighten. Decisions made now could lock in the SES to out of date rules, data and management solutions just as the agreements, projects and contracts signed 20–80 years ago have locked in present day management in both case areas (hydro-power concession periods; water rights allocations; urban growth; spatial planning).

Another related challenge is matching the scales at which hydro-climatic expertise and knowledge is generated, deployed and communicated within the scales where adaptation actions are implemented and climate impacts are experienced. A challenge in the Swiss case, is matching the level of expertise with the local level at which water is mainly governed and managed. In the Chilean case, the main issue is the level at which decisions and plans on water use are made; this tends to be presidential or ministerial (based heavily on neo-liberal economic prescriptions), while technical and operational experts are relegated to less prominent and end of pipe roles in the planning and management process.

The secularism, independence and role of technical and operational experts become even more important in adaptation issues due to the negative repercussions of maladaptation and the limited time frame which is now available for mitigation and adaptation to climate impacts. However, at the same time, political leadership is required in relation to climate change to ensure that the bigger, more complex issues are taken on board. This challenge manifests itself in both cases but very differently. In the Chilean case, regional technical and operational experts are seen to be handcuffed by central level politicians and central planning ministries, limiting their ability to apply their expertise, data and knowledge to the problems at hand. In Switzerland, the issue is the inverse of the top down challenge, with the federal and regional administrations unable to foster watershed based plans that would more integrally integrate ecosystem and climate concerns in water resources management.

14.1.3 Networks

Networks are important to both reactive coping (in terms of relationships and levels of trust between different water stakeholders) as well as longer term adaptability, in terms of the ability of actors to engage in the 'wider decision environment that will affect their longer-term resilience' (Tompkins and Adger 2004). The connectivity between actors is influenced by levels of trust, modes of negotiation and incentives for cooperation, all vital since connectivity alone does not lead to a willingness to cooperate during extreme climate stress. Knowledge networks are also vital for the integration of scientific data and information into long term planning and decision making processes, as well as for time sensitive access to monitoring data requisite for managing extreme events such as drought and floods.

Networks disseminate and share information and data as well as build or erode agreement and cooperation within institutions responsible for assessment and monitoring. Universities have been cited as important venues for dialogue and debate in order to facilitate learning across different sets of stakeholders (Garmestani and Benson 2010). Collaboration and information sharing across different actors and levels elucidates the extensive and pervasive challenge of getting stakeholders to cooperate and collaborate either formally or informally and the need for balance in power, authority, agency and autonomy across different sectors and levels of governance for effective coordination and collaboration to long term complex challenges as well as mobilising for ad-hoc extreme events.

There are challenges and impediments in both case areas to the effectiveness of existent networks for challenges relating to climate change. In the Chilean case, while there is a willingness to cooperate on single projects for shared benefits that constitute hard infrastructural adaptations (e.g. reservoirs and wells), connections between different actors tend to be based on financial or economic incentives alone, with no other glue binding actors together (i.e. basin planning for a stable and sustainable system is lacking). The development of the *Mesa Tecnica de Aconcagua* in relation to the Aconcagua Project provides a platform for those in favour of the project to share information and present supporting findings to the DGA and other stakeholders in the basin. Elsewhere in Chile, *Mesa del Agua* have been set up as watershed boards, in a set of pilot projects developed by the DGA in the past decade (Bio Bio, Huasco, Copiapo[1]). However, these institutions have failed to incorporate the full suite of watershed stakeholders, reducing their ability to effectively build cooperation across divergent views but instead the opposing viewpoints in the Aconcagua Project and related groundwater management issues are as deeply entrenched as ever.

In the Swiss case, the networks that do exist tend to be sector specific, but based more on intentions of knowledge and expertise development than on specific projects. The TRC is perhaps one area where participation has taken a consultative form, in that the implementation plan was presented to the COREPILs post facto, and approval or commentary requested on a seemingly done deal, to the chagrin of agricultural stakeholders who stand to lose land as a consequence of the enlargement (NZZ 2009). An earlier inclusion of affected stakeholders into the implementation process through communication and information networks, as they exist for water provision and other mountain water challenges, may have allowed a better understanding for the benefits that such an enlargement could bring in the long run, rather than the short term implications of land loss.

Across these different administrative or spatial scales, too strong a commitment and concentration of governance actions, rules or autonomy at one level, whether it be higher or lower, can be seen to hamper the response at another level, eroding the fine balance that could enable more coherent adaptation strategies. The role of incentives and trust building in *networks* highlights the importance of balancing out mismatches in authority, autonomy and agency (see Part III, Chap. 12) to ensure that diverse stakeholders across the complex system have the right incentives to move collectively towards more integrated and adaptive approaches. Moreover, building more effective and functional networks across these administrative and sector scales is particularly relevant to water institutions because of the imbalances of natural and economic resources between upstream and downstream water users (especially notable in the Chilean case in the disagreements between the different Juntas).

The importance of balancing lower and higher levels of governance authority is matched by other recent research, reinforcing the empirical evidence that bottom-up governance and decentralisation is not as vital a characteristic for adaptive and

[1] MOP, Unidad Técnica, Programa de Manejo de Recursos Hídricos a Nivel de Cuencas Hidrográficas (PMRH), proyecto MOP-BM, volumen 1, informe, Santiago, 5 de febrero de 2001 (Dourojeanni 2010).

integrative management approaches as earlier theory had suggested (Huntjens et al. 2010). The approaches characterised as transformative or adaptive in the decentralised Swiss model were driven by top-down policy and legislative frameworks. However, these top down frameworks were in part informed by the strong voice afforded to environmental organisations within the Swiss direct democratic governance model (see discussion in Part II). On the other hand, the Chilean case represents a centralised model of governance where the lowest governance level is the user or rights owners, who through the Water Code are granted a high level of autonomy in the management of water resources. In both cases the autonomy of the lowest level of governance (in Chile this level is private, and in Switzerland it is public) limits the ability to proactively build solutions for broader more complex issues in water resources management.

Hence, evidence from the Swiss and Chilean cases reinforce the finding in Huntjens et al. (2010, 2011), that fine tuning the balance between bottom-up and top-down approaches may be more important than proposing the more simple solution of promoting bottom-up and decentralised governance for managing water issues. Public authorities at higher or lower levels, whether in a centralised or decentralised system, have an important role to play in conflict resolution, cooperation building and facilitation, priority and standard setting as well as certain levels of information generation and provision (Huntjens et al. 2010).

The importance of trust building for cooperation has been highlighted in a number of studies by Elinor Ostrom and her collaborators (Poteete et al. 2010), in the investigation of collective action for cooperative solutions to resource management challenges. It is the mix of the design principles relation to the availability of knowledge on short and long term impacts with the ability to share that knowledge equitably between actors that can in effect have more influence on cooperation and trust generation than top down policy or rule setting. Moreover, the Chilean case reinforces the evidence that in the absence of trust or respect for government, top down rule setting can also increase the challenges for enforcement and implementation (Ostrom 2010). Interestingly, in the Swiss case, stakeholders in the agricultural sector were not only aware of the research by Netting (1981) and Ostrom (1990), but also expressly pointed out that it was in the interest of the canton to foster elements of the common property systems that had managed the Suonen/Bisses systems for centuries, to ensure collective action and responsibility for irrigation and watercourses was maintained at the local level.

The TRC, as the example of a transformative outcome, aligns different regime and knowledge indicators for the development of a management approach that takes into account both anticipatory and reactive adaptive capacity development and mobilisation. The aim is not only to enhance longer term resilience of the flood prone areas of the Rhône valley, but also to develop information and knowledge networks that would take better account of climate change related increases in flow and limit their damage through flexible buffers (e.g. evacuation corridors, buffer zones). While non state actors, such as environmental organisations, played an integral role in shaping the legislative baselines of the project, the participation of affected actors in the canton itself has been based mainly on consultation through commissions on the implementation of the project (COPIL/COREPIL).

A more innovative approach that involved the co-production of knowledge across multiple levels and stakeholders to develop the project, could help build cooperation across currently disenfranchised stakeholders (Huntjens et al. 2010; Olsson et al. 2006). Investing this time (earlier on in the project) has shown, as have other studies, a need to develop understanding, learning and thus foster cooperation across stakeholders when dealing with uncertainty and change, whether related to climate change or other variables (Stubbs and Lemon 2001).

14.2 Speeds and Scales of Change

While balancing flexibility and predictability is important to address the challenges between structure and autonomy across administrative scales, it is equally important to address adaptation to and preparedness for different scales of change as shown in Fig. 14.3 below. A community or system's adaptiveness to local climate conditions may not imply an ability to cope with changes or impacts at different speeds or scales, as is evident across both cases. The adaptive actions associated with historical variability, drought and scarcity are limited in terms of upscaling to face more complex challenges. Furthermore, in the Swiss case, perceptions of being well prepared for tougher climatic conditions in the Valais (in comparison to other areas of Switzerland) appear to lull sectors such as agriculture in particular into a false sense of security that managing climate impacts will not require alternative solutions or management approaches.

This is in keeping with other findings that suggest that adaptation to certain stress conditions (drought/rain shadow effects) within one set of parameters (historical variability) does not imply long-term adaptability to conditions whose persistence and impacts will be more pervasive (Folke et al. 2010). It also reflects empirical evidence from other studies of river basins that suggest experience of one type of extreme can the limit preparations for another form of extreme (Huntjens et al. 2010). Similarly, highly optimised tolerance theory (HOT) posits that systems that tend to become very robust to frequent kinds of disturbance may become fragile in relation to infrequent events (Carson and Doyle 2000). While in the flooding events in the Swiss case, high-impact low-frequency events are seen to have elicited a longer-term adaptive response to changing conditions at multiple levels, a transition which the TRC is on the cusp of. However, in other cases, events facilitate immediate adaptive behaviour, but fail to translate these smaller transformations into more permanently adaptive regimes, such as the response to increasing drought conditions in the Chilean case, local level responses to flooding events in the Swiss case (i.e. the backlash against the TRC), and the response to drought periods such as 2003 in the Swiss case.

Therefore, it reaffirms findings from other studies of the potential for short sharp shocks, such as trigger or focussing events, to become windows of opportunity for transition to a new water course management philosophy (Folke 2006; Herrfahrdt-Pähle 2010; Olsson et al. 2006). This in turn can enable more proactive adaptation to longer term creeping system changes, such as gradual changes in mean precipitation, changing seasonality and decreasing flows from glacier and snow melt. Tompkins and Adger (2004) also recognise this dichtomy in adaptation, both gradual and anticipatory as well as to single significant extreme events or shocks. But they add that both forms of adaptation should 'involve encouraging the evolution of new institutions that are sensitive to the resilience of the ecosystems they are managing and knowledgeable about the specific nature of the risks of climate change' (Tompkins and Adger 2004, p 10).

However the presence of capacity to adapt to one off events may not engender the mobilisation of capacity to respond to more gradual yet in the long run significant forms of change, as seen by the dichotomy of responses to inter-annual droughts in the Chilean case and the drier climate in the Swiss case in comparison to the more complex, inter-related and anticipatory changes to climate impacts on seasonality, variability and availability.

In the Swiss case, while the windows of opportunity that the flooding events opened were capitalised on, the rapid fading of the memory of those impacts highlights the importance of knowledge (information and communication) indicators, to ensure that both individual and institutional memory is maintained. Integrating opposing stakeholders into the tight communication and knowledge networks could be one means of finding more cohesive stakeholder acceptance of the implementation plan in a faster manner than the current top down communication and participation strategy that is in place through the newsletters and COREPIL. In the Chilean case, the high impact recent drought events, potentially, are providing a window of opportunity for a heightened level of self-questioning and stakeholder cooperation and collaboration to move beyond just the technical engineering solutions to security and supply challenges, but to also better enable the institutional setting to cope with increased drought impacts so that the resilience of the SES does not further degrade.

The informality of the Chilean approach grants freedom and autonomy to the user level to quickly react and find solutions to smaller issues (tourno). However, these changes and coping techniques (increased groundwater abstraction) have the potential for longer term degradation of the resilience of the ecosystem. The Chilean system is also characterised by high levels of mistrust between sectors and institutions that not only hamper the implementation of reactive adaptations but disincentivise collaboration across sectors and levels for solutions to the larger more complex issues. A focus could therefore be on enhancing the reactive elements of the system, but paying closer attention to how elements in the knowledge network indicators could improve proactive adaptation and lead to positive rather than negative transformation (more on this further down).

14.3 Navigating Structural Tensions and Trade-Offs Across Multiple Governance Scales

As discussed in Part I, in a number of studies, different iterations of flexibility or predictability are taken as indicators of adaptive capacity. While these approaches have theortetical support, the research presented in this book suggests that it might be more useful to utilise these two concepts as core tensions in developing adaptive capacity, rather than just another element of adaptive capacity. Thus, flexibility and predictability become a guiding tension through which to measure and balance adaptive planning (see Fig. 14.3). The next step therefore is to develop and propose a method for navigating this tension, in order to minimise the trade-off between the development and mobilisation of proactive and reactive adaptive capacity.

Striking the appropriate balance in the governance arrangement to develop flexible yet robust adaptive responses will present a constant, but evident challenge for policy and decision makers. Managing this paradox is key for decision makers to grapple with the challenge of how to develop an SES to be simultaneously well prepared and adapted (high proactive adaptive capacity, e.g., long-term and iterative planning, integration of uncertainty and climate change impacts) but also quick to respond (high reactive adaptive capacity, e.g., quick innovations and transformations in response to specific events) to the different scales of change. Building

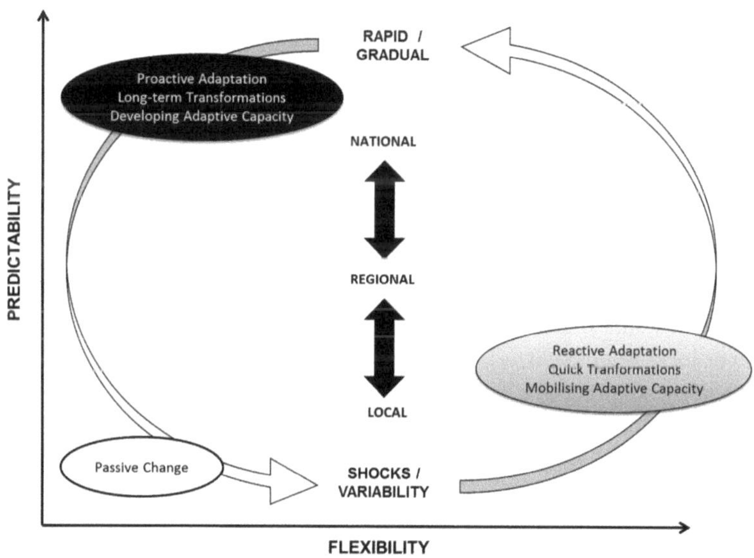

Fig. 14.3 Enhancing proactive and reactive adaptive capacity by balancing predictability and flexibility across different scales of governance (national, regional, local) and change (gradual and rapid)

adaptive capacity with 'regulated flexibility' through local preparedness and planning, while providing the necessary support, guidance, and resources at higher scales represents a challenging but initial step in the right direction to address the trade-offs in developing reactive and proactive adaptive capacity.

An area that policy and law makers should focus on is the challenge of how to best utilise legal provisions and regulations to guide and encourage adaptive behaviour without handcuffing water managers and stakeholders to codified rules which may be out of date in future years. In pursuit of this goal, more attention could be paid to how best to utilise elements of procedural law that provide structure while building in flexible instruments that provide a timeframe and process for review and the establishment of new goals that fit the present day reality, rather than the reality when the law was origionally crafted and passed.

In the Swiss case, the implementation plan of the TRC has provided for a period of review every 10–20 years, to ensure that the plan is constantly updated to be appropriate to the best available science. Lessons could be drawn from the provisions in the implementation plan that provide for this type of structured process of review in areas of contract and administrative law that govern hydropower concessions and irrigation prioritisation. Furthermore, lessons could be drawn for the Chilean case, where the rule of the Water Code and supposed legal certainty pertaining to water rights are major challenges for adaptation in the water governance system.

Legislation and property rights concerning water resources could be subject to provisions that allow for 10 year review processes of the underlying data upon which the assumptions for the validity of those provisions are made. While this still allows for goals and normative principles to be set in stable legal structures, since stationarity of the system cannot be assumed, greater flexibility for experimentation at lower governance levels could be provided for by enhancing the networks that already are in place (Cosens 2010). Establishing or strengthening the requisite institutional channels (formal forums and planning processes and informal networks with multi-purpose incentives) for collaborating amongst stakeholders and facilitating information exchange could also help address this particular challenge.

Both the cases highlight the challenges of integrating shifting hydrogical baselines into substantive law. Autonomy and strong property rights at the user level (Chile) or local level (Switzerland) can fragment adaptive responses, and present a major barrier to proactive and integrated planning and management of water resources for more complex challenges. However, referring back to Part I, jurists have highlighted that rights are an area of the law constantly re-negotiated and subject to cultural frames of reference. The Chilean water rights system has itself changed twice in the twentieth century. While this in itself has left the system quite broken and unclear, it does suggest that the current impasse over the Water Code and associated protection of constitutional water rights might not be as fixed and impenetrable as has been assumed.

Competing interests and non-integrated priority setting are two of the biggest challenges in developing more transformative and sustainable adaptive solutions in both cases. In order to balance competing interests at different political levels and across the different sectors, decision makers should aim to simultaneously invest in bottom-up (community adaptation planning, integration of climate impacts into longer-term planning, and adaptive capacity assessments) and top-down efforts (national and regional level technical, strategic and financial support systems, long-term planning requirements, investment in shared scientific and adaptation databases, mechanisms for cross-region, cross-sector learning) as an initial step for joining up segregated and contradictory policy priorities across water stakeholders. Table 14.1 builds upon these ideas, by presenting a multi-scale framework to address the challenge and tensions implicit in adaptive capacity through more practical institutional mechanisms.

Table 14.1 draws on and develops from the framework set out in IPCC (2001) and Tompkins and Adger (2005, p 566). Proactive approaches relate to taking the longer term view through a number of approaches including planning process and guidelines, policy and legal frameworks that represent long-term and iterative processes that can integrate new information as it manifests. Reactive approaches relate to flexible mechanisms and networks that can rapidly respond with quick innovations and transformations to minimise short and long term damage from specific events.

At the national or federal level, a focus on both vertical and horizontal integration has been suggested. From a proactive perspective, efforts could be directed to providing stability in change, partnering ministries or federal administrative bodies to set more integrative policies on the basis of sound environmental and climate information. This process could be enabled by formalising knowledge relationships with appropriate bodies; in some contexts this might be intergovernmental bodies, in other NGOs or in other research institutes and universities. More formal inter-disciplinary partnerships for policy setting would allow for a broader mix of information and knowledge (beyond traditional disciplines of lawyers and engineers) to inform the development or revision of legislation and regulation.

One evident challenge is that while policy should inform legislative developments, in governance contexts such as Chile and Switzerland, this can be a time-consuming and in some cases fruitless task. In Chile, the constricted and dogmatic nature of political dialogue on the Water Code and water resources reform limits the scope for addressing climate challenges through formal legislative change and reform. In the shorter term, it is worth focussing on the more dynamic elements of the system, i.e. informal elements and those that relate to knowledge and network indicators to foster approaches that are better equipped for quickly dealing with the challenges relating to climate change.

In Switzerland, evidence shows how federal policy making does filter into federal and cantonal legilsation. However, the timescales over which these policy priorities trickle down into actual rules at the canton and local level[2] can take years or decades, and even then, the autonomy of the communes can impede the effective

[2] http://faunavs.ch/?p=subject&tag=21&action=detail&id=12; http://www.vs.ch/Navig/legislation.asp?Language=fr

Table 14.1 Multi-scale framework to address tensions and trade-offs in developing adaptive capacity

Scale	Proactive	Reactive
National Federal and central governments National institutions and organisations Federal and national legislation & regulation	Robust national and legislative frameworks that merge aspects of water law, land use law, environmental law together with energy, mining and other sectoral laws that impact on water resources. Framework laws that set long-term planning requirements, but integrate flexibile mechanisms that allow for resetting goals as new information is uncovered. National and regional level technical, strategic and financial support systems, investment in shared scientific and adaptation databases, mechanisms for cross-region, cross-sector learning. Develop mandate of NGOs and learning networks as covenors and partnership developers. Time specific planning – i.e. distil complex model/scientific outcomes to temporally relevant pieces that are digestible for solution building at lower levels. Blue skies thinking: policy making process of identifying form of behavioural change needed to develop adaptive capacity.	Development of catastrophe management capacity – both financial and technical (i.e. conflict resolution; technical and knowledge assistance) to deploy at lower levels. Development of multi-scale governance networks that devolve decision making capacity in times of extreme events to lower levels of government administration. Develop mechanisms and institutional channels to provide a process for reviewing, revising and establishing new goals and priorities in the wake of extremes and disasters. Incorporation of best available technology and science/data into crisis management institutional frameworks to better reflect shifting hydro-climatic baseline on which these decisions are made.

Region Regional governments and administrations Basin organisations Regional institutions and organisations	Develop scientific networks to better connect science and management – focus on universities, NGOs, research institutes and science professionals knowledge sharing platforms and linkages with government planning, decision makers and water managers. Institutional mechanisms to translate national goals into the basin context – investing in, and where appropriate formalising, bridging organisations to link in top down and bottom up priorities, issues and solutions. Investment in regional collaboration to integrate lessons-learnt, knowledge and information across stakeholders and sectors. Create ownership of the planning process – investing in communication and education programmes at both regional and local level, rather than just consultation and participation.	Financial and technical preparedness plans so that federal or central level help can be rapidly coordinated to local levels in extreme periods. Investment in monitoring capacity and maintenance of up to date and relevant data. Enhance availability and accessibility through multiple communication channels and networks in the lead up to and during extreme events. Development of cross-sector or agency review panels – to ensure best available knowledge (from top down and bottom up sources) is integrated into rights, procedures and plans. Development of expedient and accessible conflict resolution vehicles in periods of heightened stress (i.e. developing institutional redundancy).
Local Municipality, commune levels User groups	Integration of climate impacts into longer-term planning, and adaptive capacity assessments through community adaptation planning. Particpation in planning process from a climate knowledge perspective, rather than just a presentation of final plans for approval. Development of local campaigns and activities to improve knowledge of climate impacts as well as variety of transformative approaches to adaptation – e.g. school and university competitions, advertising campaigns etc. Initiate dialogue on evolving property rights to better correspond with shifting hydro-climatic baselines. Development of access to bridging organisations with cross-scale and multi-disciplinary/sector foci for fostering trust horizontally and vertically.	Better matching of levels of autonomy with resources capacity or access to capable networks to enable better mobilisation of plans of action in extreme periods amongst rights holders or municipalities. Access to financial and technical support – with more balanced focus on both these aspects, not just financial access. Development of local networks that are connected through looser horizontal structures rather than rigid hierarchies to create ownership and faster response.

implementation of high level priorities or even provisions in the law. Furthermore, other studies have suggested that formal legal rules are more irrelevant than lawyers would tend to expect during extreme periods, due to their short advance notice period (Hurlbert 2009; IISD 2006). During these periods, more flexible and quickly accessible institutions are needed that can respond to stakeholder needs over periods of days, weeks or months.

However, this is not to discard the importance of addressing the challenges in the legal and policy framework, as these are core drivers of the developments of knowledge and network elements of the governance system. But to suggest that these longer term challenges should be seen in the context of proactive capacity building, while the quick wins in knowledge partnerships and conflict resolution mechanisms can be tackled, now, to develop capacity that can be better mobilised in the next extreme event. From a reactive perspective, while quick reactive capacity can best be mobilised at the local and regional levels, there are governance actions that, concurrently, can be implemented at the national level that can enable this process. National and federal actors should acknowledge that more extremes are likely to increase the need for larger financial support and enhanced financing mechanisms to support regional and local coping efforts, quickly and efficiently.

Craig (2009) suggests that lawmakers should think more creatively about means of restructuring legal safeguards so that public authorities have more flexibility to deal with climate change impacts. His suggestions include 'general planning requirements coupled with abbreviated administrative procedures for specific implementation decisions, periodic rather than continual judicial review for rationality, the ability to rely on post-decisional evaluations rather than pre-decisional justifications, and/or increased emergency authorities in order to achieve true capacity for adaptive management in the face of climate change impacts to resources and ecosystems' (p 55). Ruhl (2009) also suggests that in the interests of the law becoming more adaptive, those that shape the law should emphasise a shift from a preservation focus to one of 'transitionalism', in order to better allow for frequent reconfigurations that take into account trans-policy linkages and trade-offs across scales.

At the regional level the focus has been set to vertical integration, since the role of institutions and actors at this level can provide valuable linkages between top down and bottom up actions, in order to build trust, provide support, both financially and technically, and develop consensus between local needs and realities and national priorities. Establishing intermediaries and formalising bridging organisations, such as NGOs and universities could enhance the role of regional level institutions and actors in trust building across and between higher and lower levels of governance. Integrating scenario planning and analysis at this level would raise understanding of climate related uncertainty and provide a potential setting for collaborative knowledge networks between regional or national research institutions that could enable greater capacity and openness to learning at lower levels as well.

The networks developed for longer term partnerships could also improve monitoring and information flows during extreme periods. Furthermore, during crisis periods, higher levels provide critical support functions when local capacity may fail or be inadequate. Therefore, the regional level focus to develop multiple mechanisms to provide this short term support function to assist local level capacity in coping with non-linear dynamics in SESs, would include investing in monitoring systems, leading indicators, scenario planning and communication and information flows (Langlet 2010).

The focus at the local level is on horizontal integration, namely connecting different sectoral actors and communities together to build cooperation for resolving long term complex challenges but also develop networks and knowledge that can be quickly mobilised to react in times of crisis, that may be redundant during 'normal times'. The call for a rapid evolution of property rights, needed in conjunction with climate adaptation (Ruhl 2009), is perhaps best addressed at the local level, where individual or company stakeholders own and negotiate water rights or use rights, rather than at higher levels of governance and then implemented at local levels. This is likely to be a complex and emotive process, but jurists and water rights owners (farmers, companies, utilities, municipalities) need to develop stronger partnerships to develop innovative solutions to resolve the challenges relating to the mismatch between the current and future hydrological realities and the obsolete baselines upon which their rights were formulated.

The development of bridging organisations (e.g. local assessment teams) that comprise multi-sector actors in the SES, could provide the requisite arena for trust-building, learning, conflict resolution and adaptive co-management and that would provide a dual role in facilitating proactive preparatory capacity as well as arenas for mobilising joint responses in crisis time that are not dependant on higher levels. Additionally, autonomy alone is meaningless without the requisite agency, plus access to financial mechanisms. Therefore enabling access to and development of financial and technical capacity are equally important. In turn, this requires regional and national levels to have the capacity and resources to assist the local level.

While policy setting and at the national level should still remain an adaptation priority for higher levels of government, until the constrictive elements of present legislation and regulation are transformed, the limits of their impact in developing capacity to manage the impacts of climate change at the local level should still be recognised. Table 14.1, therefore focusses on the mix of regime, knowledge and network based approaches and mechanisms that are invaluable complements to legislative provisions and fixed rules in meeting climate related challenges. Most importantly, in the absense of governments being able to effectively integrate water related policy priorities and legislative processes at the national level, focussing on the mechanisms in Table 14.1 could enable water stakeholders themselves to cross scales and sectors to develop a more joined up approach at the basin level for maximum benefits in climate change adaptation.

Some studies (Garmestani and Benson 2010; Herrfahrdt-Pähle 2010) have shown how these different scale, specific foci can be couched in the Panarchy model (Gunderson and Holling 2002; Chapin et al. 2009). These studies apply the Panarchy model to institutional change, thus matching institutions and governance actions to the appropriate level. Garmestani and Benson (2010) suggest matching up the institutional foci at higher governance levels to the phase of growth and accumulation (the foreloop phase of r to K) characterised by slow and incremental transition. Conversely, faster changes at the lower governance levels should be matched up with the rapid phase of reorganisation that leads to renewal (backloop phase from Omega to Alpha). Between these different levels, bridging organisations and networks are vital to maintain open lines of communication, financial and operational support, as well as provide an arena for the accumulation and application of scientific information.

Traditionally, there has been a weighted focus on the legislative and infrastructural frameworks that structure water resources management, that is typical of what has traditionally been a sector dominated by mathematically minded technicians and engineers (Huntjens et al. 2010; Ingram 2011; Pahl-Wostl 2007). Engineers, mathematicians and economists stereotypically conform to the worldview that problems can conform to neat mathematical models. In other words, they tend to treat clouds as if they were clocks (Pearce 2002a). Complex systems however, tend to defy neat stereotypes, which is a partial explanation for the on-going challenge of meeting related challenges with paradigms and panaceas (Ingram 2011; Meinzen-Dick 2007; Ostrom 2007). It may be more useful to combine more nuanced indicators with a multi-scale framework that focuses not only on the rule based elements of the SES but also on the way in which information and knowledge is developed, shared and applied aims to embrace the complexities implied in developing and mobilising adaptive capacity, rather than ignore or constrict them. For more discussion of this latter issues, please refer to Chap. 15 on coping with and communicating uncertainties.

14.4 Conclusion

Emergent themes through the course of analysis presented in Part III revealed the tension in generating different forms of responses to different speeds or scales of change and across different spatial scales. The structural challenge of mobilising flexible fast responses in periods of drought or flooding was seen in juxtaposition with the corresponding need for a more predictable structure to guide longer term adaptation planning. These emergent contradictions in adaptive capacity were matched with concepts of proactive and reactive adaptive capacity to set out a means of navigating the structural tensions inherent in adaptive capacity. Analysis has shown that proactive adaptive capacity could be associated with predictability and guidance at higher levels, while reactive capacity could be enabled through flexibility and autonomy at lower governance levels. In turn, transformational responses are related to building longer term resilience in the SES, and are linked to proactive and preparatory adaptation. Persistent Adaptive actions provide smaller scale processes of change for quick and flexible reactions to events as and when they occur, to maintain the resilience of the system in the face of surprise.

Furthermore, this tension is influenced by different levels of governance and scales of change. A framework was presented in Table 14.1 as a means of navigating this core tension across spatial and temporal scales through more nuanced indicators that address both reactive and proactive adaptive capacity. In coping with shifts in variability and increasingly recurrent extremes, institutions across the case areas showed varying degrees of ability to mobilise for different kinds of shocks. The development of the framework set out in the table and its underlying discussion aims to contribute new and more nuanced insights into means of developing both proactive and reactive adaptive capacity that contribute to both the growing body of literature and practitioners alike. The structure proposed in Table 14.1 could be used to develop adaptive capacity assessments that take these multi-scale challenges into account, and help guide decision makers and water managers to develop adaptation solutions that take both facets of adaptive capacity into account.

It is suggested that developing the capacity to both adapt proactively and mobilise reactively to different scales or speeds of change frames adaptive capacity in a way that focuses it both on the accommodation of uncertainty, as well as the short and long term transformational potential within a governance system. Focussing on the transformational potential of adaptive capacity should be about maintaining options and choices where possible and recognising that passive, steady state, command and control approaches have tended to cut off options when the ambient climate changes. For example dykes can only go so high, reservoirs can spill over or dry up if flows exceed or deplete beyond the parameters for which they were constructed. Pinpointing the elements of the governance system that enable more persistent and transformative adaptive responses is a means to developing adaptive capacity in order to create rather than minimise future water resources options. The indicators developed and presented in Part III, and the multi-scale framework presented in this chapter presents an approach that could be further developed to enable short term reactive capacity (e.g. crisis management, coping abilities) that would be more consistent with more proactive strategies.

The approach aligns reactive and proactive in one framework so that short term strategies would not counteract longer term proactive approaches that seek to maintain the resilience of the SES rather than exacerbate underlying challenges that potentially limit adaptation to greater magnitudes of climate change in the future. Water managers and adaptation planners would be well advised to pay closer attention to these different aspects of developing and mobilising adaptive capacity, to ensure that fostering one set of responses at one level, does not detract from or counteract effects for another form at a different level, thereby limiting either short term reactive capacity or longer term proactive capacity, both of which are equally important.

References

Carson J, Doyle J (2000) Highly optimized tolerance: robustness and design in complex systems. Phys Rev Lett 84(11):2529–2532

Chapin FS, Folke C, Kofinas GP (2009) A framework for understanding change. In: Chapin FS, Kofinas GP, Folke C (eds) Principles of ecosystem stewardship resilience-based natural resource management in a changing world. Springer, New York

Cosens B (2010) Resilience and administrative law in transboundary river governance. Paper presented at the Law for Social-Ecological Resilience Conference, Stockholm Environmental Law and Policy Centre and Stockholm Resilience Centre, Stockholm, 17–19 November 2010

Craig RK (2009) 'Stationariy is dead' – long life transformation: five principles for climate change adaptation Law. Florida State University, Tallahassee

Dourojeanni A (2010) Organizaciones de Gestion de Recursos Hidricos por Cuencas: Segunda Parte. Fundacion Chile [online]. http://www.pnuma.org/agua-miaac/CODIA%20ORGANISMOS%20DE%20CUENCAS/MATERIAL%20ADICIONAL/PRESENTACIONES/PONENTES/Tema%206%20Gestion%20Transfronteriza/5%20CURSO%20BRASIL%20GIRH%20ORGANIZACIONES%20INDICADORES.pdf

Ebbesson J (2010) The rule of law in governance of complex socio-ecological changes. Glob Environ Chang 20(3):414–422. doi:DOI 10.1016/j.gloenvcha.2009.10.009

Engle NL, Johns OR, Lemos MC, Nelson DR (2011) Integrated and adaptive management of water resources: tensions, legacies, and the next best thing. Ecol Soc 16(1):19 [online]. http://www.ecologyandsociety.org/vol16/iss11/art19/

Folke C (2006) Resilience: the emergence of a perspective for social-ecological systems analyses. Glob Environ Chang 16(3):253–267

Folke C, Carpenter SR, Walker B, Scheffer M, Chapin T, Rockström J (2010) Resilience thinking: integrating resilience, adaptability and transformability. Ecol Soc 14(4):20 [online]. http://www.ecologyandsociety.org/vol15/iss24/art20/

Garmestani AS, Benson MH (2010) Actualizing panarchy within environmental policy: mechanisms for tweaking institutional hierarchies to mimic the social social-ecological systems they manage. Paper presented at the Law for Social-Ecological Resilience, Stockholm University, Stockholm, 17–19 November 2010

Gunderson LH, Holling CS (2002) Panarchy: understanding transformations in human and natural systems. Island Press, Washington, DC

Herrfahrdt-Pähle E (2010) The transformation towards adaptive water governance regimes in the context of climate change. Universität Osnabrück, Osnabrück

Huntjens P, Pahl-Wostl C, Grin J (2010) Climate change adaptation in European river basins. Reg Environ Chang 10:263–284

Huntjens P, Pahl-Wostl C, Rihoux B, Schlüter M, Flachner Z, Neto S, Koskova R, Dickens C, Nabide Kiti I (2011) Adaptive water management and policy learning in a changing climate: a formal comparative analysis of eight water management regimes in Europe, Africa and Asia. Environ Policy Gov 21:145–163

Hurlbert M (2009) The adaptation of water law to climate change. Int J Clim Ch Strateg Manag 1(3):230–240

IISD (2006) Designing policies in a world of uncertainty, change and surprise: adaptive policy-making for agriculture and water resources in the face of climate change. International Development Research Centre & The Energy and Resources Institute, Winnipeg/Manitoba/New Delhi

Ingram H (2011) Beyond universal remedies for good water governance: a political and contextual approach. In: Garrido A, Ingram H (eds) Water, Food and Sustainability. Routledge, New York

IPCC (2001) Climate Change 2001: impacts, adaptation, and vulnerability technical summary. A report of Working Group II of the Intergovernmental Panel on Climate Change. WMO/UNEP, New York

Langlet D (2010) Concluding panel debate and closure. Paper presented at the Law for Social-Ecological Resilience, Stockholm University, Stockholm, 17–19 November 2010

Meinzen-Dick R (2007) Beyond panaceas in water institutions. Proc Natl Acad Sci U S A 104(39):15200–15205

Netting RM (1981) Balancing on an Alp: ecological change and continuity in a Swiss Mountain Community. Cambridge University Press, Cambridge

NZZ (2009) Verliert die Landwirtschaft zu viel Boden?: Zunehmender Widerstand gegen die dritte Rhonekorrektion. Neu Zurcher Zeitung, 25 April 2009 [online]. Available at: http://www.nzz.ch/nachrichten/politik/international/verliert_die_landwirtschaft_zu_viel_boden_1.2457442.html

Olsson P, Gunderson LH, Carpenter SR, Ryan P, Lebel L, Folke C, Holling CS (2006) Shooting the rapids: navigating transitions to adaptive governance of social-ecological systems. Ecol Soc 11(1):18. [online] http://www.ecologyandsociety.org/vol11/iss11/art18/

Ostrom E (1990) Governing the commons: the evolution of institutions for collective action. Cambridge University Press, Cambridge

Ostrom E (2007) A diagnostic approach for going beyond panaceas. Proc Natl Acad Sci U S A 104(39):419–422

Ostrom E (2010) A multi-scale approach to coping with climate change and other collective action problems. Solutions 1(2):27–36. Online: http://www.thesolutionsjournal.com/node/565

Pahl-Wostl C (2007) Transitions towards adaptive management of water facing climate and global change. Water Resour Manag 21(1):49–62. doi:10.1007/s11269-006-9040-4

Pearce WB (2002a) Systems: schools of thought and traditions of practice. The Fielding Graduate Institute, Santa Barbara

Poteete A, Janssen M, Ostom E (2010) Working together: collective action, the commons, and multiple methors in practice. Princeton University Press, Princeton

Ruhl JB (2009) Climate change adaptation and the structural transformation of environmental law. Environ Law 20(November):343

Stubbs M, Lemon M (2001) Learning to network and networking to learn: facilitating the process of adaptive management in a local response to the UK's National Air Quality Strategy. Environ Manag 27:321–334

Tompkins EL, Adger WN (2004) Does adaptive management of natural resources enhance resilience to climate change? Ecology Soc 9(2):10 [online]. http://www.ecologyandsociety.org/vol19/iss12/art10

Tompkins EL, Adger WN (2005) Defining response capacity to enhance climate change policy. Environ Sci Policy 8:562–571

Chapter 15
Coping with and Communicating Uncertainty

Abstract Understanding and communicating uncertain and complex dynamics is one of the biggest challenges to mobilising action in the face of climate change as well as a wide range of other environmental issues. This short chapter provides some reflections that draw on a wide range of different disciplines as to how to potential communicate uncertainties. It is suggested that applying a diverse range of insights on motivating action in the face of complex science and uncertain futures could enhance the cohesiveness of cross-sector action and ingenuity of the approaches to climate change adaptation at the watershed level.

Keywords Inter-disciplinary approaches to managing uncertainty • Communicating uncertainty • Behavioural economics • Scientific complexity • Climate change adaptation

15.1 A Wicked Challenge

Understanding the challenges of communicating uncertain and complex dynamics is a core issue in terms of how to generate action on climate change, not just for water governance, but for earth system governance in the anthropocene. Attitudes to risk and uncertainty vary according to cultural, social and psychological values, which therefore influence the way in which climate change adaptation is viewed at the country level (Tompkins and Adger 2005), as well as how water governance systems respond to climate change threats. Environmental and resource problems have been termed by some researchers as particularly complex problems, indeed as 'wicked' problems (Rittel and Webber 1973). The phrase was originally coined by Rittel and Weber in relation to challenges of social and public policy making, where it is difficult to apply rational scientific or expert based solutions to problems which are not only complex to describe, but are subject to a large number of divergent lay opinions. Wicked problems therefore require wicked solutions (Roberts 2000).

M. Hill, *Climate Change and Water Governance: Adaptive Capacity in Chile and Switzerland*, Advances in Global Change Research 54,
DOI 10.1007/978-94-007-5796-7_15, © Springer Science+Business Media Dordrecht 2013

Environmental issues and uncertainty relating to climate change typify the challenge of resolving complex problems (Balint et al. 2011). Not only are there many divergent opinions in relation to climate change itself, the relationship with anthropogenic carbon emissions, but high levels of uncertainty and lengthy time scales encompass the current level of scientifically generated projections for its future development. Moreover, the complexity and increasingly political contentiousness of the climate models that define policy communication of climate change impacts further complicate the challenges relating to action on climate change (Akerlof et al. 2012). Therefore, the plethora of issues concerning the communication of climate related issues is a direct challenge for the many different facets of water governance research that posit information, transparency, knowledge and data to be crucial for effective, adaptive and integrative water resource management (Engle 2010; Huntjens et al. 2011; Iza and Stein 2009; Ostrom 2007; Pahl-Wostl et al. 2007b; UNECE 2009).

15.2 Finding the Tools

Taking these issues discussed above into account, finding the right tools and developing a better understanding for how to deal with uncertainty in water relevant decision making is a key issue for enhancing adaptive capacity. Lessons can be learnt from studies related to the issues of wicked problems and the challenges of generating solutions to complex unqualified problems over long time periods. Developing adaptation options that do not lock in decision makers to a limited set of responses, despite the lack of information and high uncertainty that surrounds impacts, is vital to ensure that choices for alternative solutions are maintained as local impacts of climate change become more apparent (Keeney and McDaniels 2001).

Decision support systems, such as one proposed by Keeney and McDaniels (2001) employ shorter time frames to evaluate climate change policies in order to take better account of the 'pervasive and overwhelming uncertainties about climate change impacts' (p 992). Using timeframes of less than 20 years, they suggest, would support the selection of policies based more on near term consequences, making the challenge of addressing climate change seem more manageable, while also enabling a process of learning to make better decisions for future climate impacts. The approach also focuses on developing management capacity through learning supported by adaptive management discourse as well as institutional economists such as Nordhaus (1994), where strategies are revised as new information becomes available. Nordhaus proposes an 'act then learn' process for decision making under uncertainty that allows for the incorporation of new information into decision making, as and when it becomes available.

While these studies have focussed on climate change policy, with a focus on mitigation (Tompkins and Adger 2005), they nevertheless provide expedient insights for the field of adaptation and the challenge in decision making for enhancing adaptive capacity under the uncertainty of climate change impacts (Kane and Yohe 2000; Smit et al. 2000). Indeed, the learning based approaches that mitigation focussed studies have espoused (Brunner 1996; Keeney and McDaniels 2001; Nordhaus 1994) embrace many of the learning based elements promoted in the resilience based disciplines of adaptive management, adaptive co-management and adaptive governance (Folke et al. 2002; Pahl-Wostl et al. 2007a; Huntjens et al. 2011).

In addition to lessons that can be drawn from other discourses in the social sciences that advocate that decision making on climate change issues should better account for uncertainty and climate impacts, a great deal of value and insight could also be added to the debate by looking towards other disciplines that rest at the intersection of technology, economics and psychology. The fields of behavioural economics and commercial marketing are increasingly being employed within the wider sustainability discourse to better understand motivational behaviour and the psychology of decision making as it relates to economic or other choices (Sutherland 2011). The challenge of overcoming apathy in developing adaptive capacity was identified in both the Swiss and Chilean cases, in relation to actors' perceived inability to cope with the long term impacts of climate change. Not only would a focus on the shorter term and mid-range, impacts help generate manageable options to mid-term climate impacts, but motivational tools from the fields of advertising, marketing and business (Sutherland 2011; Wales 2011) might also be employed to stimulate adaptation planning.

The cross-disciplinary field of behavioural economics (Jackson 2005; Pomykala 2005) inherently recognises a broader set of human limitations and potential in the evaluation and framing of choices than traditionally recognised by neoclassical economists, with its total focus on the utility function (Rabin 1998). As Mullainathan and Thaler (2001) summarised, humans deviate from the standard economic model in three key ways. The 'bounded rationality reflects the limited cognitive abilities that constrain human problem solving. Bounded willpower captures the fact that people sometimes make choices that are not in their long-run interest. Bounded self-interest incorporates the comforting fact that humans are often willing to sacrifice their own interests to help others' (Mullainathan and Thaler 2001, p 1). Increasingly, such empirical findings on motivational factors for choice building are being applied by NGOs in the field of advocating for sustainability, particularly in relation to sustainable consumption, climate change mitigation and environmental protection (Crompton 2008; Dawnay and Shah 2005; Jackson 2005; Kaplan 2000; Wheatley and Frieze 2006).

New Economics Foundation reports have suggested that policy makers should better leverage the role that other people's behaviour plays in changing a particular form of behaviour (Dawnay and Shah 2005), as they draw from the work of Malcolm Gladwell (2000) on 'mavens', 'connectors' and 'salesmen'. Each category relates to a small group of people that can have significant impacts on changing the behaviour of individuals' behaviours.[1] In other areas, their report proposes means of taking

[1] Mavens are people who have such expert knowledge that you would take their advice if given it (and Mavens enjoy giving it for free). The Connectors have many connections, so information they have has the potential to be distributed to a large number of people. The Salesmen are people with the power to persuade us to change our behaviour. Policy-makers may find it useful to focus their efforts to create behaviour change on these specific types of people who will help promote wider change.

into account the intrinsic motivations that people have to do the right thing (Frey et al. 2004), the difference between willingness to pay versus willingness to accept (Kahneman et al. 1991), and people's computational biases (i.e. that people are not rational in how they compute decisions as to whether or not do something) including the influence of how a problem is framed (Frederick et al. 2002) into policy formation to better incentivise behavioural change and acceptance of new policies (Dawnay and Shah 2005).

Challenges in acceptance and adherence to top down rules were identified as significant issue for the TRC in the Swiss case, while difficulty in reaching agreement in the Aconcagua Project and commitment breaking in the Turno were issues in the Chilean case. Studies have suggested that publicly written commitments are stronger than verbal ones (McKenzie-Mohr and Smith 1999), while elsewhere the importance of providing for participation in rule-making helps improve acceptance of and happiness with policy outcomes as people find it motivating rather than demotivating (Kaplan 2000). It is interesting however that in the Swiss case, the participative TRC has even so experienced difficulty in persuading agricultural stakeholders to be on board a project that they perceive as unnecessary in the extent of its ecological provisions. Behavioural economics would suggest that since immediate losses are stronger incentives than long term rewards, in trying to incentivise stakeholders to embrace the ecological parameters of the project, it should avoid presenting farmers with the immediate losses they will encounter (i.e. their land). The farmers' unwillingness to accept the loss of their land could lead to a far higher evaluation of that piece of land than may have originally been expected (Pearce 2002).

In the Swiss case, commercial marketing techniques that challenge the incumbent paradigm could be integrated into the communications plans at the canton level, to improve understanding and acceptance for the more transformational elements of the TRC. Equally, pin pointing mavens, connectors and salesmen outside the canton administration and the environmental organisations to champion the project could contribute to a speedier participative process by spreading digestible information about the project through mediums other than workshops, websites and newsletters. A similar set of techniques could be used in the Chilean area to reduce the divisions between the different actors involved in the Aconcagua Project as well.

While for scientists it is clear that climate related challenges rely on complex data, complex systems and complex solutions, stakeholder acceptance of adaptation actions might be better based on simplified dilutions of these challenges (both causes and effects). A key point for the specific communication of uncertainty in climate change impacts is the evidence that too much information is counter-productive (Dawnay and Shah 2005). Therefore, while design principles and discourses on adaptive and integrated water management stress the importance of information and knowledge in effective water resources management, lessons from behavioural economics could help policy makers and water managers find the appropriate balance in the provision of targeted data and information to construct viable choices under climatic uncertainty. Linking this information and knowledge with personal motivations to 'do the right thing' could also help shift apathy regarding stakeholders' limited ability to adapt to climate impacts (i.e. forced change) to motivations for engaging in a collaborative process for resolving local and regional challenges (i.e. intentionality). The learning and information related elements in the previous chapter relate to how the different governance levels can contribute to finding this balance to address the different informational needs for proactive and reactive adaptive capacities. These steps could enhance the cohesiveness of cross-sector action and ingenuity of the approaches at the watershed level.

References

Akerlof K, Rowan KE, Fitzgerald D, Cedeno AY (2012) Communication of climate projections in US media amid politicization of model science. Nat Clim Chang. Advance online publication. http://www.nature.com/nclimate/journal/vaop/ncurrent/abs/nclimate1542.html#supplementary-information

Balint PJ, Stewardt RE, Desai A, Walters LC (2011) Wicked environmental problems: managing uncertainty and conflict. Island Press, Washington, DC

Brunner RD (1996) Policy and global change research: a modest proposal. Clim Chang 32(2):121–147

Crompton T (2008) Weathercocks and signposts: the environment movement at a crossroads. World Wildlife Fund, UK

Dawnay E, Shah H (2005) Extending the "rational man" model of human behaviour: seven key principles. Briefing note for the Environment Agency by NEF. New Economics Foundation, London

Engle NL (2010) Adaptation to extreme droughts in Arizona, Georgia, and South Carolina: evaluating adaptive capacity and innovative planning and management approaches for states and their community water systems. University of Michigan, Ann Arbor

Folke C, Carpenter S, Elmqvist T, Gunderson L, Holling CS, Walker B (2002) Resilience and sustainable development: building adaptive capacity in a world of transformations. Ambio 31(5):437–440

Frederick S, Loewenstein G, O'Donoghue T (2002) Time discounting and time preference: a critical review. J Econ Lit 40(2):351–401

Frey BS, Benz M, Stutzer A (2004) Introducing procedural utility: not only what, but also how matters. J Inst Theor Econ 160:377–401

Gladwell M (2000) The tipping point. Little, Brown and Company, Boston

Huntjens P, Pahl-Wostl C, Rihoux B, Schlüter M, Flachner Z, Neto S, Koskova R, Dickens C, Nabide Kiti I (2011) Adaptive water management and policy learning in a changing climate: a formal comparative analysis of eight water management regimes in Europe, Africa and Asia. Environmen Policy Gov 21:145–163

Iza A, Stein R (2009) RULE – reforming water governance. IUCN, Gland

Jackson T (2005) Motivating sustainable consumption: a review of evidence on consumer behaviour and behavioural change. Report to the Sustainable Development Research Network. Centre for Environmental Strategy. University of Surrey, Guildford, UK

Kahneman D, Knetsch JL, Thaler RH (1991) Anomalies: the endowment effect, loss aversion, and status quo bias. J Econ Perspect 5(1):193–206

Kane SM, Yohe GW (2000) Societal adaptation to climate variability and change: an introduction. Clim Chang 45(1):1–4

Kaplan S (2000) Human nature and environmentally responsible behavior. J Soc Issues 56(3):491–508

Keeney RL, McDaniels TL (2001) A framework to guide thinking and analysis regarding climate change policies. Risk Anal 21:989–1000

McKenzie-Mohr D, Smith W (1999) Fostering sustainable behavior: an introduction to community-based social marketing. New Society Publishers, Gabriola Island, Canada

Mullainathan S, Thaler R (2001) Behavioral economics. In: Smelser NJ, Baltes PB (eds) International encyclopedia of the social and behavioral sciences. Elsevier, Amsterdam/New York.

Nordhaus WD (1994) Managing the commons: the economics of climate change. MIT Press, Cambridge, MA

Ostrom E (2007) A diagnostic approach for going beyond panaceas. Proc Natl Acad Sci USA 104(39):419–422

Pahl-Wostl C, Craps M, Dewulf A, Mostert E, Tabara D, Taillieu T (2007a) Social learning and water resources management. Ecol Soc 12(5):5 [online]. URL: http://www.ecologyandsociety.org/vol12/iss12/art15/

Pahl-Wostl C, Kabat P, Möltgen J (eds) (2007b) Adaptive and integrated water management. Coping with complexity and uncertainty. Springer, Berlin

Pearce D (2002) The role of 'Property Rights' in determining economic values for environmental costs and benefits. Report to the Environment Agency, Bristol, UK

Pomykala J (2005) Behavioural economics: a crash course. http://www.altruists.org/static/files/A%20Page%20on%20Behavioural%20Economics.htm

Rabin M (1998) Psychology and economics. J Econ Lit 36(1):11–46

Rittel H, Webber M (1973) Dilemmas in a general theory of planning. Policy Sci 4:155–169

Roberts NC (2000) Wicked problems and network approaches to resolution. Int Public Manag Rev 1(1)

Smit B, Burton I, Klein R, Wandel J (2000) An anatomy of adaptation to climate change and variability. Clim Chang 45:223–251

Sutherland R (2011) The new sweet spot – and how to find it. Paper presented at the TEDx WWF, Geneva, Switzerland, 13 Oct 2011

Tompkins EL, Adger WN (2005) Defining response capacity to enhance climate change policy. Environ Sci Policy 8:562–571

UNECE (2009) Guidance on water and adaptation to climate change. United Nations Economic Commission for Europe, Geneva

Wales A (2011) The water-food-energy nexus – why everything you consume is connected. Paper presented at the TEDx WWF, Geneva, Switzerland, 13 Oct 2011

Wheatley M, Frieze D (2006) Using emergence to take social innovation to scale. Berkana Institute, Spokane, WA

Chapter 16
Addressing the Challenge of Institutional Infrastructure in a Technically Focussed World

Abstract In this concluding chapter of the book, the key results are recapitulated and implications of the findings for policy, practice and theory are reflected upon. In addressing these mounting challenges, focussing on how to transition and transform to more sustainable water governance and management paradigms, is a crucial piece of the puzzle that includes technical and hard infrastructural adaptation, but should not be limited to it. It is high time that the social and institutional infrastructure that defines the decision making environment for technical and physical infrastructural adaptation is paid equal attention. Water governance regimes do need to be both adaptable to amalgamating pressures as climate change develops but also structured to foster elements of a system that allows for more holistic and sustainable adaptation to take place. However, beyond certain tipping points, there are state changes to which adaptation and the ability to cope may be virtually impossible. Thus, policies and institutions focussing on mitigation and adaptation should become better integrated in order to take better advantage of potentially valuable synergies, and ensure the avoidance of mal-adaptation that might in turn increase climate change drivers.

Keywords Rhône, Canton Valais, Switzerland • Aconcagua, Region V, Chile • Transformation of water governance • Resilience based climate change adaptation • Social and institutional infrastructure

16.1 Overview

This book hopes to have contributed new insights into governance related determinants of adaptive capacity through the empirical investigation of extreme events across two highly contrasting case areas. The research questions posed, aimed to identify key components of adaptive capacity that could be empirically observed in the case areas in response to extreme events; the way in which different governance

M. Hill, *Climate Change and Water Governance: Adaptive Capacity in Chile and Switzerland*, Advances in Global Change Research 54,
DOI 10.1007/978-94-007-5796-7_16, © Springer Science+Business Media Dordrecht 2013

regimes (and mechanisms within these regimes) facilitated adaptive capacity in the water sector; and the key tensions in building adaptive capacity across different contexts and scales. In answering these questions, primary objectives were to contribute to the conceptualisation and operationalisation of adaptive capacity; better understand how the governance context and elements within those frameworks contribute to an enabling environment for adaptive capacity; better understand the challenges in generating adaptive capacity across temporal and spatial scales; and generate a framing of adaptive capacity that could better serve policy and decision makers, to guide them through the complex choices in mobilising adaptive capacity.

In order to meet these goals, a set of determinants of adaptive capacity were identified and developed, drawing on the body of literature relating to adaptive capacity across the interconnected disciplines of adaptation, vulnerability, resilience, adaptive governance and adaptive management, as discussed in Part I. These determinants were used to explore adaptive capacity in relation to extreme hydrological events, through semi structured interviews and questionnaires. Determinants related to the way actors used and shared information and knowledge, accessed and utilised networks, the levels at which decisions were made, how different interested actors were integrated into planning and decision making, the experience and expertise of water managers and decision makers, the access to financial and human resources as well as the role of individual or institutional leadership. In addition, the governance assessment allowed for a deeper exploration of the legal, regulatory and policy framework connected to water, climate and extremes, within which and with which these governance determinants interacted.

16.2 Summary of Findings

16.2.1 Governance

The aim of the indicator based governance assessment was to develop a better understanding of the institutional framework within which climate change impacts will take place. The assessment was a key contribution to the deliverables of the ACQWA project, and both the summary and full assessments provide a rich and detailed picture of the governance framework in relation to IWRM and highlighted the core challenges in each case area to the development and implementation of an IWRM based approach.

In the Swiss case, despite the fulfilment of accountability, transparency and participation indicators, the assessment suggested that there is a significant gap between the conceptual strands of IWRM in federal laws and policies and their translation at the regional and local levels. The complex institutional framework, legislative provisions and levels of sovereignty which govern water resources in the Canton Valais implied a lack of coordination and long term planning amongst the different politico-administrative levels and sectoral groups. These challenges are linked to concerns that the ramifications of climate change and expanding water uses are not adequately reflected in the current governance framework.

In the Chilean case, greater challenges persist across the governance indicators, in particular in relation to transparency and accountability. While water governance at the political level is driven through a centralised approach, water management happens in the private sphere and is driven by private interests. Despite the strong codified nature of water governance through the Water Code, the weakness of enforcement and capacity in the DGA means that provisions relating to protection of aquatic ecosystems can effectively be ignored at the basin level. The market focus on water management has meant that public institutions responsible for water rights management or water and environmental issues have very limited capacity to address water issues.

16.2.2 Adaptive Capacity

Part III presented the outcomes of the analysis relating to adaptive capacity in relation to the extreme events employed across the different case events (as detailed in Part II). Chapter 10 presented the different institutional and governance mechanisms that were mobilised, drawn on, or relevant to preparing for or navigating the case events in each case area. The adaptive mechanisms presented took into account both proactive and preparatory adaptive actions as well as reactive and autonomous adaptive actions pertaining to drought and flooding situations. Across the two cases areas a broad mix of adaptive mechanisms were recorded, ranging from historical coping techniques to legal prescriptions for prioritising uses in periods of scarcity to more radical policy reform. Chapter 10 then categorised these adaptive responses according to the concepts of transformative adaptation, persistent adaptation and passive change in order to establish linkages between the governance mechanisms that allowed for more sustainable and resilient approaches compared to those that fostered responses that might not build adaptive capacity or even degrade resilience in the face of increasing stresses and uncertainty. Higher concentrations of transformative and persistent adaptive responses were seen in the Swiss case area than in the Chilean case area.

Chapter 11 presented the different perceptions of stakeholders, as to the elements of the governance system that supported or hindered effective water management or its ability to cope with climate variability and impacts. Across both cases and levels of governance, inter-jurisdictional issues and lack of information and data were common barriers. However, in Chile issues relating to trust, enforcement and institutional capacity were significant preoccupations. In Switzerland, stakeholders across all levels concentrated more on issues of local autonomy, including challenges and strategies related to the decentralised mode of governance, that is a barrier to the integration of water governance. Common bridges related to flexibility and autonomy at local or user levels, use of financial incentives as a means of addressing intra-jurisdictional challenges to promote efficiency in Chile, or enhancing ecological and social benefits in Switzerland. Stakeholders across both case areas, though predominantly at regional and federal or national levels, cited the importance of research networks and knowledge partnerships in developing their understanding of the challenges and solutions to climate change impacts.

16.2.2.1 Findings 1

The initial research question sought to better understand how governance regimes and mechanisms within these regimes can facilitate adaptive capacity in the water sector. Adaptive outcomes from the Swiss cases correspond with more transformative and adaptive actions and management approaches as well as a more positive correlation with the adaptive capacity indicators. On the other hand, the adaptive outcomes in the Chilean case correspond with less transformative outcomes as well as a less positive correlation to the adaptive capacity indicators. However, despite the different governance modes of the two cases, both share common challenges in the development and mobilisation of proactive and reactive adaptive capacity, perhaps partly since both models ascribe a similar level if not type of autonomy to the local level.

The synthesis presented in Chap. 13 highlights the tension between the rigid and inflexible legislative context in the Chilean case, with the higher levels of autonomy at the user levels, which frustrates and constricts the ability of water managers and the owners of use rights to adapt in a more proactive manner to hydrological changes and stresses in the basin. While reactive coping techniques can be quickly called on through the networks and traditions that exist, more long term preparations and transformative approaches for meeting the mounting challenges are blocked by lack of trust and cooperation, lack of agency at regional operational levels and lack of accessible and appropriate information on water resources. Indicators from the Swiss case suggests that the highly networked layers of governance allow knowledge and learning to be transferred vertically across different levels of capacity but has greater challenges with horizontal integration (which mirror the challenges associated with the implementation of IWRM). Despite being characterised by more transformative and persistent adaptation, the table highlights that the areas of rule making and division of responsibility remain a challenge.

16.2.2.2 Findings 2

The second core research questions related to identifying and better understanding the key tensions implicit in building adaptive capacity across different contexts and scales. An emergent theme in the analytical process was the underlying tension of balancing predictability, guidance and certainty from higher levels of governance with flexibility and autonomy of users and rights holders at lower scales. The literature recognises that clarity in rules and legal certainty is fundamental for accountability in water governance (see Part I), while the empirical findings across both cases highlighted the issues from locking ownership and use rights into codified norms that are based on out of date hydrological data and patterns. This is a challenge that is further heightened in times of stress in the case areas, during which there is an intensified involvement of central or regional government agencies. Conversely, there is a need to recognise local individualities and needs, which can go unconsidered at higher levels of administration, participative processes used to address this dichotomy can also stall agreements on projects and frustrate multiple stakeholders, especially if not matched with requisite knowledge and information assets.

Across Governance/Spatial Scale

While proactive adaptive capacity can be associated with predictability and guidance at higher levels, reactive capacity is enabled through flexibility and autonomy at lower governance levels. One of the major challenges in climate change adaptation is therefore navigating this balance between fostering the flexibility needed to deal with an increase in the likelihood of complex and unexpected changes from climate change while maintaining the certainty and guidance for longer term preparedness through legislative, regulatory and policy frameworks. In the Swiss case, a strong level of autonomy at the local level challenges transformative policy priorities and federal and cantonal legislation at higher levels. In Chile however, the autonomy at the user level (rather than municipality) is coupled with a lack of guidance or regulating force pulling actors in the same direction.

Across Temporal Scale/Scale and Speed of Change

While balancing flexibility and predictability is important for addressing the challenges between structure and autonomy across administrative scales, it is equally important to address adaptation to and preparedness for different scales of change. A community or system's adaptiveness to local climate conditions may not imply an ability to cope with changes or impacts at different speeds or scales, as is evident across both cases. The adaptive actions associated with historical variability, drought and scarcity are limited in terms of upscaling to face more complex challenges. Furthermore, in the Swiss case, perceptions of being well prepared to tougher climatic conditions in the Valais (in comparison to other areas of Switzerland) appear to lull sectors, such as agriculture in particular, into a false sense of security that managing climate impacts will not require alternative solutions or management approaches.

In the Swiss case, while the windows of opportunity that the flooding events opened were capitalised on, the rapid fading of the memory of those impacts highlights the importance of knowledge indicators, to ensure that both individual and institutional memory is maintained. Extending and integrating opposing stakeholders into the tight communication and knowledge networks could be one means of addressing the current challenges. In contrast, perceptions of being well prepared for tougher climatic conditions in the Valais appears to have meant that alternative solutions or management approaches are not currently being considered. In the Chilean case, the high impact recent drought events are potentially providing a window of opportunity for a heightened level of self-questioning and stakeholder collaboration to move beyond technical engineering solutions to security and supply challenges, but to also better enable the institutional setting to cope with increased drought impacts so that the resilience of the SES does not further degrade. While the freedom and autonomy at the user level allows actors to quickly react and find solutions to smaller issues, many of these coping techniques have the potential for longer term degradation of the resilience of the ecosystem.

16.3 Assessing Adaptive Capacity

The literature on adaptive capacity has progressed and grown exponentially in the preceding decade, but significant challenges remain in developing rigorous, measurable and transferable indicators of adaptive capacity. In this attempt to develop a more empirically grounded and multi-pronged approach to understanding and assessing adaptive capacity, some conclusions can be drawn not only on enduring challenges, but also on how different indicators might be prioritised for different contexts and scales. The research presented in this book employed a tailored traffic light (black to white) scheme to highlight positive, neutral, and negative fulfilment of adaptive capacity indicators as well as forms of adaptive outcomes. While this approach can clearly communicate those areas of the governance system that are contributing positively or negatively to adaptive capacity within the case areas, it perhaps provides less information on which indicators are the most important within a certain context or across a range of contexts.

In terms of identifying certain indicators as being more or less important for developing and mobilising adaptive capacity, some caution is called for so that contextual sensitivities remain at the forefront of analysis and interpretation. However, some conclusions can be drawn to denote priority indicators across the different areas at different stages of diagnosis. Broadly, in reflecting on the varying levels of pertinence of the indicators and operationalised sub-indicators, it seems that those indicators that relate more to the good governance discourse (namely, ownership, accountability, responsibility, integration under the *Regime* category) could be termed as baseline requirements for navigating to more sustainable water management. However, as discussed in Chap. 13, it is these areas of the governance system that can in fact represent the most intractable part of the problem. On the other hand, those indicators that emerged through both the investigation and analytical process (based more heavily on adaptation and resilience related discourses), related more heavily to the *Knowledge* and *Network* categories and represent areas of the governance system that perhaps allow greater scope and flexibility for amendment and revision as new information or new conditions emerge.

Decisions on which particular aspects of these indicators to focus on would need to be made based upon local and expert analysis and judgement. Other studies that have taken a traffic light approach have also emphasised the importance of context when weighting adaptive capacity indicators or criteria (Gupta et al. 2010), underlining the need for any decisions taken on such an indicator tool to be embedded with contextual meaning as well as local insights and understanding. Therefore, such judgements would not only take into account the performance of the system according to the indicators, but also an understanding of which positive or negative indicators could best contribute to resolving either baseline (sustainable water management) or climate related (uncertainty, extreme events based) issues within the particular case. While future assessments could therefore still include the full suite of indicators for the three overarching categories, specific indicators and sub-indicators may be focussed on in more depth according to the particular needs of individual cases (whether basins, institutions, and local, regional or national case areas).

In the Chilean case, *Regime* related indicators are highlighted as the area of the governance system detracting from adaptive capacity the most. However, regime related challenges in the system are also perhaps the most intractable to resolve. So, pinpointing areas of *Knowledge* and *Network* related indicators (e.g. strengthening and better enabling expertise and knowledge systems already in place, and developing stronger networks that go across public and private spheres around these information systems) that could make more of a difference, faster, could provide not only baseline but also climate related benefits. In Switzerland, performance across the indicators is more positive, but again *Regime* related indicators of ownership and effectiveness are highlighted as detracting most from adaptive capacity. The role that *Network* related indicators have to play (especially trust building in the collaboration indicator) in remediating challenges in effective deployment of federal and cantonal provisions for more transformative adaptation measures, delineates the potentially reinforcing nature of these different categories as highlighted in other studies (Gupta et al. 2010).

One of the defining aspects of two case areas is the presence or absence of trust. In Switzerland trust is seen to broadly function, despite certain intra-jurisdictional challenges, while in Chile levels of both institutional and individual trust are identified as low (view collaboration indicators in the *Network* category). However, tackling issues of mistrust is particularly challenging, as well as dependant on whether relationships need to be mended between individuals or different institutional contexts or levels. This is an area of research that warrants further investigation, perhaps drawing on discourses relating to the fields of psychology and behavioural sciences.

Another notably important aspect of the indicators across both cases relates to matching expertise and knowledge with a willingness to learn (used in this study as an input in the perceptions indicators in the *Knowledge* category) and from there a conversion from learning to action (used in this study as an output in terms of the learning aspects of the adaptive outcomes). It highlights the challenge touched upon in Sect. 6.5 about how to effectively communicate uncertainty for action, and the issue of understanding at which point actors have enough information to start acting upon it. This is a major issue for the both the climate change adaptation as well as the broader mitigation debate.

Finally, while significant developments have been made in understanding adaptive capacity and factors that determine its presence since the 2001 IPCC determinants, a heavier focus has been placed on categorical (vulnerability, resilience, adaptive capacity, coping, etc.) rather than analytical (criteria, determinants, indicators, etc.) definitions. Perhaps due to the interdisciplinary nature of research on adaptive capacity, the body of work on adaptation related fields has concentrated on clarity around defining the different terms and concepts in this field, but perhaps less on ensuring conceptual clarity in terms of determinants, indicators, and criteria, whose use are often undefined within the literature.

However, there has been a significant and growing body of work on environmental, sustainability and ecological management criteria in different fields of the political, social and bio-physical sciences (OECD 1997; Slocombe 1998), on which the adaptation and adaptive capacity community could not only draw, but also strive for a better level of convergence and consistency in analytical definitions. However, in the quest for more measurable criteria of adaptive capacity indicators, the importance of context should continue to be prioritised. Researchers should ensure that in searching for broader analytical clarity and applicability in the development of indicators and measurable criteria of adaptive capacity, the nuance of specific contexts and priorities is not lost.

16.4 Contributions and Ways Forward

The outcomes of analysis and development of a multi-level framework for navigating the core tensions in adaptive capacity generate a framing that pays closer attention to the challenges inherent in policy construction across the complex spatial and temporal scales within which climate impacts will unfold. For instance, mobilising adaptive capacity to respond to variability at the local level also requires a longer term commitment to preparing for shocks and uncertainties at higher levels of governance. Couching these challenges in terms of the scales of governance and change across which they play out, can break down seemingly insurmountable challenges to more manageable and addressable issues as presented in Chap. 13 (Figs. 13.1 and 13.2).

Other studies (Tompkins and Adger 2004, 2005) have established the different forms of adaptation, as being both proactive (planning for future climate change, developing adaptive capacity) and reactive (autonomous reaction to events, mobilising adaptive capacity). Furthermore, evidence suggests that the ability to respond and absorb shocks reactively depends also on the proactive development of strategies and plans that not only enhance the resilience of the SES but that can be mobilised, as and when shocks occur (i.e. reactively). Concurrently, multiple reactions to climate events can open a window of opportunity to develop plans and policies that would enhance proactive adaptive capacity. Water managers and decision makers therefore need to focus on developing more proactive and long term preparedness to climate change and uncertainty in addition to flexible reactive approaches. While the proactive adaptive capacity can be linked with predictability and guidance at higher levels, it is suggested that reactive capacity is linked with flexibility at the local level. Therefore, meeting both aspects of adaptive capacity is about balancing out these two elements, in order to avoid the trade-off between predictability and flexibility across scales of governance (as presented in Chap. 14).

The governance assessment presents a detailed framework in which to understand how the governance system of each case area can assist in the implementation of IWRM. However, given the breadth of criticism against the reliance on panaceas to resolve water governance challenges (Meinzen-Dick 2007; Ostrom 2007; Ingram 2011) and the broad challenges of implementing IWRM (Engle et al. 2011; Medema et al. 2008), the indicators employed in the present governance assessment may not be adequate to provide insights into how the governance system may build and mobilise proactive and reactive adaptive capacity requisite for responding to different forms of shock at different scales. Therefore, while the assessment of legal provisions, case law and policy according to indicators of accountability, transparency, participation and those that relate to IWRM provide a valuable baseline from which to assess elements of the water governance system that may help or hinder adaptive capacity, they are limited in their ability to comprehensively account for adaptive capacity in the face of potentially increasing climatic uncertainty.

Adaptive capacity was conceptualised through its role in the transformation potential of a system to a more stable and sustainable state as a means to absorb future shocks and uncertainty. This framing of adaptive capacity assigns importance to both the capacity needed for proactive adaptation for preparing for future threats and uncertain shocks in relation to climate change impacts, as well as reactive adaptation for flexibly responding to unanticipated shocks from climate variability. This conceptualisation of adaptive capacity was explored in relation to a number of case extreme events through a set of determinants, which allowed for the development of a better understanding of the governance mechanisms associated with managing fast and slow hydrological change, as well as guidance and flexibility at different governance scales.

From an operational perspective, the cross-scale (of both governance and magnitudes of change) analysis enabled the development of a synthesis tool (Chap. 13) to guide decision makers on where resources could be best used to address elements of the governance system that hinder adaptive capacity and where to foster elements that enable adaptive capacity. Equally, it enabled the development of a multi-scale framework (Chap. 14) to address the challenges and tensions implicit in adaptive capacity through more practical institutional foci at each different scale.

Part 1 discussed the challenges in assigning causation and establishing linkages between more adaptive and integrative approaches and successful management of climatic events and adaptation. The methodological approach followed aimed to recognise and take into account these challenges by exploring adaptive capacity in relation to specific extreme events and categorising the governance mechanisms associated with those preparing for, managing and reacting to those hydrological extremes according to a resilience and institutional learning based framework of SES change before further characterisation and operationalisation of adaptive capacity indicators.

Furthermore, the analytical definitions employed within the study recognised the inability to exclusively link pressures relating to climate change from other economic, environmental or developmental pressures. This process aimed to establish clearer connections between such approaches and the mobilisation of adaptive capacity before, during or after a climatic event, while recognising the complex interactions between different pressures and the overarching challenges of causation, for which there is a need for longer and systematic monitoring and assessment of strategies in preparation and mobilisation phases as climate events happen over longer periods of time. However, the approach taken in this study has not only presented a method to assess how governance approaches can assist in the implementation of adaptive and integrative water governance approaches, but actually how successful elements of those governance mechanisms are at dealing with uncertainty through different kinds of hydro-climatic events as and when they come.

Some recent studies of adaptive capacity (Keskitalo et al. 2010) have still focussed on the broad vulnerability based determinants[1] of adaptive capacity (Smit and Wandel 2006; Yohe and Tol 2002) to provide insights into adaptive capacity in developed country contexts. However, while such aggregate determinants can still provide useful insights at a more aggregate level, they lack the nuance that can affect the development and mobilisation of adaptive capacity across the different scales addressed in the research presented. For instance, public perception of climate change and its impacts were found to be astute across stakeholders interviewed in both case areas, but despite high awareness and acceptance of climate change impacts, a strong sense of apathy was also prevalent in actors' edibility to adapt to large scale climate change, which impacted planning and preparation for smaller scales of change.

Furthermore, this book hopes to have provided valuable insights into new case areas to the growing body of literature on adaptive capacity. It is equally important to build the body of evidence on adaptive capacity as well as the mix of cases from a range of developed, emerging and developing contexts to better under what drives adaptive behaviour and enables different social systems to cope with and successfully adapt to climate related threats (Keskitalo et al. 2010). Just as Elinor Ostrom and collaborators have compiled a strong body of evidence to challenge conventional wisdom on the tragedy of the commons, collective action dilemmas and the social trap discourse (Ostrom 2010), a similar concerted effort to compiling case

[1] The range of technological options, structure of institutions, stock of human and social capital, access to risk-spreading procedures, ability of decision makers to manage information, public perception of causes of change and likely impacts (Adger et al. 2007).

studies focussing on adaptive capacity development and mobilisation, would contribute robust empirical evidence to inform policy and decision making at multiple scales. Such efforts have begun with projects such as NeWater, but further efforts to extend such a database with cases from other individual projects could serve the interests of both research and policy well.

16.5 Policy Recommendations

Despite calls for greater integration of climate change and water in research and policy, institutional barriers have persisted, whether in separate energy and environment lobbies, environmental and water lobbies, or climate and water lobbies. Increased dialogue across these disciplines is crucial to an adequate response to climate impacts, yet despite the flow of water across sector interests, institutions and disciplines; reality shows that the compartmentalisation of different interests still hinders an adequate understanding and resolution of water resource challenges. Because water is an issue that crosses multiple jurisdictional and institutional boundaries, water adaptation issues have the potential to act as a useful platform from which to initiate dialogue across political and economic lines.

In both cases, but Chile in particular, as is commonplace in other regions (Matthews et al. 2011), water related climate change adaptation policy is approached from a technical and infrastructural perspective across separate ministries. This study suggests that a focus on the institutional aspects of adaptation should come before that on the infrastructural adaptation planning, as without a robust and functioning social and institutional infrastructure, money and resources could be significantly wasted on infrastructural investments that may not resolve the key issues concerning water allocation and water resources management (see the challenges in implementing the TRC and Aconcagua Project). Policy makers should therefore focus on not only reinforcing and strengthening monitoring frameworks for bio-physical indicators but also supplementing them with developments in monitoring indicators in relation to the social and governance infrastructure in which decisions on infrastructure are made.

The country specific recommendations below aim to avoid the panacea trap, by drawing on the complex dynamics presented and explored early in Part IV and also taking into account the local physical and governance peculiarities. The challenges for Chile and Switzerland are very different, but commonalities can be found. Across both case areas, there is a need for better integration across sector policy and legislation. Both cases still represent the silo approach to water governance, where water is split across multiple ministries or administrative bodies. A top-down recognition that water runs through the different economic (mining, industry, energy, agriculture) and security (natural hazards, domestic water supply) priorities of government, should be matched by a concerted collaborative effort across ministries and administrative bodies to remove contradictory policy approaches and priorities as well as legislative provisions in order to put an end to the mixed signals that are sent out to the different stakeholders at lower governance levels that impact food-energy-water nexus.

In order to achieve this, both cases are likely to rely on a mix of leadership coupled with facilitative intermediaries that can provide a creative and flexible space to resolve contradictory and in some cases ideological battles over water resources governance. This is not likely to be an easy or speedy process, which calls for an intermediary focus on developing the more flexible and adaptive elements of the governance system as per the discussion in Part IV. How this process of policy integration would be achieved would vary considerably across the Swiss and Chilean cases, as what works in Switzerland, would probably not function in the Chilean governance and cultural context.

Intermediaries and bridging organisations have emerged organically in the Swiss case (e.g. Wasser Agenda 21, Mountain Water Network, WWF, ProNatura, Sector Associations, etc.) where non-governmental actors have a stronger and more organised voice in the political process. Additionally, the role of universities and research institutes are generally seen as neutral and often play a positive role in the development of alternative management approaches. Alternatively, in Chile, the establishment of bridging organisations and intermediaries might instead be developed from international organisations and research bodies, such as the World Bank, UN-ECLAC and OECD that inform and generate dialogue horizontally across ministries and vertically with regional actors.

Findings from the adaptive outcomes and adaptive capacity indicator assessment indicate that the Swiss case manifests a higher adaptive capacity, in particular through its ability to build knowledge networks and plan for future challenges. However, the challenges of such a decentralised, devolved and participative governance system can also challenge the ability for actors at one governance level (in this case at the regional level) to take advantage of a window of opportunity to transition to more transformative approaches in the TRC. This is not to suggest that the participative approach should not be followed, since the direct democratic and participative process does have its role to play in building longer term legitimacy, ownership and public and personal accountability by ensuring that a multitude of voices are taken into account.

However, presently, regional and national actors also begrudgingly acknowledge the role of participation in obstructing proactive and potentially transformative adaptation. This is not meant as a call to abandon such participation, but rather for policy makers to pay closer attention to the process, stage and arena in which stakeholders are engaged in that process as discussed in Chap. 13.

This fine balance that exists in Swiss politics between rule of local autonomy, decentralised principles of implementation and hands-off federal principles of cooperation, is being challenged not only by the increasing costs associated with more frequent extreme events, but also by the diminishing ability for local municipalities to manage those events and cover the costs of damage in their aftermath. The increasing reliance on canton and federal technical and financial assistance is both a challenge for higher levels of government in Switzerland, as well as an opportunity. The Federal and Valais governments both recognise the many challenges that the mountain communities of the Valais face from this set of environmental and socio-economic transitions that are presently taking shape.

Recently, Federal Councillor Doris Leuthard (Head of the Department of the Environment, Transport, Energy and Communications, DETEC) commented at the Swiss Forum for Sustainable Development that 'Switzerland's metropolitan, urban, rural, tourist and Alpine spaces all have different strengths and functions. They make Switzerland what it is. It is therefore important to think, plan and act in terms of functional spaces. And that is something which demands greater international cooperation' (DuPasquier 2011). Her comments speak to the on-going challenge to find the balance between preservation and development in Swiss governance particularities, in finding the balance between strengthening traditions that worked in the past, while recognising the aspects that must change in order to face the complex multi-scale challenges that local communities are increasingly coming up against.

At the local level this is in part being tackled through the enhancing and expanding networks across different communes and more intermittently outside the canton and that represents a transition of these alpine communities from closed off units (within which there was cooperation and common property) to a more integrated region for adapting to more complex (i.e. global scale) challenges. It mirrors evidence on developing social resilience from other case areas, such as the case of Trinidad and Tobago in Tompkins and Adger (2004). These societies or communities that have a high dependence on resources vulnerable to climate change not only show a tendency to spread their risk (i.e. integrating more villages into one water provision area), but also to extend their 'spaces of engagement to enable them to find a wider support network' (Tompkins and Adger 2004). These spaces of engagement and communities of practice provide an opportune arena for accurate knowledge concerning the development of precipitation, drought and scarcity situations (currently being developed across different research consortia) to be applied to the negotiation and development of different sets of water agreements, including use prioritisation and integrated land and flood management.

An area that policy makers at the federal level could perhaps focus on would therefore be to better link their own agenda and developing strategies with not only the regional but also the local level communities. This need not imply more technical or financial support, and these routes could also provide an arena for increased connectedness given the increasing reliance on state assistance and finances. Rather, more creative strategies to link up with, but not impose top down control, with the communities of practice that are being formed across local and regional levels might allow a more cohesive, proactive and cooperative approach for preparing for windows of opportunity, so that when periods of stress arise in the future the groundwork has already been made to adopt more transformational approaches. Investing a portion of financial capacity and time in enhancing cooperation across these different scales and sectors could contribute to speedier and smoother passage of the increasingly costly protection projects anticipated over the coming decade (Meier 2011).

Chile is characterised by complex contradictions, a centralist neo-liberal state, where the rule of the Water Code meant most interviewees consulted the Code directly during the interview, yet enforcement of it is negligible. Proponents of the

Chilean market model speak to the high degree of flexibility and autonomy it provides to the water rights users to resolve management issues, yet this same autonomy is seen by some experts as a major barrier to building collaborative solutions to the complex water challenges that water rights holders now face in overexploited and increasingly drought prone basins. An initial step for water managers and decision makers in the Chilean case would be to maximise the institutional assets they have to hand (e.g. Juntas de Vigilancia and Canalistas) in order to move from the status quo of individual flexibility to cohesive flexibility. One policy focus could be to incentivise user groups to formalise their user based organisations to take on a more legitimate governance role (to develop practices and ideas for more effective monitoring, enforcement, collaboration, conflict resolution) for cooperation across the basin (but beyond dam building).

This could perhaps draw from lessons learnt in the transformation of the coastal marine resources, where new scientific information on stock depletion was taken up by pre-existing social networks of fishermen, informing their ideas and practices and finally connecting to political leadership to generate a governance transformation once the window of opportunity manifested (Gelcich et al. 2010). In the current situation of the Aconcagua Basin, pre-existing networks exist across the basin (despite the challenges for some sections to formalise their Junta de Vigilancia), that could provide a critical intermediary between the user level and the regional ministerial bodies.

The current DGA have already expressed a critical focus on improving access to accurate and up to date information on water rights and the status of hydrological resources. Beyond this fundamental priority, serious considerations and dialogue need to happen around the unsustainable mismatch between current rights allocation and hydrological projections for Chile. This may seem like an impossible task, but an initial step could be for regional DGA and MMA hydro-climatic experts to begin a dialogue with local Juntas and Canal Associations where latest monitoring, observation and modelling studies are presented to rights holders. A more open exchange on information from the regional operational arm of government could help to build trust across the different sectors and levels during 'normal periods', which might in turn allow for stakeholders to network more collaboratively for extreme periods.

A worrying sign in the wrong direction, however, was the fact that the CNR had directed resources away from training programmes on efficiency (which were supposed to have been effective) in order to pay closer attention to improving transparency of the water rights. While in itself this is also an important policy focus, reframing irrigation efficiency training in the context of future uncertainty and scarcity, rather than just agricultural expansion and profit motivation would be a very useful role for the CNR to play alongside the DGA's focus on monitoring and information.

Currently, adaptation concepts in Chile tend to be technically and hard path oriented (dams, canal repairs, groundwater wells, irrigation) infrastructural projects with little attention being paid to either natural infrastructural assets (e.g. groundwater recharge, floodplains, wetlands) for enhancing resilience of the SES (Smith and

Barchiesi 2009) or institutional infrastructure to enhance the ability of the governance system to agree on and implement efficient, equitable and sustainable solutions to mounting challenges. Turning attention to incentivising better connections between economic actors in the basin might allow actors to agree on alternative market based solutions to restore degraded ecosystems and thus enhance natural infrastructure, as has been practised in USA basins (Harmon 2010).

Finally, Tompkins and Adger (2005, p 568) note that 'learning by doing requires decision-makers to accept that they make mistakes and bad decisions…if this acceptance is not present, then learning cannot happen'. Just as the focus is on technical adaptation, the majority of literature on the Chilean water model is concerned with whether or not the existence of a market can be identified (Thobani 1995), despite more recent focus by practitioners and academics on the effectiveness of the model (Bauer 2004; Dourojeanni and Jouravlev 1999). Optimistically, there is a growing interest in the government as well as NGO communities as to what precisely the market is and is not effective for.

Government researchers and analysts, as well as Chilean and World Bank academics need to move beyond the dogma of the market and place greater emphasis on this latter question, in relation to the accumulating stresses in the energy-water-food and environment nexus, to be able to better understand the linkages water adaptation can provide for sustainability, prosperity and resilience. Furthermore, findings in this book showing the limitations of the model in developing and mobilising adaptive capacity to shocks at different scales should be taken as a reinforcement of other studies that have warned against selling the neoliberal market panacea to other emerging and developing countries as an attractive alternative approach to water resources governance (Bauer 2004).

16.6 Final Thoughts

Across the world, impacts from climate change are being increasingly experienced through either too much or too little water, at times in some areas in close succession to one another (e.g. 2010 floods that followed the severe drought years in the Murray Darling Basin). While these impacts are likely to intensify through shifts in climate, water governance challenges do not stem from climate change alone, but are subject to a mix of interrelated political, environmental, technical and socio-economic pressures. In addressing these mounting challenges, focussing on how to transition and transform to more sustainable water governance and management paradigms, is a crucial piece of the puzzle that includes technical and hard infrastructural adaptation, but should not be limited to it. For too long, water related issues have resided in the kingdom of engineers and economists. Rigidity, constraint, and structured rules have persisted, constraining both the physical and governance systems.

As we move into a new period where uncertainty, unpredictability and complexity (from environmental and social drivers) mount, we can no longer expect to treat clouds as if they were clocks (Pearce 2002; Sutherland 2011). While clocks are ordered and predictable, clouds are typified by change and variation. In approaching the issues of adapting to climate related challenges in the water sector, we require a more balanced mix of solutions, that will incorporate both the perspective of clocks (engineering, technical solutions, dyke and dam building, irrigation technologies; legal frameworks with no mechanisms for review as hydrological baselines shift) as well as of clouds (accounting for complex adaptive and inter-related systems through multi-scale flexible policies and legislation that pay closer attention to how actors formulate, share and act on knowledge and information).

In designing institutional and governance responses for enhanced adaptive capacity, closer attention should be paid not only to the scales of governance at which particular policies should be fostered at, but also to the different speeds and magnitude of change for which they can be mobilised. Taking these issues into account in institutional and policy design, could guide governance reforms to allow for the generation of responses that attempt to accommodate uncertainty, rather than stop uncertainty. Equally, it would assist in refocusing the adaptation discussion beyond the confines of technical, efficiency and infrastructural responses to impacts at purely local or national levels.

Water governance regimes need to be both adaptable to amalgamating pressures as climate change develops but also structured to foster elements of a system that allows for more holistic and sustainable adaptation to take place. Intensified partnership and collaboration is needed not just across different scales of governance but also across the different sector rivalries within basins or watersheds themselves. In this respect, some heartening lessons can be drawn from multi-party solutions to watershed protection that have been developed through public-private partnerships such as the Water Futures Project (SABMiller 2009), in which companies collaborate with other stakeholders to protect the watersheds upon which they rely (Wales 2011) in the face of mounting challenges from over-abstraction and climate change impacts.

The focus of this book has been on adaptation and the governance frameworks that allow for greater adaptability in the face of escalating pressures within river basins due to the potential degree of warming in which the climate system is now locked in. However, it must also be acknowledged, that beyond certain tipping points, there are state changes to which adaptation and the ability to cope may be virtually impossible. The more catastrophic levels of climate change (rapid and significant sea level rise) are likely to impact on resources and ecosystem services to such an extent that it would have the potential to push the most adaptive governance system past its ability to absorb such a level of shock and disturbance. Therefore, despite the focus on the ability to adapt in this book, the necessity of mitigating the more extreme levels of temperature rise must also remain a global priority. In turn, policies and institutions focussing on mitigation and adaptation should become better integrated in order to take better advantage of potentially valuable synergies, and ensure the avoidance of mal-adaptation that might in turn increase climate change drivers.

Finally, there are some lessons that the adaptation community (both researchers and practitioners) can perhaps learn from the natural world with which it is preoccupied. Individual, institutional and political willingness to change and adjust to change has a fundamental role to play in transitioning to more adaptable governance systems that can manage water sustainably in the world of the anthropocene. Creativity is likely to play a large role in enabling innovation in governance transitions. In the field of Zoology, researchers have pinpointed the importance of play in problem-solving and conflict resolution in bonobo apes society (Behncke Izquierdo 2011). 'Play is our adaptive wildcard, it helps us adapt to an increasingly complex and challenging world through greater creativity and cooperation...in order to successfully adapt to a changing world, we need to play' (Behncke Izquierdo 2011). The focus on play for these zoologists is in part captured by the call for more flexible and iterative integration of knowledge, information and learning in the field of adaptive governance to enhance the generation of more innovative responses for increasingly complex problems.

Designers, in the fields of architecture, products and services have increasingly been drawing inspiration from the ecology (McDonough and Braungart 2002). Some designers have taken inspiration from ecology for application to institutional design to challenge the human search for hierarchical structure by drawing lessons from alternative modes of connection (e.g. the non-hierarchical fungal mat connections between Aspen Trees in the US) (Fulton Suri 2011). In a similar manner to resilience theory, it suggests that lessons can be learnt from natural processes to organise social structures in a way that builds cohesive action. Fulton Suri (2011) presses the need for institutional design to learn from and mirror the inter-connectedness of the natural world in the very institutions that we construct to manage it. Increased monitoring and observation of governance and intuitional indicators could enable policy makers to better account for the interconnections in complex systems, and thus foster governance and institutional frameworks that can accommodate the climate related challenges of increasingly unpredictable and indeterminate uncertainties.

References

Adger WN, Agrawala S, Mirza MMQ, Conde C, O'Brien K, Pulhin J, Pulwarty R, Smit B, Takahashi K (2007) Assessment of adaptation practices, options, constraints and capacity. In: Parry ML, Canziani OF, Palutikof JP, van der Linden PJ, Hanson CE (eds) Climate change 2007: impacts, adaptation and vulnerability. Contribution of working group II to the fourth assessment report of the Intergovernmental Panel on Climate Change. Cambridge University Press, Cambridge

Bauer C (2004) Siren song: Chilean water law as a model for international reform. Resources for the Future, Washington, DC

Behncke Izquierdo I (2011) Evolution's gift of play, from bonobo apes to humans. Paper presented at the TED 2011, Longbeach, CA, March 2011

Dourojeanni A, Jouravlev AS (1999) El Código de Aguas de Chile: entre la ideología y la realidad. vol LC/R.1897. Comisión Económica para América Latina y el Caribe, Santiago de Chile

DuPasquier A (2011) Promoting an alliance between town, country and the mountains. Bundesamt für Raumentwicklung, 10 Oct 2011

Engle NL, Johns OR, Lemos MC, Nelson DR (2011) Integrated and adaptive management of water resources: tensions, legacies, and the next best thing. Ecol Soc 16(1):19 [online]. URL: http://www.ecologyandsociety.org/vol16/iss11/art19/

Fulton Suri J (2011) What nature can teach us about design. Paper presented at the TEDx WWF. Geneva, Switzerland, 13 Oct 2011

Gelcich S, Hughes T, Olsson P, Folke C, Defeo O, Fernández M, Foale S, Gunderson LH, Rodríguez-Sickert C, Scheffer M, Steneck RS, Castilla JC (2010) Navigating transformations in governance of Chilean marine coastal resources. Proc Natl Acad Sci USA 107(39):16794–16799

Gupta J, Termeer C, Klostermann J, Meijerink S, Van den Brink M, Jong P, Nooteboom S, Bergsma E (2010) The adaptive capacity wheel: a method to assess the inherent characteristics of institutions to enable the adaptive capacity of society. Environ Sci Policy 13(6):459–471. doi:10.1016/j.envsci.2010.05.006

Harmon R (2010) How the market can keep streams flowing. Paper presented at the TEDx Rainier. Seattle, WA, 10 Oct 2010

Ingram H (2011) Beyond universal remedies for good water governance: a political and contextual approach. In: Garrido A, Ingram H (eds) Water for Food in a Changing World. Routledge, New York, pp 241–261

Keskitalo CH, Dannevig H, Hovelsrud G, West JJ, Gerger Swartling Å (2010) Adaptive capacity determinants in developed states: examples from the Nordic countries and Russia. Reg Environ Chang [Online]. doi:10.1007/s10113-010-0182-9

Matthews JH, Wickel BAJ, Freeman S (2011) Converging currents in climate-relevant conservation: water, infrastructure, and institutions. PLoS Biol 9(9):e1001159. doi:10.1371/journal.pbio.1001159

McDonough W, Braungart M (2002) Cradle to cradle. North Point Press, New York

Medema W, McIntosh BS, Jeffrey PJ (2008) From premise to practice: a critical assessment of integrated water resources management. Ecol Soc 13(2):29 [online]. URL: http://www.ecologyandsociety.org/vol13/iss22/art29/

Meier D (2011) Kosten für Hochwasserschutz steigen massiv. Deutsche Radio Schweiz, 13 Oct 2011

Meinzen-Dick R (2007) Beyond panaceas in water institutions. Proc Natl Acad Sci USA 104(39):15200–15205

OECD (1997) Better understanding our cities: the role of urban indicators. Organisation for Economic Co-operation and Development, Paris

Ostrom E (2007) A diagnostic approach for going beyond panaceas. Proc Natl Acad Sci USA 104(39):419–422

Ostrom E (2010) A multi-scale approach to coping with climate change and other collective action problems. Solutions 1(2):27–36 [Online]. http://www.thesolutionsjournal.com/node/565

Pearce WB (2002) Systems: schools of thought and traditions of practice. The Fielding Graduate Institute, Santa Barbara

SABMiller (2009) Water futures. http://www.sabmiller.com/index.asp?pageid=1772. Accessed 15 Oct 2011

Slocombe DS (1998) Defining goals and criteria for ecosystem-based management. Environ Manage 22(4):483–493

Smit B, Wandel J (2006) Adaptation, adaptive capacity and vulnerability. Glob Environ Chang 16(3):282–292

Smith DM, Barchiesi S (2009) Environment as infrastructure – resilience to climate change impacts on water through investments in nature IUCN. Gland, Switzerland

Sutherland R (2011) The new sweet spot – and how to find it paper presented at the TEDx WWF. Geneva, Switzerland, 13 Oct 2011

Thobani M (1995) Tradable property rights to water: how to improve water use and resolve the water conflicts. Finance & Private Sector Development Note no. 34. World Bank, Washington, DC

Tompkins EL, Adger WN (2004) Does adaptive management of natural resources enhance resilience to climate change? Ecol Soc 9(2):10 [online]. URL: http://www.ecologyandsociety.org/vol19/iss12/art10

Tompkins EL, Adger WN (2005) Defining response capacity to enhance climate change policy. Environ Sci Policy 8:562–571

Wales A (2011) The water-food-energy nexus – why everything you consume is connected. Paper presented at the TEDx WWF. Geneva, Switzerland, 13 Oct 2011

Yohe G, Tol RSJ (2002) Indicators for social and economic coping capacity – moving toward a working definition of adaptive capacity. Glob Environ Chang 12(1):25–40

MIX
Papier aus verantwortungsvollen Quellen
Paper from responsible sources
FSC® C105338

If you have any concerns about our products,
you can contact us on
ProductSafety@springernature.com

In case Publisher is established outside the EU,
the EU authorized representative is:
Springer Nature Customer Service Center GmbH
Europaplatz 3, 69115 Heidelberg, Germany

Printed by Libri Plureos GmbH
in Hamburg, Germany